Organometallic Reactions and Syntheses

Volume 6

ADVISORY BOARD

CONTRIBUTORS TO THIS VOLUME

MICHAEL F. FARONA

University of Akron
Akron, Ohio

RUSSELL N. GRIMES

University of Virginia
Charlottesville, Virginia

C. U. PITTMAN, JR.

University of Alabama
University, Alabama

A Continuation Order Plan is available for this series. A continuation order will bring delivery of each new volume immediately upon publication. Volumes are billed only upon actual shipment. For further information please contact the publisher.

Organometallic Reactions and Syntheses

Volume 6

Edited by

ERNEST I. BECKER

University of Massachusetts
Boston, Massachusetts

and

MINORU TSUTSUI

Texas A & M University
College Station, Texas

PLENUM PRESS · NEW YORK AND LONDON

Library of Congress Cataloging in Publication Data

Becker, Ernest I   1918-
   Organometallic reactions.

   Includes Bibliographies.
   Vol. 6 published by Plenum Press under title:
Organometallic reactions and syntheses.
    1. Organometallic compounds–Collected works. I. Tsutsui, Minoru, 1918-   joint
author. II. Title.
QD411.B36   547'.05   74-92108
ISBN 0-306-39906-7 (v. 6)

© 1977 Plenum Press, New York
A Division of Plenum Publishing Corporation
227 West 17th Street, New York, N.Y. 10011

Printed in the United States of America

Preface

The primary literature on organometallic chemistry has undergone phenomenal growth. The number of papers published from 1951 to 1976 is about equal to all prior literature. Together with this intense activity there has developed a complexity in the literature. Thus, specialized texts and teaching texts, a review journal, an advanced series, and a research journal have all appeared during this period. The present series also reflects this growth and recognizes that many categories of organometallic compounds now have numerous representatives in the literature.

The purpose of *Organometallic Reactions and Syntheses* is to provide complete chapters on selected categories of organometallic compounds, describing the methods by which they have been synthesized and the reactions they undergo. The emphasis is on the experimental aspects, although structures of compounds and mechanisms of reactions are discussed briefly and referenced. Tables of all of the compounds prepared in the category under consideration and detailed directions for specific types make these chapters particularly helpful to the preparative chemist. While the specific directions have not been referenced in the same way as are those in *Organic Syntheses* and *Inorganic Syntheses*, the personal experiences of the authors often lend special merit to the procedures and enable the reader to avoid many of the pitfalls frequently encountered in selecting an experimental procedure from the literature.

We acknowledge a debt of gratitude to the contributing authors whose dedication and skill in preparing the manuscripts cannot adequately be rewarded. It has been gratifying to note that virtually all invitations to contribute have been accepted at once. We also owe thanks to the publisher for encouragement and even the "gentle prod" when necessary to see these volumes to their completion.

Ernest I. Becker
Minoru Tsutsui

v

Contents

Chapter 2

Reactions of Metallocarboranes

RUSSELL N. GRIMES

Chapter 3
Homogeneous Catalysis by Arene Group-VIB Tricarbonyls

MICHAEL F. FARONA

Chapter 1

Vinyl Polymerization of Organic Monomers Containing Transition Metals

Charles U. PITTMAN, JR.

I. INTRODUCTION AND SCOPE

In this chapter the reactivity toward vinyl homopolymerization and copolymerization of organic monomers containing transition metals is discussed. Vinylferrocene (1) is an excellent example and the most thoroughly studied member of this unique class of monomers. Other model examples include vinylcyclopentadienylmanganesetricarbonyl (2), styrenetricarbonylchromium (3), and trans-bis(tributylphosphine) (4-styryl)palladium chloride (4). Each of these monomers undergoes vinyl addition polymerization, and each contains a transition metal in the monomer which is retained in the polymer. The first vinyl polymerization of an organometallic derivative was the radical-initiated homopolymerization of vinylferrocene by Arimoto and Haven[1,2] in 1955. They also showed that vinylferrocene could be copolymerized with methyl acrylate, styrene, and chloroprene. Despite this success, organometallic vinyl monomers were virtually neglected for the next decade in comparison with the vast attention given to the preparation of condensation polymers[3,4] and coordination polymers[5] of transition-metal derivatives.

Recently, studies of vinyl derivatives of transition-metal compounds have appeared which allow a more quantitative description of their reactivity

CHARLES U. PITTMAN, JR. ● Department of Chemistry, University of Alabama, University, Alabama.

$$
\begin{array}{ccc}
\underset{\text{Fe}}{\text{(cp)}}\!-\!CH{=}CH_2 & \underset{Mn(CO)_3}{\text{(cp)}}\!-\!CH{=}CH_2 & \underset{Cr(CO)_3}{\text{(arene)}}\!-\!CH{=}CH_2 \\
(1) & (2) & (3)
\end{array}
$$

$$
\begin{array}{c}
CH{=}CH_2 \\
\text{(arene)} \\
Bu_3P\!-\!Pd\!-\!PBu_3 \\
| \\
Cl \\
(4)
\end{array}
$$

in homo- and copolymerization. It is of fundamental interest to understand how organometal functions affect vinyl polymerization behavior. In particular, it is of interest to see how monomers such as (1–4) behave in terms of the quantitative copolymerization scheme pioneered by Mayo, Lewis and Walling[6-8] and the Q–e scheme of Price.[8,9] This knowledge will help to predict the copolymerization behavior of organometallic monomers with classic organic vinyl monomers, and it also will provide a qualitative picture of the electronic effects that organometallic functions impart. Thus, in addition to reviewing the literature, the correlation of existing reactivity data is tentatively interpreted in this chapter to provide a framework in which to view organometallic monomers.

    Transition-metal-containing monomers offer the polymer chemist many unique options. Since transition metals often exist in more than one stable oxidation state, mixed-oxidation polymers can be envisioned. Polymers with catalytic properties can be constructed. The thermal or photochemical degradation of organometallic polymers might be designed to free metals or metal oxides within films or crosslinked polymers to produce systems with new properties. With these opportunities coexist many pitfalls. The transition-metal–ligand system may undergo undesirable reactions with radical, anionic, cationic, or Ziegler–Natta catalysts which would preclude polymerization. Many organometallic compounds react with air or moisture or are thermally unstable. The fields of transition metal and polymer chemistry are seldom bridged by the same chemists. These and other problems exist. Even where such polymers have been prepared, detailed property studies still lag behind synthetic efforts.[10,11]

II. POLYMERIZATION AND COPOLYMERIZATION OF VINYL MONOMERS CONTAINING FERROCENE

A. Homopolymerization

Vinylferrocene has been homopolymerized using radical, cationic, and Ziegler–Natta initiators. Peroxide initiators are not suitable. Many research groups[1,12-15] have demonstrated that vinylferrocene is oxidized by peroxides to ferricenium ion and that peroxide decomposition is catalyzed in this system. Azo initiators have been used most often both in bulk and in solution systems.[1,2,12-21] Persulfate has been used in emulsion systems but only oils were formed.[1,2] In solution, fairly low-molecular-weight polymers are formed with azobisisobutyronitrile (AIBN)[15-19,22,23] but higher molecular weights are achieved in bulk polymerization.[13,15] Table I summarizes some representative molecular weights which have been obtained.

Using cationic or Ziegler catalyst systems, only very low-molecular-weight polymers have been found.[1,24-25] Despite the high tendency of the cyclopentadienyl rings of ferrocene to undergo electrophilic attack, only vinyl propagation was observed. Simionescu et al.[25] reported that $Et_2AlCl/M(acac)_2$ and $Et_3Al/M(acac)_2$—where M = Ni, Cu, and V(O)—gave polymers which were partially methanol soluble.

Baldwin and Johnson[14] reported that the homopolymerization of vinylferrocene [VF] and styrene were quite similar, and they claimed the rate of polymerization followed the normal bimolecular termination mechanism found with most vinyl monomers, i.e., $r_p = k[VF]^1[In]^{1/2}$ for AIBN initiation at 70° in benzene solution. Kinetics were followed by dilatometry and nmr. However, Pittman et al.[16,22] showed that solution polymerizations in benzene gave low yields unless multiple initiator additions were employed. In addition, the molecular weights resulting from solution polymerizations were low $\overline{M}_n = 4000$–6000 and the distribution was narrow ($\overline{M}_w/\overline{M}_n = 1.2$–$1.5$ using 1 g VF per 1 ml benzene).[16,20] Furthermore, the molecular weight was not increased on reducing the initiator concentration, and the degree of polymerization increased with an increase in monomer concentration reaching its highest level in bulk polymerizations.[16] Cassidy et al.[13] obtained degrees of polymerization as high as 226 in bulk polymerizations. These relationships are very unusual for vinyl polymerization and unlike the behavior of styrene.

An extensive study of the kinetics of vinylferrocene homopolymerization in benzene has been performed by George and Hayes.[15,19] They conclusively demonstrated that the $r_p = k[VF]^{1.2}[AIBN]^{1.1}$ in benzene at 60°, using AIBN as the initiator. This rate law requires a monomolecular termination process. The termination step apparently involved an intramolecular electron transfer from iron to the chain-propagating radical, followed by trapping of the anionic chain end and the formation of a Fe(III) high-spin complex at the chain end.

TABLE I
Representative Number-Average Molecular Weights Obtained in
Polyvinylferrocene Homopolymerizations

Solvent temperature	Vinylferrocene, mol/liter	Initiator	Initiator, mol/liter	Conversion, %	M_n	Ref.
Benzene, 60°	1.0	AIBN	1.0×10^{-2}	1.1	6,100	15
Benzene, 60°	1.0	AIBN	1.0×10^{-2}	15.0	5,000	15
Benzene, 60°	2.0	AIBN	1.0×10^{-2}	10.0	8,800	15
Benzene, 60°	3.0	AIBN	1.0×10^{-2}	1.9	5,800	15
Benzene, 60°	3.0	AIBN	1.0×10^{-2}	7.6	11,000	15
Benzene, 60°	4.0	AIBN	1.0×10^{-2}	2.1	8,500	15
Benzene, 60°	4.0	AIBN	1.0×10^{-2}	7.0	12,800	15
Benzene, 60°	4.0	AIBN	1.8×10^{-2}	3.1	6,800	15
Benzene, 60°	4.0	AIBN	2.0×10^{-2}	3.2	7,900	15
Benzene, 80°	5.3	AIBN	6.4×10^{-2}	25	4,600	16
Benzene, 80°	5.3	AIBN	1.2×10^{-1}	69	10,000	16
Benzene, 70°	5.3	AIBN	1.2×10^{-1a}	60	4,200	16
Dioxane, 60°	1.0	AIBN	5.0×10^{-2}		5,800	18
Dioxane, 60°	2.5	AIBN	4.9×10^{-2}		9,500	18
Dioxane, 60°	3.0	AIBN	5.0×10^{-2}		10,800	18
Dioxane, 60°	2.0	AIBN	1.0×10^{-2}		11,500	18
Dioxane, 60°	2.0	AIBN	2.5×10^{-2}		11,100	18
None, 80°	bulk	AIBN	0.74^b	21.6	7,300	16
None, 80°	bulk	AIBN	1.02^b			
			1.00	46.8	6,300	16
None, 70°	bulk	AIBN	2.00^b	70.6	23,200	16
None, 60°	bulk	AIBN	2.00^b	57.6	36,800	16
Toluene, 0°	1.0	$BF_3 \cdot OEt_2$	0.05	52	2,200	24
CH_2Cl_2, 0°	1.0	$BF_3 \cdot OEt_2$	0.05	35	1,500	24
CH_2Cl_2, 0°	1.0	$TiCl_4$	0.05	63	1,100	24
CH_2Cl_2, 0°	1.0	$Et_2AlCl \cdot t$-BuCl	0.05	28	3,500	24

[a] Dual addition of initiator at 3-hr intervals.
[b] Weight percent of AIBN.

This process is shown in Scheme 1. The conversion of ferrocenium groups to a high-spin Fe(III) species is not unexpected since Golding and Orgel showed that organic solutions of ferricenium salts decompose to give Fe(III) compounds.[26] Each termination event produced a high-spin Fe(III) complex.[15,19,27] This was supported by spectral evidence.[15,27] Magnetic susceptibility measurements confirmed that the polymers did not contain ferromagnetic impurities.[15,27] The nmr spectra of polyvinylferrocene exhibited pronounced paramagnetic broadening which could not be removed by treatment with ascorbic acid.[15,27] Since ascorbic acid readily reduces ferrocenium ions to ferrocene, this finding is consistent with a high-spin paramagnetic Fe(III) species.

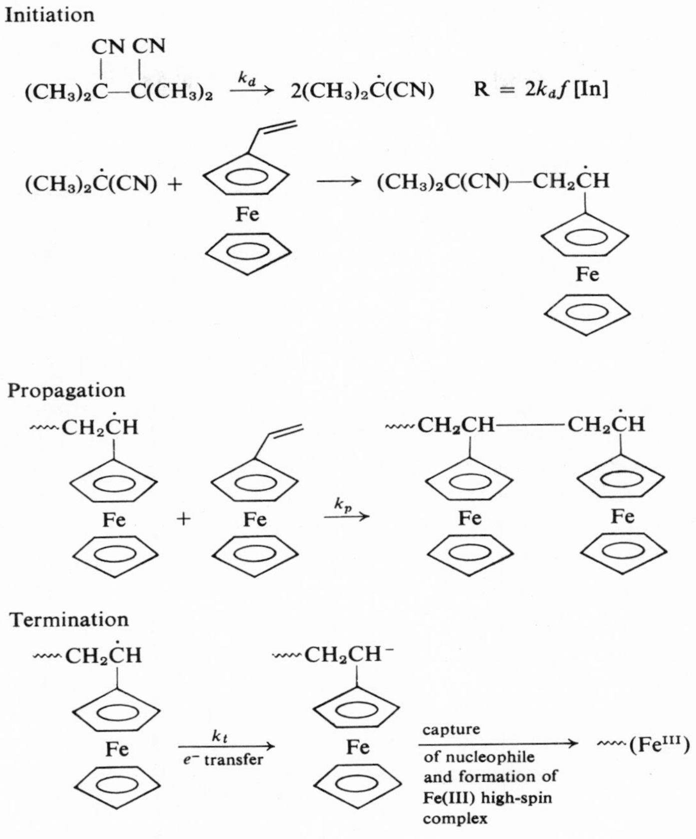

Initiation

$$R = 2k_d f \, [\text{In}]$$

Propagation

Termination

$$Rp = (2k_p k_d f/k_t) \, [\text{M}] \, [\text{In}]$$

SCHEME 1

The Mössbauer spectra of high-molecular-weight bulk polymerized polyvinylferrocene exhibits a clean doublet ($Q = 2.4$ mm sec^{-1}) consistent with unoxidized ferrocene groups.[22] However, the Mössbauer spectra of the lower-molecular-weight polymers, produced in benzene, exhibited a peak at 0.14 mm sec^{-1}, which is not consistent with either ferrocene or ferrocenium ion spectra.[15,27] This peak was temperature dependent. The area of this peak varied from 3 to 14% of the total iron content and its intensity was larger for low-molecular-weight polymers.[27] The lower-molecular-weight polymers are the ones with the higher percentage of end groups. The relative intensity of this peak correlated well with the degree of polymerization,[27] which strongly implies that about one such Fe(III) species exists for each polymer chain present. It is just these low-molecular-weight polymers which have a higher mole fraction of chain ends. Finally, a broad line esr spectrum was observed: $g = 2.06$ and 425 gauss band width in benzene and 600 gauss width in the solid.[15,27] The intensity of this band was greatest for the lower-molecular-weight polymers.

The first-order termination step in vinylferrocene homopolymerizations in benzene requires that the degree of polymerization follow the relation $\overline{DP} = k_p[M]/k_t$. This explains why the degree of polymerization was found to be approximately proportional to the monomer concentration and independent of initiator concentration.

The low molecular weights found in solution homopolymerizations and the fact that the degree of polymerization (DP) was not strictly linear with monomer concentration suggested that chain transfer to monomer and polymer could not be neglected in vinylferrocene polymerizations. Further, Pittman et al.[16] found in viscosity–molecular-weight studies, that as the molecular weights increased in bulk polymerizations, the resulting polymers became increasingly branched. Gel-permeation chromatography (gpc) demonstrated that polymers produced in bulk also gave binodal distributions.[16] Low-molecular-weight fractions ($5000–23,000 = \overline{M}_n$) were readily fractionated and the values of K and a in $|\eta| = K(M)^a$ were 6.64×10^{-4} and 0.49 in benzene, and 7.20×10^{-5} and 0.72 in THF, respectively.[16] The high-molecular-weight modes could not be fractionated.[16] For example, it proved impossible to fractionate a sample with $\overline{M}_n = 1.5 \times 10^5$ and $\overline{M}_w = 6.3 \times 10^5$.[16] The low viscosities observed for the high-molecular-weight nodes suggest a particularly dense, highly branched polymer.[16] Branching would be expected if chain transfer to polymer was pronounced. This supposition was confirmed by George and Hayes.[15] The value of the chain-transfer constant to monomer, C_m, was 8×10^{-3} for vinylferrocene versus 0.06×10^{-3} for styrene at 60°.[15] As the concentration of polymer in solution increases (this is quite important in high-conversion bulk polymerizations) the polymer increasingly acts as a chain-transfer agent leading to highly branched soluble and perhaps even to crosslinked insoluble polymers. Gel-permeation chromotography studies of lower-molecular-weight fractions are best interpreted by concluding that polyvinylferrocene is even branched at low conversions.[27]

The high chain transfer to monomer suggests that the following process should be important[27]:

$$\sim\!\!\text{CH}_2-\overset{\bullet}{\text{CH}} \ + \ \text{CH}\!\!=\!\!\text{CH}_2 \ \longrightarrow \ \sim\!\!\text{CH}_2-\text{CH}_2 \ + \ \overset{\bullet}{\text{C}}\!\!=\!\!\text{CH}_2$$
$$\quad\quad\ \ \underset{\text{Fc}}{|} \quad\quad\quad \underset{\text{Fc}}{|} \quad\quad\quad\quad\quad\quad \underset{\text{Fc}}{|} \quad\quad\quad \underset{\text{Fc}}{|}$$

$$\text{CH}_2\!\!=\!\!\overset{\bullet}{\text{C}} \ + \ \text{CH}_2\!\!=\!\!\text{CH} \ \longrightarrow \ \text{CH}_2\!\!=\!\!\text{C}\!\!\left(\!\!\text{CH}_2-\text{CH}\!\!\right)\!\!-\!\!\text{CH}_2-\overset{\bullet}{\text{CH}}$$
$$\quad\quad \underset{\text{Fc}}{|} \quad\quad\quad \underset{\text{Fc}}{|} \quad\quad\quad\quad \underset{\text{Fc}}{|} \quad \underset{\text{Fc}}{|} \quad\quad \underset{\text{Fc}}{|}$$

A growing ferrocene radical (Fc) could abstract a vinyl hydrogen. The resulting vinyl radical would then initiate a new polymer chain, but this would result in a vinylidene end group. George and Hayes argued that ir bands at 920 and 850 cm^{-1} are most consistent with terminal vinylidene groups.[27] However, cyclopentadienyl ring out-of-plane C–H deformations are intense in the 810–850 cm^{-1} region and specific assignments are at best tentative.

For most vinyl monomers, the rate of polymerization, r_p, is proportional to $k_p/k_t^{1/2}$, but for vinylferrocene (in benzene) $r_p \propto k_p/k_t$. Thus, the value of k_p/k_t was estimated to be 33 dm^2/mol for vinylferrocene assuming $f = 0.75$.[15] Typically, values of k_p lie between 10^2 and 2 $\times$ 10^3 for vinyl monomers. Thus, if k_p for vinylferrocene is in this range, then k_t will be between 3 and 60 sec^{-1}. The activation energy, E_0, for this polymerization was also remarkably high (139 kJ/mol) versus about 80 kJ/mol for the majority of organic monomers.[15] Consider the equation

$$E_0 = E_p + E_d - E_t$$

where E_p, E_d, and E_t are the activation energies for propagation, initiation, and termination, respectively. For AIBN in benzene at 60°, $E_d = 127$ kJ/mol and by using an estimate of $E_t \approx 8$ kJ/mol the value of E_p was estimated to be about 20 kJ/mol for vinylferrocene.[15]

In contrast to homopolymerizations in benzene, the kinetics of vinyl-ferrocene polymerizations in dioxane, initiated by AIBN, were first order in monomer and half order in AIBN at 60°.[28] The equation $R_p = k$ [VF]$^{0.97}$ [AIBN]$^{0.42}$ held. Furthermore, the value of E_0 was 84 kJ/mol between 50° and 80°. Thus, the solvent apparently makes a remarkable difference. Chain transfer to monomer remained very important in dioxane ($C_m = 1.2 \times 10^{-2}$ from molecular weight studies).[28] This value of C_m was close to that observed in benzene where chain transfer to polymer was important. This value sets an upper theoretical limit of 18,000 on $\overline{M}_n$ for polyvinylferrocene produced under the conditions used here. It is now apparent why solution polymerizations always gave low molecular weights. At [VF] = 1–2 mol/dm^3 and [AIBN]=1–10 $\times$ 10^{-2} mol/dm^3, the values of $\overline{M}_n$ ranged from 5790 to 11,500. Chain transfer to solvent was negligible. The value of $k_p/k_t^{1/2} = 2.36 \times 10^{-2}dm^{1.5}$/mol$^{1/2}sec^{1/2}$ which is similar to that of styrene (3.41 $\times$ 10^{-2}).[28]

The kinetics in dioxane support the normal mechanism for vinyl mono-mers, including bimolecular termination by recombination or disproportion-ation. Mössbauer and esr studies supported this conclusion.[28] Unlike the polymers produced in benzene, no trace of Fe(III) Mössbauer bands or esr signals were observed.

Dioxane has no π-electron system and should act as an inert diluent. If benzene assisted the chain-end electron transfer (proposed to account for the monomolecular termination mechanism in benzene) via complexation through its π-system, then the difference in kinetics between the two solvents could be understood. Rate and $\overline{DP}$ studies of styrene polymerizations in the presence of small amounts of vinylferrocene were then performed in benzene and bulk.[29,30] These studies provided evidence that a specific interaction between styrene and vinylferrocene occurred in bulk. However, benzene–vinylferrocene interactions predominated in benzene. The physical nature of this interaction is unknown, since it is currently supported only by kinetic arguments and the observation that the molar extinction coefficient for ferrocene units in vinylferrocene–styrene copolymers is greater than that in polyferrocene alone.[29]

Vinylferrocene and its homopolymer are stable in solution except when stored in chlorinated solvents.[16] In CCl_4, yellow polyvinylferrocene solutions become greenish and eventually precipitate. This process is markedly faster as the incident light intensity increases and is due to the formation of ferricenium salts and further decomposition.[16]

Vinylferrocene can be polymerized in halogenated solvents without initiator.[31] The electron-transfer path has been invoked to explain the results.

Low-molecular-weight polymers with binodal distributions are often obtained. Also, substantial amounts of chlorine (3%) are found in the polymers due to the high chain transfer to solvent.[31] Ferricenium salt sites are formed in these polymers also.

Polyvinylferrocene exhibits a very high glass-transition temperature (Tg).[22] Differential scanning calorimetry shows two transitions, the first at 170–190° and the second at 225–247°.[22,32] The softening point is 280–290°.[13] Assignment of the T_g is difficult because thermal decomposition under nitrogen is fairly rapid at 240° and the polymers studied have had low molecular weights. The 225–247° transition is probably the correct glass transition based on recent evidence[32]; polymers of very high molecular weights would have even higher T_g values.

B. Copolymerization of Vinylferrocene

The first copolymers of vinylferrocene were reported by Arimoto and Haven.[1] Methyl acrylate and vinylferrocene (80:20) were copolymerized (80°, AIBN initiation) to give a solid copolymer melting from 190° to 210°, containing 10.3% Fe.[1] Similarly, styrene and vinylferrocene (80:20) copolymerized to give a product melting from 158° to 165° with 4.3% Fe. Emulsion copolymerization with a large molar excess of chloroprene with small amounts of vinylferrocene, initiated by persulfate, gave elastomers.[1] It was not until the extensive investigations of Pittman and Aso [12,20,24,32–34] that the reactivity of vinylferrocene toward copolymerization was understood.

Lai, Rounsefell, and Pittman[12] copolymerized vinylferrocene with styrene, methyl acrylate, methyl methacrylate, and acrylonitrile in benzene at 70° using AIBN initiation. Copolymerization readily occurred except with vinyl acetate or isoprene. Subsequently, copolymerization studies were carried out with N-vinylpyrrolidone,[32] N-vinylcarbazole[34] maleic anhydride,[33] and even vinylcyclopentadienyltricarbonylmanganese[50] to provide a wide range of comonomer reactivities with which to examine vinylferrocene reactivity.

Before these copolymerizations are discussed, a brief introduction to copolymerization kinetics and relative reactivity ratios will be given and the Q–e scheme will be provided as a framework. In a copolymerization, where the assumptions are made that the chain length and the penultimate monomer units do not affect reactivity, four propagation steps must be considered:

$$\text{\textasciitilde}M_1^{\cdot} + M_1 \xrightarrow{k_{11}} M_1 - M_1^{\cdot}$$

$$\text{\textasciitilde}M_1^{\cdot} + M_2 \xrightarrow{k_{12}} M_1 - M_2^{\cdot}$$

$$\text{\textasciitilde}M_2^{\cdot} + M_2 \xrightarrow{k_{22}} M_2 - M_2^{\cdot}$$

$$\text{\textasciitilde}M_2^{\cdot} + M_1 \xrightarrow{k_{21}} M_2 - M_1^{\cdot}$$

Applying the usual steady-state conditions, the relative rate of incorporation of M_1 and M_2 into the copolymer is given by the following copolymer equation [8]:

$$\frac{d[M_1]}{d[M_2]} = \left(\frac{[M_1^0]}{[M_2^0]}\right)\left(\frac{r_1[M_1^0] + [M_2^0]}{r_2[M_2^0] + [M_1^0]}\right)$$

where $r_1 = k_{11}/k_{12}$ and $r_2 = k_{22}/k_{21}$. A knowledge of r_1 and r_2 permits one to calculate the composition of a copolymer of M_1 and M_2 from any initial monomer–feed ratio (M_1^0/M_2^0) at any percent conversion to polymer using the integrated form of the copolymer equation.[6,8]

The values of r_1 and r_2 are evaluated by running many copolymerizations at known feed ratios and anlyzing the M_1/M_2 ratio in the polymer. These composition–conversion data are then fitted to best values of r_1 and r_2. Fitting these data should be performed by a nonlinear least-squares method.[35,36] Linear methods, such as the Fineman–Ross technique,[37] are mathematically incorrect and can, in some cases, lead to large errors.[36]

The best method to evaluate the reactivity ratios is to use the integrated form of the copolymer equation (to account for monomer drift), employing a nonlinear least-squares fitting procedure with an objective mathematical criterion of confidence.[35,36] Furthermore, polymerizations to relatively low conversions should be performed at the optimized initial M_1^0/M_2^0 ratios. Such a method has been developed and reviewed by Tidwell and Mortimer.[35,36] Since the optimum M_1^0/M_2^0 ratio cannot be determined unless r_1 and r_2 are known, and the optimum M_1^0/M_2^0 ratios should be used (derived from these crude r_1 and r_2 values) to do further experiments[36]; r_1 and r_2 are first computed and the procedure is repeated.

Accurate values of the reactivity ratios are also necessary in order to evaluate monomer reactivity in terms of the $Q–e$ scheme. This scheme is a semiquantitative empirical attempt to combine the effects of resonance stabilization and polarity on the reactivities of monomers and their radicals[8,9]: Q is a resonance stabilization parameter and e is a parameter related to the polar character of the vinyl group.[8,9] The scheme has been extensively reviewed[9,38]; it is sufficient to show here how the values of Q and e for the two monomers (M_1 and M_2) are computed from the reactivity ratios. The following equation expresses the relationship of Q and e to the reactivity ratios:

$$r_1 = (Q_1/Q_2) \exp - e_1(e_1 - e_2)$$
$$r_2 = (Q_2/Q_1) \exp - e_2(e_2 - e_1)$$
$$r_1 r_2 = \exp - (e_1 - e_2)^2$$

The values of e for a large number of monomers are tabulated in the *Polymer Handbook*.[39] A few are listed here to provide a basis of comparison with organometallic monomers. A large negative value of e signifies a monomer is electron rich in radical copolymerizations, while the more positive e becomes the more electron deficient the monomer becomes. Careful copolymerization studies with standard organic monomers permit the classification of organometallic monomers according to the $Q–e$ scheme wherever the terminal copolymerization model holds.

This classification has now been done for vinylferrocene. Extensive

$$e = -1.96 \qquad e = -1.37 \qquad e = -0.80$$

$$e = +0.39 \qquad e = +1.20 \qquad e = +2.25$$

TABLE II

Reactivity of Vinylferrocene, M_1 in Copolymerizations with Organic Monomers, M_2 According to the Q–e Scheme

M_2	e_2	r_1	r_2	e_1	Reference
Vinylcyclopentadienyl-tricarbonylmanganese	−1.99	0.49	0.44	−2.6	50
N-Vinylcarbazole	−1.40	0.60	0.20	−2.4	34
p-N,N-Dimethylaminostyrene	−1.37	0.15	3.8	−2.2	48
1,3-Butadiene	−1.05	0.3	3.5	−1.05	51[a]
	−1.05	0.14	3.97	−1.80	48[b]
	−1.40	0.14	3.97	−2.1	48[b]
N-Vinyl-2-pyrrolidone	−0.90	0.67	0.33	−2.1	34
Styrene	−0.80	0.08	2.70	−2.04	12
	−0.80	0.07	2.63	−2.1	12[b]
	−0.80	0.2±.1	4±1	−1.3	20[c]
Methyl methacrylate	+0.40	0.52	1.22	−0.29	12
	+0.40	0.56	1.25	−0.20	12[b]
Methyl acrylate	+0.58	0.82	0.63	−0.21	12
	+0.58	0.73	0.61	−0.32	12[b]
Acrylonitrile	+1.20	0.15	0.16	−0.73	12[d]
	+1.20	0.11	0.17	−0.81	12[b]
Diethyl fumarate	+1.25	Alternating copolymer			49
Fumaronitrile	+1.96	Alternating copolymer			49
Maleic anhydride	+2.25	0.02	0.19	−0.11	33[e]

[a] Calculated from nine nonoptimized data points using the differential form of copolymer equation in a linear least-squares fitting mode.
[b] Calculated from data in Ref. 51 (or 12) using the integrated form of the copolymer equation in the nonlinear least-squares fitting mode.
[c] Nonoptimized experimental design used.
[d] $r_1 \cdot r_2 = 0.024$, high alternation tendency.
[e] $r_1 \cdot r_2 = 0.003$, essentially alternating copolymer.

copolymerization studies are summarized in Table II. Here, the comonomers were chosen to span a wide range of e values from very electron rich to very electron deficient. The e values for the comonomers M_2 are listed in the table. It is particularly striking that vinylferrocene exhibits one of the largest negative e values ever reported. From styrene copolymerizations an $e_1 = -2.1$ for vinylferrocene is obtained! This value is derived from the data in Ref. 12, where the positive value of e had inadvertently been cited. The value of -2.1 may be compared to the following series of e values: styrene, -0.80; N,N-dimethylaminostyrene, -1.37; 1,1-dianisylethylene, -1.96. Thus, vinylferrocene is an exceptionally electron-rich monomer in free-radical copolymerizations. This can be further emphasized by looking at the values of $1/r_2$ in the following series of copolymerizations where M_2 is styrene:

Monomer 1	Fe (ferrocenyl vinyl)		CH_3O– (p-methoxystyrene)
$1/r_2$	~ 0.36		0.86

(styrene)		O_2N– (p-nitrostyrene)
1.0		5.26

Compared to a phenyl group, the ferrocene nucleus exhibits an extra-ordinary ability to stabilize an adjacent positive charge.[40-43] For example, σ^+ of p-ferrocene (determined from p-ferrocenylphenylethyl chloride solvolysis studies) is -0.71, and σ^+ for an α-ferrocenyl substituent is -1.4.[42] Under conditions of the weaker electron demand found in dissociation of p-ferrocenylphenylbenzoic acid the value of $\sigma_p = 0.18$[44] which is similar to that of a methyl group. Other available evidence suggests that a ferrocenyl substituent destabilizes radical anions[45] and radicals[46,47] in comparison with the effect of a phenyl ring.

The second striking feature in Table II is that the Q–e scheme appears to break down when electron-attracting comonomers are employed. This phenomenon is similar to a sudden change of slope in a Hammett σ–ρ plot. It could signify a change in mechanism. Since the Q–e scheme is defined in terms of the terminal copolymerization model,[8] the Q–e scheme no longer is appropriate to apply and has no meaning if another process is operating.

If a fraction of the monomer units entering the copolymer were due to a growing chain adding to a charge-transfer complex, this fraction of the polymer will always give a $1:1$ M_1/M_2 ratio:

$$\sim\sim M_2^{\cdot} + [M_1 \rightarrow M_2] \longrightarrow \sim\sim M_2 — M_1 — M_2^{\cdot}$$

If this were the only mode of propagation, an alternating copolymer would result. It can be seen that with acrylonitrile and maleic anhydride the alternation tendency is strong. However, a tendency toward alternation could also be due to a polar stabilization of the transition states for $^{\delta+}M_1^\bullet$---$^{\delta-}M_2$ and $^{\delta-}M_2^\bullet$---$^{\delta+}M_1$. This would lead to larger values of k_{12} (versus k_{11}) and k_{21} (versus k_{22}) within the framework of the terminal copolymerization model. If this were occurring without other complications, one might expect the $Q–e$ scheme to hold over the full range of comonomers employed. As can be seen in Table II, the $Q–e$ scheme does not hold when M_2 is electron deficient.

The copolymerizations of vinylferrocene with maleic anhydride[33] ($e = +2.25$), fumaronitrile[49] ($e_2 = +1.96$), and diethyl fumarate[49] ($e_2 = +1.25$), deserve special mention. Each gives 1:1 copolymers. At initial vinylferrocene to maleic anhydride ratios of 54:46 to 94:6 essentially alternating copolymers were formed.[33] The addition of $ZnCl_2$ to the media did not further affect copolymer composition. Both fumaronitrile and diethyl fumarate gave 1:1 alternating copolymers with vinylferrocene over the range of $M_1:M_2$ ratios from 0.1:0.9 to 0.9:0.1 in benzene.[49] The molecular weights were below $\overline{M}_n = 8000$ in every case. The rate was shown to be first order in AIBN in the fumaronitrile copolymerizations, again demonstrating a

monomolecular termination step was operating. The initial rates were studied as a function of $[M_1]:[M_2]$ and were found to maximize at a 1:1 ratio.[49] Some evidence was provided for the existence of vinylferrocene–fumaronitrile (or diethyl fumarate) charge transfer complexes from the uv–visible difference spectra which exhibited weak absorptions at 380 nm and 370 nm, respectively.[49]

The great importance of (a) using a nonlinear least-squares[35,36] fitting of r_1 and r_2 to experimental results, and (b) using the integrated form of the copolymer equation is illustrated in Table II for vinylferrocene–butadiene copolymerizations. George and Hayes[51] reported nine copolymerization experiments to conversions as high as 14%. Using the linear Fineman–Ross technique,[37] they calculated $r_1 = 0.3$ and $r_2 = 3.5$ from which was computed $e = -1.05$ for vinylferrocene. Applying the correct nonlinear approach and accounting for the composition drift by using the integral form of the copolymer equation, it is seen that $r_1 = 0.14$ and $r_2 = 3.97$ from which e for vinylferrocene is -1.80 (or -2.1 if -1.40 is used for e_2). It is clear that proper treatment of available data is quite important when dealing with the Q–e scheme. Furthermore, applying the optimized experimental design criteria described by Tidwell and Mortimer,[35,36] one sees that only two of the nine experiments in the vinylferrocene–butadiene copolymerizations performed by George and Hayes were done in the optimized regions of monomer feed ratios (i.e., $M_1^0/M_2^0 = 63/37$ and $92/8$).

From vinylferrocene's high negative value of e, one would predict it would polymerize readily with cationic initiators such as $Et_2AlCl \cdot t\text{-BuCl}$ or $BF_3 \cdot OEt_2$. This was found to be the case, but only low-molecular-weight polymers ($\bar{M}_n < 4000$) were formed.[24,25] In copolymerizations with styrene, catalyzed by $BF_3 \cdot OEt_2$ in CH_2Cl_2, styrene was incorporated only when the styrene content of the monomer feed exceeded 90%.[24] Clearly, vinylferrocene's reactivity is higher than that of styrene. Using an initial styrene/vinylferrocene ratio of $9:1$ only 58% styrene was found in the resulting copolymer.[24] In toluene, the percentage styrene in the feed had to exceed 80% to get a copolymer.

Vinylferrocene (M_1) copolymerizes with vinyl isobutyl ether in toluene at $0°$ in the presence of catalytic amounts of $BF_3 \cdot OEt_2$.[24] The reactivity ratios were $r_1 = 0.1$ and $r_2 = 9.7$ which lead to values of $Q_1 = -0.4$ and $e_1 = -1.3$, respectively. The cationic reactivity of vinylferrocene is less than that of vinyl isobutyl ether, which appears to be in conflict with the fact that solvolysis of ferrocenylmethyl chloride is 7.6 times faster than that of CH_3OCH_2Cl. The smaller reactivity of vinylferrocene was therefore attributed to steric crowding as a consequence of vinylferrocenes formed.

Solution copolymerizations of vinylferrocene and butadiene, catalyzed by metal acetylacetonates, gave copolymers with ratios of butadiene to vinylferrocene of 20–$25:1$.[53] They were brown viscous liquids or gums with broad molecular-weight distributions and both cis- and trans-1,4- and 1,2-butadiene repeat units. Free-radical copolymerizations with butadiene in 1,4-dioxane at $60°$ proceeds readily to give low molecular weight (5000–9000) gums.[51] These low molecular weights are in agreement with the high chain-transfer constant exhibited by vinylferrocene in dioxane ($C_m = 2 \times 10^{-3}$)[15]. The molecular weight distributions were broad and it was inferred from gel-permeation chromatography studies that significant branching had occurred.[51] A regular increase in the glass-transition temperature was observed as the mole fraction of vinylferrocene in the polymer increased. This fits the simple equation:

$$T_g \text{ (copolymer AB)} = (T_g A \times \text{mole fraction A}) + (T_g B \times \text{mole fraction B})$$

For copolymers containing 0.033 and 0.353 mole fractions of vinylferrocene, the values of T_g were $-74°$ and $18°$, respectively.[51] Linseed-oil films have been made which chemically incorporate vinylferrocene from 0.1 to 5% by weight.[54] The vinyl group apparently crosslinks into the film structure during air drying.

C. Isopropenylferrocene Polymerization

Isopropenylferrocene bears the same relation to vinylferrocene that α-methylstyrene does to styrene. α-Methylstyrene has a very low ceiling temperature, above which it cannot be homopolymerized. Thus, it is not surprising that all reported attempts to homopolymerize isopropenylferrocene have failed. Reed reported that no polymer was obtained in toluene using 0.1, 0.5, or 1.0% levels of either AIBN or benzoyl peroxide.[55] However, copolymerization did occur with styrene or methyl methacrylate.[55]

The rate of methyl methacrylate polymerization sharply decreased on the addition of small amounts of isopropenylferrocene.[55] Intrinsic viscosities of these polymers also dropped when isopropenylferrocene was added. Only small amounts of isopropenylferrocene entered the copolymers compared to the amount charged. A similar effect was seen in the isopropenylferrocene–styrene copolymerizations, although it was not as severe. Thus, isopropenylferrocene does resemble α-methylstyrene, which copolymerizes with styrene and methyl methacrylate under radical initiation.[56]

Isopropenylferrocene appears to be a monomer capable of reacting with radical intermediates, but one that produces a chain-end radical. This chain-

Copolymers

end radical either has a low reactivity or readily undergoes chain transfer with monomer. The large steric effects and the resonance interaction with the ferrocene moiety suggests that the radical would have a low reactivity. It would be interesting to see if free-radical initiation at low temperatures, with nonoxidizing radicals, would result in polymerization.

D. Acrylic Ferrocene Monomers

A series of acrylic ferrocene monomers have been prepared and their homo- and copolymerization behavior studied.[57-62] These include ferrocenylmethyl acrylate, FMA,[22,57-59,61] ferrocenylmethyl methacrylate, FMMA,[57-59,62] ferrocenylethyl acrylate, FEA,[59,60] and ferrocenylethyl methacrylate, FEMA.[59,60] The synthesis of each starts from the commercially available N,N-dimethylaminomethylferrocene; Scheme 2 summarizes these routes.

The kinetics of the benzene solution homopolymerizations of FMA and FMMA were studied in benzene.[57] Over a concentration range of 0.25 to 0.79 mol/liter of FMA and 8.4×10^{-3} to 3.54×10^{-2} mol/liter AIBN, the homopolymerization of FMA was first order in monomer and half order in initiator concentration. The activation energy was 18.7 kcal/mol. Similarly, the homopolymerization of FMMA in benzene was first order in monomer and half order in AIBN concentration. These results suggest that a normal bimolecular termination step, characteristic of most vinyl polymerizations, is occurring. However, the observed activation energy was 32.7 kcal/mol between 60° and 75°. Both of these monomers warrant more detailed kinetic studies in view of George et al.[15] on vinylferrocene polymerization kinetics.

Ferrocenylethyl acrylate and ferrocenylethyl methacrylate were homopolymerized in benzene, using AIBN initiation at temperatures from 60° to 90°.[59] In both cases, the rate was first order in monomer and half order in initiator concentrations. Thus, all four acrylic ferrocene monomers follow $r_p = k[M][AIBN]^{1/2}$ and terminate via bimolecular processes. The activation energies were 21.7 and 18.6 kcal/mol for the homopolymerizations of FEA and FEMA, respectively. The extra carbon between the acrylate function and the ferrocene ring make FEA and FEMA and their polymers more resistant to the facile S_N1 solvolysis reactions that FMA and FMMA can undergo. Solvolysis reactions of the latter, are promoted by the great stability of α-ferrocenyl carbonium ions.

High monomer concentrations result in higher polymer yields in FEA and FEMA solution homopolymerizations.[59] Sample molecular weights, yields, and conditions are listed in Table III. Using crudely fractionated samples, viscosity–molecular-weight studies led to the following values of K

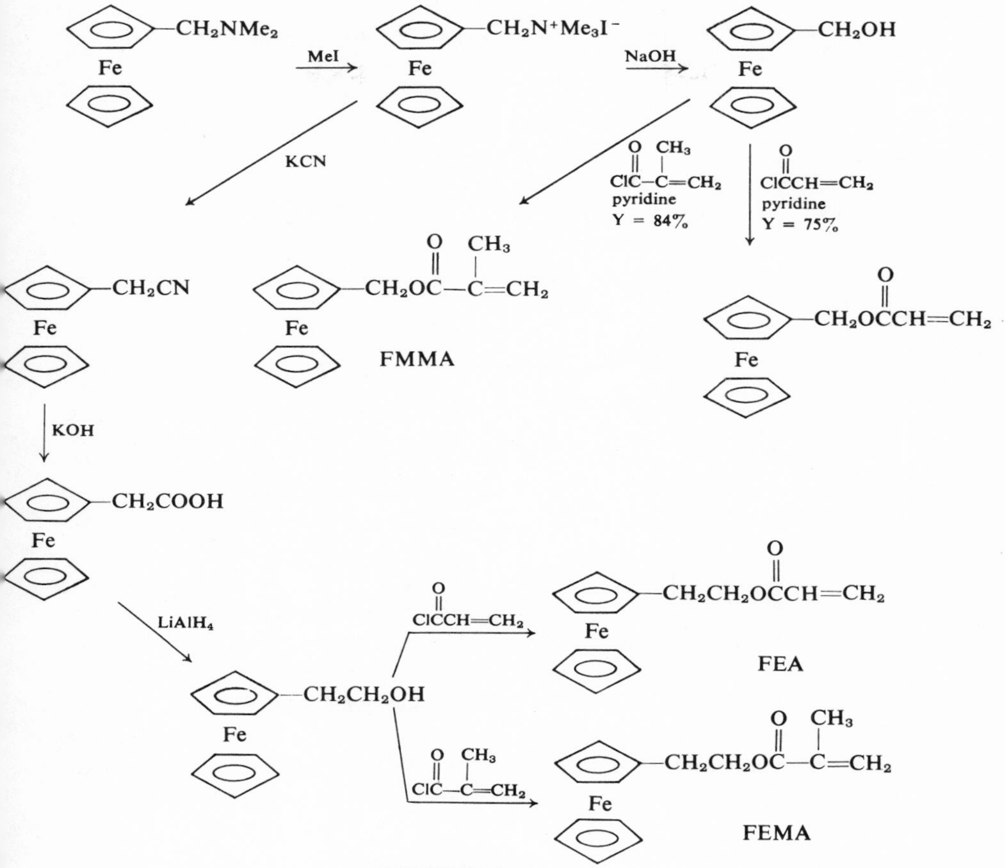

SCHEME 2

and a in the Mark–Houwink equation: $K = 4.7 \times 10^{-3}$ and $a = 0.70$ for poly(ferrocenylethyl acrylate) and $K = 3.12 \times 10^{-3}$ and $a = 0.76$ for poly(ferrocenylethyl methacrylate).[59] The polymers were yellow-orange and could be cast into brittle films. Tentative T_g values were assigned from differential scanning calorimetry (DSC) studies of these four acrylic monomers, but measurement of molar volumes is needed to confirm the assign-

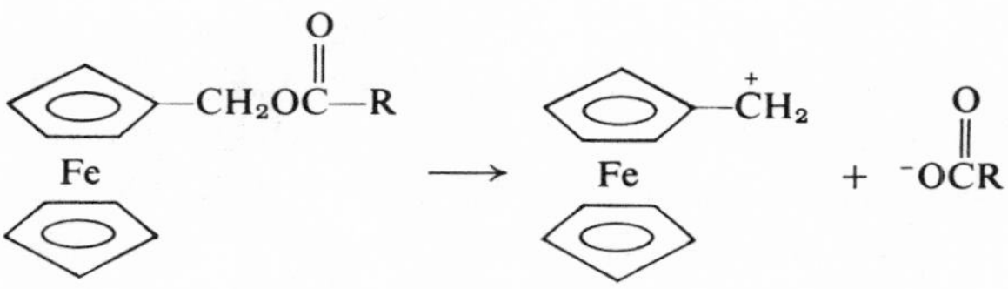

TABLE III

Sample Homopolymerizations of Ferrocenylethyl Acrylate and Ferrocenylethyl Methacrylate in Benzene Solutions

Monomer, g	+	Benzene, ml	Temp., °C	Time, hr	Yield, %	$\overline{M}_n$	$\overline{M}_w$
Ferrocenylethyl acrylate							
2.3		20	70	23	24	27×10^3	69×10^3
10.1		5	70	90	86	36×10^3	81×10^3
2.0		10	60	10	32	15×10^3	36×10^3
Ferrocenylethyl methacrylate							
2.4		20	70	8	41	83×10^3	29×10^4
2.1		10	70	6	74	64×10^3	17×10^4
2.4		20	75	8	55	35×10^3	11×10^4
8.0		6	80	12	89	24×10^3	65×10^3
5.0		10	56	12	81	40×10^3	12×10^4
3.0		10	70	12	87	21×10^3	50×10^3

ment. The T_g values are 197–210° poly(ferrocenylmethyl acrylate),[22] 185–195° poly(ferrocenylmethyl methacrylate),[57] 160° poly(ferrocenylethyl acrylate),[22] and 210–215° for poly(ferrocenylethyl methacrylate).[59]

Each of these ferrocene-containing acrylates was oxidized to Fe(II)Fe(III) mixed-oxidation-state polymers on treatment with oxidizing agents such as dichlorodicyanoquinone (DDQ) or orthochloranil.[22,57,59] The ready oxidation of ferrocene to ferrocenium took place as shown in the equation below:

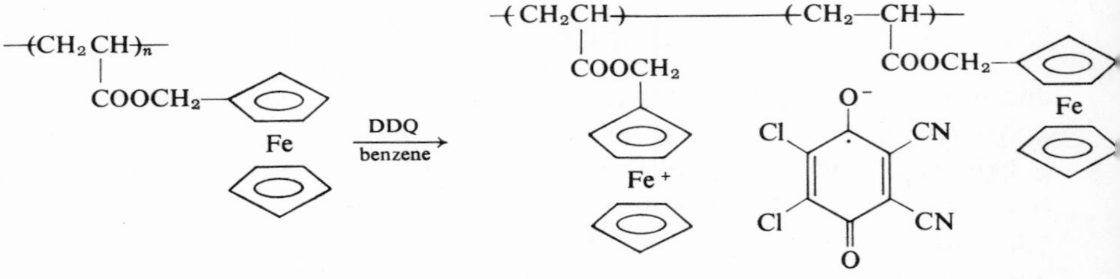

For each mole of oxidizing agent (i.e., DDQ) incorporated into the polymer, a ferrocene moiety was oxidized. This was shown conclusively by elemental analysis and Mössbauer and infrared spectroscopy.[22] For example, the strong carbonyl stretching frequency at 1680 cm^{-1} of DDQ was absent in the polysalts. In its place the $\nu_{C=O}$ at 1580–1600 cm^{-1} of DDQ$^{\div}$ was found. Thus, all the DDQ in the polymer was reduced to DDQ$^{\div}$. Two iron species were identified in the Mössbauer spectra of the polysalts.[22] Ferrocene groups exhibited a doublet centered at $IS = 0.78$ to 0.80 mm sec^{-1} (relative to nitroprusside) with a quadrapole splitting, Q, of 2.4 mm sec^{-1}. Ferrocenium units were broad singlets at $IS = 0.79$ mm sec^{-1}. The ratio of ferrocene to ferricenium was directly measured by comparing the areas of their corresponding Mössbauer peaks.[22] The ratios determined in this way agreed with the ratio calculated from elemental analysis where each DDQ was assumed to oxidize one ferrocene group (i.e., ir evidence).

The homopolymerization rate of ferrocenylethyl acrylate was much faster than that of ferrocenylmethyl acrylate.[59] Similarly, the polymerization rate of ferrocenylethyl methacrylate was greater than ferrocenylmethyl methacrylate under equivalent conditions.[59] In both cases, interposing one extra methylene unit between the vinyl center and the very bulky ferrocene substituent resulted in rate enhancement. The ferrocene moiety has a large three-dimensional volume and would be expected to sterically retard homopolymerizations as it is moved closer to the vinyl group. This trend also appeared in the copolymerization of these acrylates with styrene, methyl acrylate, methyl methacrylate, and vinyl acetate.[58,60,61] The relative reactivity ratios (r_1, r_2) for a large number of these copolymerizations are summarized in Table IV.

Using styrene as a standard comonomer (M_2), it is apparent that ferrocenylethyl acrylate ($r_1 = 0.41$, $r_2 = 1.06$)[60] is far more reactive than ferrocenylmethyl acrylate ($r_1 = 0.02$, $r_2 = 2.3$).[58] It is clear that a ferrocenylmethyl acrylate radical resists addition to more ferrocenylmethyl acrylate and that styrene radicals have a greater preference for styrene than for ferrocenylmethyl acrylate. These preferences are not as large for ferrocenylethyl methacrylate.

A comparison of the relative reactivity of ferrocenylmethyl and ferrocenylethyl methacrylates with methyl methacrylate (M_2) is also instructive. Ferrocenylmethyl methacrylate copolymerizations exhibit a very small r_1 (0.12) and a large r_2 (3.37)[58] showing that both ferrocenylmethyl methacrylate and methyl methacrylate radicals prefer to add to methyl methacrylate. On the other hand, in ferrocenylethyl methacrylate copolymerizations with methyl methacrylate, the reactivity ratios are $r_1 = 0.20$, $r_2 = 0.65$.[60] Thus, a methyl methacrylate radical actually adds more rapidly to ferrocenylethyl methacrylate than to methyl methacrylate. Positioning the bulky ferrocene one carbon further from the double bond leads to an increase in the copolymerization activity.

Analysis of the molecular-weight distributions of ferrocenylmethylacrylate and methacrylate by GPC and viscosity studies indicated that considerable branching occurred. In some cases, polymers with $\overline{M}_w \geqslant 90 \times 10^3$ gave intrinsic viscosity values below 0.25 dl g^{-1}. Plots of $|\eta|$ versus log (molecular weight) were not linear.[61] If the concentration of the ferrocene acrylate was high, insoluble polymers often resulted.[61]

TABLE IV

Summary of the Reactivity Ratios of Acrylic Monomers Which Contain Ferrocene or η^6-(Phenyl)tricarbonylchromium Groups

Monomer 1		Styrene	Methyl acrylate, M_2	Methyl methacrylate	Acrylonitrile	Vinyl acetate
(phenyl)Cr(CO)₃–CH₂OC(=O)–	r_1	60, EA[99] 0.10	70°, EA[99] 0.56			
	r_2	0.34	0.62			
(ferrocenyl)Fe–CH₂OC(=O)–	$r_1{}^a$	70°, Bz[58] 0.02 (0.013)	70°, Bz[58] 0.14 (1 × 10⁻⁴)	70°, Bz[58] 0.08 (0.065)		70°, Bz[58] 1.44 (1.68)
	r_2	2.3 (2.8)	4.46 (4.2)	2.9 (3.28)		0.46 (0.52)
(phenyl)Cr(CO)₃–CH₂CH₂OC(=O)–	$r_1{}^b$	70°, EA[100] 0.1	70°, EA[100] 0.3		70°, EA[100] 0.6	
	r_2	0.5	0.0		0.2	
(ferrocenyl)Fe–CH₂CH₂OC(=O)–	r_1	70°, Bz[60] 0.41	60°, Bz[60] 0.76			60°, Bz[60] 3.4
	r_2	1.06	0.69			0.07

Monomer		A	B	C	D	E
Methyl acrylate	r_1	0.18	1	1.5	1.22	9.00
	r_2	0.75	1	0.3	0.15	0.1
(ferrocenylmethyl acrylate: $CH_2OC(=O)\!-\!$, CH_3, $O=C$, Fe)	$r_1^{a,c}$	70°, Bz[58] 0.03 (4 × 10⁻⁶)	70°, Bz[58] 0.08 (0.03)	70°, Bz[58] 0.12 (0.13)	80°, Bz[61] 0.30	70°, Bz[58] 1.52 (1.36)
	r_2	3.7 (4.1)	0.82 (1.1)	3.37 (3.28)	0.11	0.20 (0.17)
(ferrocenylethyl acrylate: $CH_2CH_2OC(=O)\!-\!$, CH_3, $O=C$, Fe)	r_1	70°, Bz[60] 0.08		60°, Bz[60] 0.20		60°, Bz[60] 8.8
	r_2	0.58		0.65		0.06
(Cr(CO)₃ phenethyl acrylate: $CH_2CH_2OC(=O)\!-\!$, CH_3, $O=C$, Cr(CO)₃)	r_1	70°, EA[102] 0.04		70°, EA[102] 0.90	70°, EA[102] 0.07	
	r_2	1.35		1.19	0.79	
Methyl methacrylate	r_1	0.46		1		20.0
	r_2	0.52		1		0.15
(Fe(CO)₃ diene methacrylate: $CH_2OC(=O)\!-\!$, CH_3, $O=C$, Fe(CO)₃)	r_1	80°, Bz[103] 0.26	80°, Bz[103] 0.30		80°, Bz[103] 0.34	80°, Bz[103] 2.0
	r_2	1.81	0.74		0.74	0.05

[a] When M_2 = maleic anhydride with FMA, r_1 = 0.61, r_2 = 0.11[61] (80°, AIBN, Bz–EtOAc); with FMMA: r_1 = 0.28, r_2 = 0.10[61] (80°, Bz–EtOAc, AIBN).

[b] When M_2 = N-vinyl-2-pyrrolidone, r_1 = 3.71, r_2 = 0.05 (Bz–EtOAc, 80°, AIBN).[61]

[c] When M_2 = 2-phenylethyl acrylate, r_1 = 3.1, r_2 = 1.7 in EA at 70°.[100]

Note: EA = ethyl acetate, Bz = benzene.

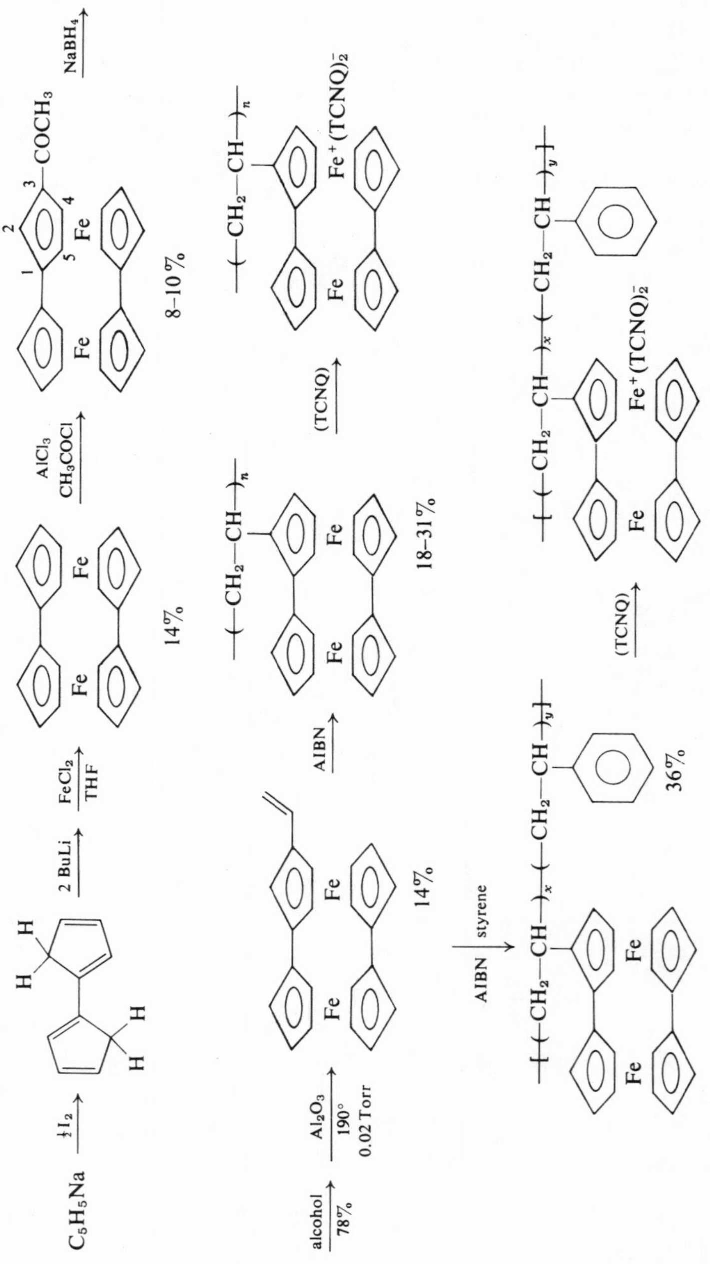
C_5H_5Na
$\frac{1}{2}I_2$
2 BuLi
$FeCl_2$
THF
14%
$AlCl_3$
CH_3COCl
$COCH_3$
Fe
8–10%
$NaBH_4$
alcohol
78%
Al_2O_3
190°
0.02 Torr
14%
AIBN
18–31%
(TCNQ)
$Fe^+(TCNQ)_2^-$
AIBN
styrene
36%

SCHEME 3

Pittman and Surynarayanan[62] reported the synthesis of 3-vinylbisful-valenediiron in 0.02% overall yield from sodium cyclopentadienide (Scheme 3). The bisfulvalenediiron group is a very large bulky function. Despite this bulk, homopolymerization was carried out in benzene using AIBN at 70°, although both the yields and molecular weights were low. Copolymerizations with styrene were also performed successfully. The homopolymer was oxidized with tetracyanoquinodimethane (TCNQ) to give a series of polymers containing the $(TCNQ)_2^-$ counterion for each monooxidized bisfulvalenediiron unit.[62] The dark conductivity of the homopolymer, in which 71% of the bisfulvalenediiron units were oxidized, was 6–9 × 10^{-3} ohm^{-1} cm^{-1}.[62]

E. Miscellaneous Ferrocene Monomers

Acryloylferrocene and *trans*-cinnamoylferrocene have been studied as potential monomers. Korshak *et al*.[63,64] found that acryloylferrocene was readily homopolymerized by radical and anionic initiators. For example, in the melt phase at 80° a homopolymer with a molecular weight of 30,000 was obtained using AIBN initiator. In benzene, AIBN initiation was reported to produce homopolymers with molecular weights of 5000–27,000. Free-radical initiation gives increasingly branched copolymers and an insoluble fraction. Lithium aluminum hydride in ether and butyllithium initiations were reported to give homopolymers with molecular weights of 18,000 and 6400, respectively. Acryloylferrocene was synthesized according to the method of Hauser.[65]

Acryloylferrocene was readily copolymerized with acrylonitrile in benzene at 80° using AIBN initiation.[63,64] Using a 1:1 ratio of monomers, a copolymer with a molecular weight of 30,000 and softening point of 230–280° was obtained. Using a 4.5:1 ratio excess of acrylonitrile gave a polymer softening at 270–300° and $\overline{DP} \simeq 75$.[21] Copolymers with styrene and methyl methacrylate have also been reported.[21]

Unlike acryloylferrocene, *trans*-cinnamoylferrocene (**5**) does not homo-polymerize.[66] It does copolymerize in bulk or solution with a variety of vinyl monomers. For example, Coleman and Rausch[66] reported copolymers of *trans*-cinnamoylferrocene with styrene, acrylonitrile, and methylmethacrylate. Cinnamoylferrocene entered the styrene copolymers readily but only a few percent was incorporated in the methyl methacrylate copolymers. Ethyl acrylate and 1,1-dihydroperfluorobutyl acrylate gave low conversions when copolymerized. *trans*-Cinnamoylferrocene did not copolymerize with maleic anhydride, vinyl acetate, or *n*-butyl vinyl ether. It gave hard rubbery co-polymers, but only low conversions, with butadiene and isoprene (softening points of 130–150°).[66]

(5)   (6)

1-Ferrocenyl-1,3-butadiene (**6**) gave low-molecular-weight oligomers ($\bar{M}_n = 1650$) in 37% conversion when homopolymerized in benzene at 70° using AIBN initiation.[66a] Kinetic studies showed the rate was first order in monomer and half order in initiator concentration.[67] At high initiator concentrations only dimeric products were obtained. Copolymerizations with styrene also gave low-molecular-weight polymers (mol wt = 1300–1950).[66a] Despite the low molecular weights, reactivity ratio values were reported ($r_1 \simeq 0.7$, $r_2 \simeq 2.3$).[66a]

Poly(1-ferrocenyl-1,3-butadiene) was successfully obtained by anionic initiation.[68] Using butyllithium in toluene, polymers with molecular weights in excess of 20,000 were obtained. By varying the monomer/initiator ratio from 4 to 52, it was shown that polydispersity increased as molecular weight increased.[68] Consistently high $\bar{M}_n$ values violated the number-average degree of polymerization relationship. These observations can be accounted for by assuming a fraction of the initiator is consumed by ring metallation which leads to a dead chain.

III. POLYMERIZATION AND COPOLYMERIZATION OF ETHYNYLFERROCENE

A surprising effort has been expended to obtain polyethynylferrocene. This monomer was first prepared[69] from acetylferrocene by a modified Vilsmeier reaction and base-catalyzed reaction of the resulting chlorovinyl derivatives.

Free radical, cationic, Ziegler, and anionic catalysts have been employed. Korshak and co-workers[70-73] conducted bulk polymerizations catalyzed by 10%-by-wt dispersed sodium which gave a 10% yield of polyethynylferrocene, $\overline{M}_n = 1700$. Using molten sodium at temperatures above 130°, molecular weights of 2500 were claimed. The same group also tried using $(Ph_3P)_2NiBr_2$ or cobalt carbonyls in THF but obtained brown powders which were probably cyclotrimers plus a poorly characterized insoluble polymer. Initiation by t-butylperoxide gave polymers claimed to have a ladder structure containing conjugated double bonds, but detailed characterization was lacking.

Soluble polyethynylferrocene ($\overline{M}_n = 1400$) resulted from initiation by dibutyl peroxide at 160°.[73] Benzoyl peroxide-initiated bulk polymerizations gave very short chains, DP = 4–8,[74] as did AIBN-initiated bulk polymerizations at 190°.[75] However, high yields (80%) of soluble polymers with molecular weights as high as 5000 could be obtained using AIBN.[75] The benzoyl peroxide-initiated polymerizations were first order in ethynylferrocene and exhibited an activation energy of 9.98 kcal mol.$^{-1}$ Simionescu[25] studied several initiator systems including benzoyl and lauroyl peroxide from 190–300°. Again, the degree of polymerization (4–6) was low, and increased yields were obtained at high initiator concentrations. Triisopropylboron gave $\overline{DP} = 13$ and some insoluble polymer while Al(iso-Bu)$_3$/V(acac)$_2$ in toluene at 50–90° gave an insoluble polymer which was not poly(ethynyl-ferrocene). Other initiators employed included (PPh$_3$)$_2$NiBr$_2$ and (PPh$_3$)$_2$Ni-(acac)$_2$ in CH$_3$Cl at 80°. Kunitake et al.[76] did not isolate polymers using AIBN initiation at 70° in bulk or in benzene solution. Ziegler catalysts of the type AlR$_3$–TiCl$_4$ gave predominantly cyclotrimers.[52] Polymers with low molecular weights (1000–3500) were achieved with AlR$_3$–Ti(OBu)$_4$ cata-lysts.[76] Sokolov found an Al(iso-Bu)$_3$/TiCl$_4$ (2:1) catalyst system, in benzene at 70°, polymerized ethynylferrocene. The resulting poly(ethynylferrocene) was stable in air to 200–250°.[77]

The exact structure of the polymers obtained in several of these studies is in doubt. Critical spectral studies were not reported in many cases. Kunitaki reported ir, nmr, and uv spectra which appear in accord with the completely conjugated structure.[76] The nmr showed only ferrocenyl and olefin protons; the uv absorption decreased continuously with increasing wavelengths. The ir spectra exhibited broad $\nu_{C=C}$ around 1600 cm^{-1} but a trace of aliphatic ν_{C-H} was observed at 2800–3000 cm^{-1}. These observations agree with those of Pittman et al. who managed to obtain some samples with no trace of aliphatic C–H absorptions.[75]

The effect of the completely conjugated backbone on electrical conduc-tivity had been one major reason for making poly(ethynylferrocene). Pittman and Sasaki[78] showed that highly purified poly(ferrocenylacetylene) was an insulator ($\sigma = 4.5 \times 10^{-14}$ ohm^{-1}cm^{-1}) with essentially the same bulk conductivity as polyvinylferrocene ($\sigma = 8 \times 10^{-15}$ ohm^{-1}cm^{-1}). Thus, extended conjugation is not effective in enhancing the conductivity of this polymer. Mixed-valence polymers were prepared by the partial oxidation of poly(ethynylferrocene) with agents such as iodine and dichlorodicyanoquin-one.[78] These partially oxidized polymers had far higher conductivities ($\sigma = 10^{-6}$–10^{-7} ohm^{-1}cm^{-1} when 40–60% of the ferrocene units had been oxidized to their corresponding ferricenium groups). The partial oxidation of poly(vinylferrocene) and poly(ferrocenylene) to Fe(II)Fe(III) polymers resulted in a similar insulator to semiconductor conversion.[78,79]

Paushkin et al. copolymerized ethynylferrocene with isoprene and obtained a rubbery material with 1.5–3.8% Fe, corresponding to 80–100 isoprene units per ferrocene moiety.[80] Copolymer molecular weights of 11×10^3 to 28×10^3 were reported. The Al(*iso*-Bu)$_3$/TiCl$_4$ catalyst system was used in toluene at 20° under nitrogen. In this system, cyclotrimerization of ethynylferrocene or isoprene homopolymerization can also occur, depending upon the conditions. The optimum conditions reported were 20% wt monomers, 10:1 isoprene–ethynylferrocene, catalyst concentration 50% based on total monomer, and an Al:Ti ratio of 2. A brief report of ethynylferrocene–chloroprene copolymerization has also appeared.[81] Thermal copolymerizations at 160–180° and pressure of 180–200 atm were performed on from 1:1

and 1:20 ratios of ethynylferrocene:chloroprene. Hydrochloric acid was liberated in each case, making definitive structural assignments difficult.[81]

The patent literature mentions the copolymerization of phenylacetylene with isobutene.[82] Thus, attempts to examine ethynylferrocene–isobutene copolymerizations are not without analogy. Paushkin[83] employed cationic initiation (BF$_3$) to make rubber-like copolymers. Best yields were obtained at high isobutene/ethynylferrocene ratios (i.e., 10:1 to 36:1). Molecular weights from 29×10^3 to 89×10^3 were reported. Values of K and a in the Mark–Houwink equation were obtained ($K = 1.52 \times 10^{-3}$, $a = 0.67$).[83] The copolymerization exhibited the normal features of a cationic mechanism. For example, the yield of polymer increased as temperature decreased giving a maximum at $-70°$. The yield decreased as the temperature was further lowered. The molecular weight increased almost linearly with a decrease in temperature from $-20°$ to $-100°$. As the proportion of ethynylferrocene in the feed increased, the yield of copolymer decreased.[83] The presence of ethynylferrocene slowed the reaction. Copolymerizations with other common organic monomers have not been discussed.

IV. POLYMERIZATION AND COPOLYMERIZATION OF VINYL ORGANOMETALLIC CARBONYL DERIVATIVES

Several metal carbonyl derivatives have now been successfully homo- and copolymerized. Monomers that have been studied include vinylcyclopentadienyltricarbonylmanganese, styrenetricarbonylchromium, η^6-(benzyl acrylate)tricarbonylchromium (BAC), η^6-(2-phenylethyl acrylate)tricarbonylchromium (PEAC), η^6-(2-phenylethyl methacrylate)tricarbonylchromium (PEMAC), η^4-(2,4-hexadiene-1-yl acrylate)tricarbonyliron (HATI), and η^4-(1,3,5-hexatrienyl)tricarbonyliron.

A. Vinylcyclopentadienyltricarbonylmanganese

The homopolymerization of vinylcyclopentadienyltricarbonylmanganese was first reported by Kozikowski and Cais.[84] Simple thermal polymerizations at 170° in bulk resulted in an orange, glassy, soluble, film-forming polymer with a softening point of 80°. Pittman *et al.*[50,85,86] demonstrated that homopolymerizations in benzene, initiated by AIBN, produced polymers having values of $\overline{M}_n$ as high as 25,000. They also performed a large number of copolymerizations with comonomers including styrene, methyl acrylate, acrylonitrile, vinyl acetate, vinylferrocene,[50] and N-vinyl-2-pyrrolidone.[32]

The copolymerization behavior of vinylcyclopentadienyltricarbonylmanganese provides the second classic example of assessing the vinyl reactivity of an organometallic monomer. The reactivity ratios for copolymerizations and the calculated e_1 values are summarized in Table V. The resulting pattern resembles that found for vinylferrocene. Vinylcyclopentadienyltricarbonylmanganese exhibits an e_1 value of -1.99 in copolymerizations with styrene (in benzene at 70°, AIBN initiation). Thus, like vinylferrocene, it is a very strongly electron-donating monomer. Furthermore, examining the Q–e scheme over the wide range of comonomers shows that large negative e_1 values are obtained when the comonomer's value of e_2 is more negative than zero. However, when electron-deficient comonomers are employed (such as acrylonitrile or methyl acrylate), sharp deviations occur in the value of e_1.

It appears that the normal terminal copolymerization model describes vinylcyclopentadienyltricarbonylmanganese copolymerizations with comonomers having e_2 values more negative than zero. However, when comono-

TABLE V

Solution Copolymerization of Vinylcyclopentadienyltricarbonylmanganese (M_1) at $70°$ [50]

Comonomer, M_2	e_2	Solvent	r_1	r_2	e_1	Molecular-weight range obtained	
						$\overline{M}_n$	$\overline{M}_w$
Vinylferrocene	−2.1	Benzene	0.44	0.49	−2.6	7,700	13,800
						9,100	15,200
N-Vinyl-2-pyrrolidone[32]	−0.90	Benzene	0.14	0.09	−2.9	5,300	7,500
						6,700	7,800
Styrene	−0.80	Benzene	0.10	2.5	−1.99	7,000	40,800
						14,500	86,000
Vinyl acetate	−0.22	Ethyl acetate	2.35	0.06	+1.62	6,000	20,500
						19,700	35,000
Methyl acrylate	+0.58	Benzene	0.19	0.47	−0.95	11,800	58,300
						14,100	76,800
Acrylonitrile	+1.20	Ethyl acetate	0.19	0.22	−0.59	12,000	22,300
						13,800	65,400

mers are electron-deficient, charge-transfer complexes are formed. These complexes begin to be incorporated into the polymer. Thus, a combination of the terminal and the charge-transfer model compete, and the Q–e scheme is no longer applicable. This interpretation is identical to that invoked for vinylferrocene.

Copolymers of cyclopentadienyltricarbonylmanganese with a wide range of M_1/M_2 compositions were made in these studies.[32,50] In general, the higher mole fraction of vinylcyclopentadienyltricarbonylmanganese in the polymer, the more brittle are the films of the copolymer. The copolymers were insulators and inhibited fungus growth in accelerated growth tests.[50] When the N-vinyl-2-pyrrolidone copolymers were heated to $210°$, bubble formation appeared.[32] Heating at $260°$ resulted in fairly rapid evolution of carbon monoxide and crosslinking occurred:

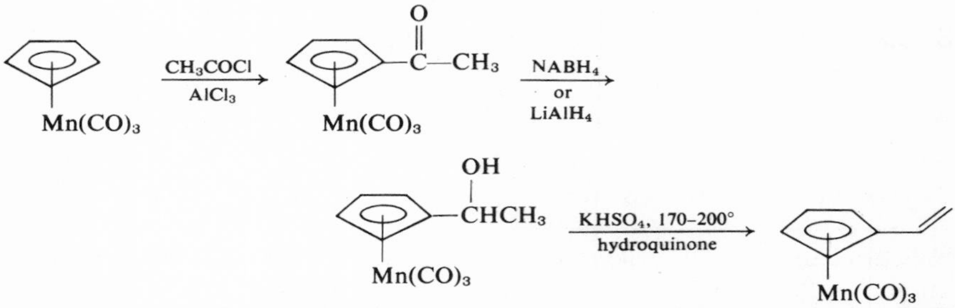

Vinylcyclopentadienyltricarbonylmanganese is prepared from cyclopentadienylmanganesetricarbonyl by acylation, hydride reduction of the carbonyl function, and dehydration.[50,84]

B. Monomers with η^6-Phenyltricarbonylchromium Groups

Styrenetricarbonylchromium was first prepared by Rausch et al. by the reaction of styrene with trisammoniatricarbonylchromium in dioxane in 50–65% yield.[86] Alternatively, benzaldehyde diethylacetal can be complexed in the same fashion and then hydrolyzed to its aldehyde. The aldehyde is converted to the monomer by a Wittig reaction[86]:

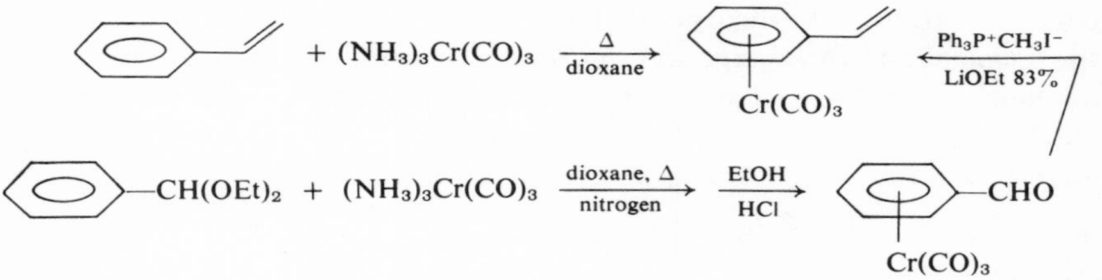

Rausch,[86] Pittman,[87] and Pauson[88] all showed that styrenetricarbonylchromium did not homopolymerize with radical initiators such as AIBN or benzoylperoxide. Since this monomer was unstable toward acid,[88] cationic polymerizations were precluded. No high-molecular-weight products were obtained with butyllithium.[88] Copolymerizations where M_2 is styrene, methyl acrylate, or vinylcyclopentadienyltricarbonylmanganese are readily carried out using AIBN in benzene at 70°.[87] In each case, the value of $r_1 = 0$. Thus a growing chain ending in a styrenetricarbonylchromium radical cannot add to another molecule of styrenetricarbonylchromium. This is consistent with the resistance of this monomer to homopolymerization. The reactivity ratios in copolymerizations with styrene (M_2) were $r_1 = 0$, $r_2 = 1.39$ and with methyl methacrylate (M_2) were $r_1 = 0$, $r_2 = 0.75$.[87]

The reason that a styrenetricarbonylchromium radical resists adding to another mole of this monomer is obscure. Steric arguments are difficult to justify since copolymerization is easily carried out with bulky vinylcyclopenta-dienyltricarbonylmanganese.[87] It is well known that η^6-complexed $Cr(CO)_3$ moieties have a strong electron-withdrawing inductive effect, along with the ability to act as strong electron-donating groups, upon centers of electron deficiency, by a direct contact or resonance interaction.[89-98] Therefore, it is possible that charge separation in the transition state occurs when the comonomer is electron rich, while with electron-deficient monomers, charge separation could exist in the reverse sense. In either case, a polar contribution could stabilize the transition state and promote the addition step.

Acrylates in the η^6-phenyltricarbonylchromium series are readily homo- and copolymerized.[99-101] η^6-(Benzyl acrylate)tricarbonylchromium, synthesized from benzyl alcohol and hexacarbonylchromium as outlined below, was homopolymerized in ethylacetate or THF using AIBN initiation at temperatures from 55–70°;[99] different yields (26–91%) have been obtained. A variety of molecular-weight ranges were prepared, further fractionated, and the viscosity–molecular-weight relation $|\eta| = KM^a$ determined in DMF.[99] The values of K and a were 3.95×10^{-3} and 0.82, respectively. Differential scanning calorimetry showed an endothermic reaction at 235–260° corresponding to the loss of carboxyl groups.[99] Unfortunately, the homopolymerization kinetics could not be obtained, due to bubble formation in the dilatometers employed.

η^6-(Benzyl acrylate)tricarbonylchromium (BAC) forms random co-polymers with styrene ($r_1 = 0.10$, $r_2 = 0.34$ where styrene is M_2) and with methyl acrylate ($r_1 = 0.56$, $r_2 = 0.63$) in ethyl acetate when initiated with AIBN.[99] Copolymerization of BAC with ferrocenylethyl acrylate takes place readily in ethyl acetate to give a polymer containing both iron and chromium.[99] The vinyl reactivity of BAC has been compared to that of other iron- and chromium-containing acrylates and methacrylates in Table IV. It can be seen that BAC is somewhat less reactive than ferrocenylethyl acrylate or methyl acrylate but more reactive than ferrocenylmethyl acrylate.

η^6-(2-Phenylethyl acrylate)tricarbonylchromium (PEAC) has been prepared, starting from 2-phenylethanol, by a route analogous to that used for BAC.[100] Several unsuccessful attempts to make the tricarbonylmolybdenum analog were reported.[100,101] The solution copolymerizations of (PEAC) (M_1) with styrene, methyl acrylate, acrylonitrile, and 2-phenylethyl acrylate have been studied in ethyl acetate at 55–85° using AIBN initiation. All these copolymers were soluble in acetone, ethyl acetate, and THF. Copolymers with large mole fractions of (PEAC) were insoluble in benzene.[100]

The molecular-weight distributions in (PEAC) copolymers were exceptionally broad. Gel permeation chromatography showed they were usually binodal with values of $\overline{M}_n$ above 10^6 for the high-molecular-weight node and $\overline{M}_n = 10^4$ to 6×10^4 for the low-molecular-weight node.[100] Undoubtedly, the high-molecular-weight nodes were highly branched. Since both r_1 and r_2 were greater than one in copolymerizations of 2-phenylethyl acrylate (PEAC) and a binodal product was obtained, fractionation studies were carried out to prove that homopolymerization of each monomer had not occurred. Analyses of both nodes showed they were copolymers of the same composition.[100] Copolymers of (PEAC) crosslinked on standing in solution for a few hours or on storage as solids for long periods.

Homopolymers of (PEAC), or its styrene copolymers, could be cast as yellow films on glass. Upon standing in room light or sunlight or upon uv irradiation with a high-pressure Hg lamp, the films turned light green. Infrared studies showed disappearance of the metal carbonyl stretching frequencies. Irradiation with wavelengths both above and below 320 nm caused this decomposition with the shorter wavelengths responsible for faster degradation. Loss of CO and formation of Cr_2O_3 took place in the films which became crosslinked in the process.[100]

Thermal decomposition of the films at 150° in the dark resulted in mixed chromium oxides imbedded in the crosslinked film.[100] This process represents a chemical method of forming polymers with metal oxides in very small particle sizes.

η^6-(2-Phenylethyl methacrylate)tricarbonylchromium[102] was prepared in a manner analogous to (BAC) and (PEAC) as outlined below. It homopolymerized readily (ethyl acetate, 70°, AIBN) to give homopolymers with values of $\overline{M}_n = 1.5 \times 10^4$.[102] Copolymerization with styrene, methyl methacrylate, and acrylonitrile was investigated in ethyl acetate at 70°.[102] Initiation with AIBN was used so that its vinyl reactivity could be compared to (BAC), (PEAC), and (FMA). The reactivity ratios of all these copolymerizations are summarized in Table IV.

Microgel formation and binodal soluble fractions were observed frequently in polymerizations carried to high conversion.[102] Exposure to light or heat (150°) resulted in decomposition of the η^6-phenyltricarbonylchromium moiety as described above.

It is of some interest to compare the vinyl reactivity of (BAC), (BEAC), and (PEMAC) with methyl acrylate, (FMA), and (FEA). This can be done by studying Table IV; (BAC) and (PEAC) are somewhat less reactive than methyl acrylate in copolymerizations with styrene, methyl acrylate, and acrylonitrile. For example, the values are $r_1 = 0.10, 0.1,$ and 0.18, respectively, when M_2 = styrene and M_1 = (BAC), (PEAC), or methyl acrylate. Ferrocenylmethyl acrylate is far less reactive. It exhibits an r_1 value of 0.02 when M_2 = styrene. In fact its r_1 values in copolymerizations with methyl acrylate (0.14), methyl methacrylate (0.08), and vinyl acetate (1.44) are all very small when compared to the corresponding r_1 values when M_1 = methyl acrylate (1, 1.5, and 9.0, respectively). Ferrocenylethyl acrylate is more reactive; (PEMAC) is a significantly less active methacrylate than methyl methacrylate as evidenced by its small r_1 values when M_2 = styrene (0.04), methyl methacrylate (0.09), and acrylonitrile (0.07).

C. η^4-(2,4-Hexadiene-1-yl acrylate)tricarbonyliron

η^4-(2,4-Hexadiene-1-yl acrylate)tricarbonyliron (HATI) is the only η^4-dienetricarbonyliron vinyl monomer which has been successfully polymerized. It has been synthesized by the reaction of 2,4-hexadienal with ironpentacarbonyl which gives the η^4-dientricarbonyl complex by displacement of two moles of carbon monoxide.[103] Sodium borohydride reduction gave the corresponding alcohol which is readily acylated by acryloyl chloride in pyridine. Broad molecular-weight distributions are obtained in solution

homopolymerizations of (HATI) carried out in benzene using AIBN as the initiator.[103] When high monomer concentrations were used, a very high-molecular-weight tailing is observed. This is due to chain transfer to polymer which gives a branched structure and, possibly, microgel formation.

η^4-(2,4-Hexadiene-1-yl acrylate)tricarbonyliron copolymerizes readily with styrene, acrylonitrile, vinyl acetate, and methyl acrylate in benzene at 80°.[103] The reactivity ratios were obtained where M_2 is styrene (r_1 = 0.26 r_2 = 1.81), methyl acrylate (r_1 = 0.30, r_2 = 0.74), acrylonitrile (r_1 = 0.34, r_2 = 0.74), and vinyl acetate (r_1 = 2.0, r_2 = 0.05).[103] These studies show that (HATI) is far more reactive than ferrocenylmethyl acrylate. Methyl acrylate

TABLE VI

Values of T_g and Molecular Weights for Sample Copolymers of
η^4-(2,4-Hexadiene-1-yl acrylate)tricarbonyliron

M_2	Mole % in polymer	$\overline{M}_n$	$\overline{M}_w$	T_g
Acrylonitrile	33.4	2.6×10^3	5.2×10^3	125
Acrylonitrile	85.5	7.1×10^3	19.7×10^3	158
Vinyl acetate	85.7	11.2×10^3	27.4×10^3	—
Vinyl acetate	19.3	9.8×10^3	46.1×10^3	—
Styrene	22.3	8.8×10^3	22.6×10^3	153
Styrene	81.4	9.3×10^3	28.0×10^3	174
Methyl acrylate	33.4	8×10^3	4×10^6	165
Methyl acrylate	90.4	8×10^3	24.5×10^3	178

copolymers also exhibited a broad high-molecular-weight tail indicative of branching.

The glass-transition temperature of the homopolymer was not established by DSC but it was observed for several copolymers as summarized in Table VI.

The thermal decomposition of two (HATI) copolymers, one with 66% acrylonitrile and the other with 88% styrene, was studied in bulk, in cumene, and diluted in KBr pellets.[103] At 200° under nitrogen an insoluble polymer was produced in each sample, and the iron carbonyl bands disappeared as the organometallic unit decomposed. The identity of the residual iron compounds was not elucidated. In air at 200°, Fe_2O_3 was deposited within the polymer as decomposition took place. The resulting polymers were highly crosslinked and completely insoluble in all solvents tested.

The η-^{4}(diene)tricarbonyliron functions, when attached to styrene polymers, can be converted in high yields to π-allyltetracarbonyliron cations in the presence of HBF_4 and CO.[104] These cation sites react readily with nucleophiles to give the 1,4-addition products of the diene moiety.

η^4-(1,3,5-Hexatrienyl)tricarbonyliron can be prepared, but it will not homopolymerize with radical or anionic initiators. Addition of small amounts of this compound to styrene inhibited styrene polymerization.

D. Titanium Acrylates

Novel acrylic monomers containing the titanocene nucleus have been prepared and polymerized. Korshak et al.[105] reported that allyloxytitanocene chloride was homopolymerized at 100° using benzoyl peroxide as the initiator. However, the molecular weight of this polymer was very low. A styrene copolymer ($\overline{M}_n = 70,000$) and a methyl methacrylate copolymer ($\overline{M}_n = 22,100$) were reported. However, the authors did not report analytical data and the $M_1 : M_2$ ratio of the copolymers was not given. The high copolymer molecular weights, relative to the homopolymers, suggest a low relative reactivity for the titanium monomer, possibly coupled with high chain-transfer activity.

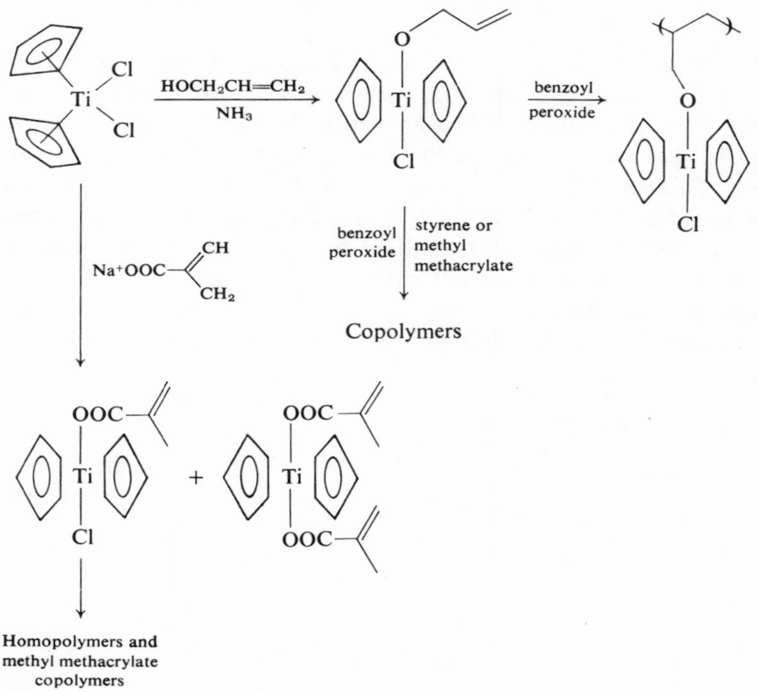

Copolymers

Homopolymers and
methyl methacrylate
copolymers

Methacryloyloxytitanocene chloride and dimethacryloyltitanocene were prepared from titanocene dichloride and sodium methacrylate.[106] Both monomers readily homopolymerized in DMF at 80° with benzoyl peroxide initiation. The dimethacrylate gave a crosslinked polymer.[106] Neither polymer exhibited impressive thermostability; *TGA* studies indicated decomposition below 220°. Solution copolymerizations with methyl acrylate gave polymers which were said to resemble plexi-glass.[106] However, detailed analytical data are not available.

V. POLYMERIZATION OF 1,1'-DISUBSTITUTED FERROCENES

The polymerization of 1,1'-divinylferrocene and 1,1'-diisopropenyl-ferrocene are unique examples of cyclopolymerizations that have now been well studied by Russian[107–109, 112, 114] and Japanese groups.[110, 111, 113] Knox

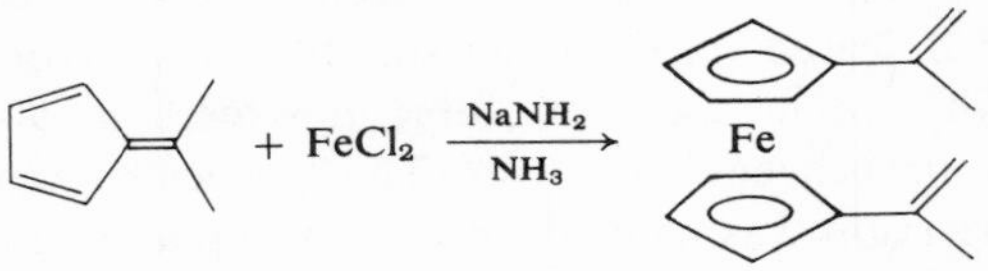

and Pauson[115] first suggested the possibility of preparing cyclolinear polymers from 1,1'-divinylferrocene. They also prepared 1,1'-diisopropenylferrocene from dimethylfulvalene and ferrous chloride in ammonia–sodium amide. Sosin *et al.*[108] employed a route from 1,1'-diacetylferrocene (shown above) which had been previously used by Riemschneider[116] and Schlögl.[117] 1,1'-Diisopropenylferrocene underwent radical initiated (benzene, AIBN, 85°)

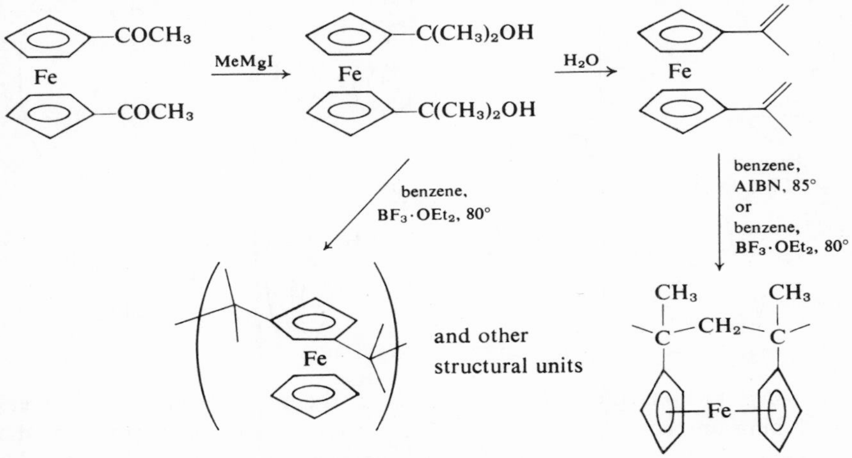

and cationic (benzene, $BF_3 \cdot OEt_2$, 80°) homopolymerization.[107] The yields and molecular weights reported were low in these radical-initiated systems (27%, 2–3 × 10³). Anionic initiation with $LiAlH_4$ or Al(iso-Bu)₃/$TiCl_4$ did not give polymers. Polymers from both radical and cationic reactions were assigned linear cyclopolymer structures.

Cationic initiation ($BF_3 \cdot OEt_2$) at 20° gave 80% yields, low molecular weights (6–7 × 10³), and a softening point of 200°. By raising the temperature to 80°, higher molecular weights (25–26 × 10³) and higher softening points (280–300°) were obtained.[107] A yellow polymer (mol wt = 25 × 10³) prepared by $BF_3 \cdot OEt_2$ initiation, which gave no esr signal, showed unusual conductivity fluctuations with temperatures. 1,1-Bis(α-hydroxyisopropyl)-ferrocene gave polymers of irregular structures (mol wt = 70–80 × 10³) when subjected to $BF_3 \cdot OEt_2$ in benzene.[107]

Sosin and Dzhashi[108] reported that 1,1'-divinylferrocene was also readily polymerized to partially soluble polymers with $BF_3 \cdot OEt_2$. The ready cationic polymerization of 1,1'-divinylferrocene and 1,1'-diisopropenylferrocene was explained, based on the great stability of α-ferrocenyl carbonium ions. 1,1'-Divinylferrocene gave insoluble polymers in 90% yields using AIBN initiation in benzene at 80°.[109] This suggested that normal vinyl addition polymerization was competing with cyclopolymerization. Kunitake et al.,[110] in a detailed study, showed that poly(1,1'-divinylferrocene) (mol wt = 10 × 10³) containing 82% cyclic and 18% linear units was formed using $BF_3 \cdot OEt_2$ at 0° in methylene chloride. The percentage of cyclic units dropped to 67 in toluene. Using AIBN at 60°, over 96% cyclic units and molecular weights of 21 × 10³ were obtained:

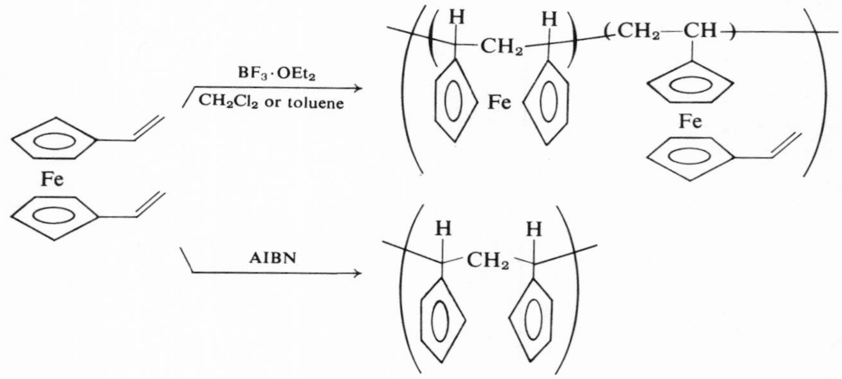

The vinyl protons are readily observed in the nmr spectra of polymers containing unreacted vinyl groups. This permitted ready assignment of the percentage of cyclic units. The infrared spectra of cyclic structures exhibited a splitting of the 820 cm^{-1} band (the out-of-plane deformation of the cyclopentadienyl-ring hydrogens) into 810 and 850 cm^{-1} bands.

Kunitake et al.,[110,111] found the rate of cationic polymerization of 1,1'-divinylferrocene was faster in CH_2Cl_2 than in toluene. The amount of benzene-soluble polymer increased as the monomer concentration decreased. The benzene-insoluble fraction was also insoluble in H_2SO_4, suggesting that some of the pendant vinyl groups had served as points of crosslinking.

The relative rates of cyclopolymerization to normal vinyl polymerization (k_1/k_2) was determined for a series of cationic initiators.[111] The initiators

were grouped into two classes. $BF_3 \cdot OEt_2$, $SnCl_4$, and $TiCl_4$ gave values of $k_1/k_2 \simeq 1$, while $AlCl_3$, $AlEtCl_2$, and $AlEt_2Cl$ gave values of 2–3. Intramolecular cyclization had a higher activation energy than intermolecular propagation. However, the frequency factor for cyclopolymerization was higher than that for intermolecular vinyl propagation by a factor of 68 when the monomer concentration was 0.5 ml^{-1} ($AlEtCl_2$, CH_2Cl_2). This frequency factor ratio for 1,1'-divinylferrocene is much lower than that found for ortho-divinylbenzene[118,119] which was 2–5 × 10^4. This difference is caused by the rotation of the cyclopentadienyl rings about iron. This rotation must be stopped in the cyclic transition state. Thus, that transition state is entropically far more unfavorable relative to the starting material 1,1'-divinylferrocene, than in the ortho-divinylbenzene case.

Solvent effects were studied in cationic polymerization of 1,1'-divinylbenzene.[111] Intramolecular cyclization of the growing chain became more favorable as the "tightness" of the ion pair decreased. As the solvent becomes more polar, solvent-separated ion pairs are formed, and the rate of cyclization increases relative to intermolecular vinyl addition (Table VII). This behavior is very similar to that observed in the cationic polymerization of ortho-divinylbenzene.[118,119] Some features of the polymerization mechanism are

TABLE VII

Solvent Effect on the Cyclopolymerization of 1,1'-Divinylferro-
cene Initiated by $BF_3 \cdot OEt_2{}^a$ 0°

Solvent	Dielectric constant	k_1/k_2
$C_6H_5CH_3$	2.4	0.67
$CHCl_3$	4.8	0.97
C_6H_5Cl	5.6	0.84
CH_2Cl_2	9.1	1.06
$C_2H_5NO_2$	28.1	1.13
CH_3CN	37.5	1.49

shown here. This mechanism does not completely represent the process since k_1/k_2 varied with initial monomer concentration. Direct ring alkylation may occur as one of the crosslinking reactions.

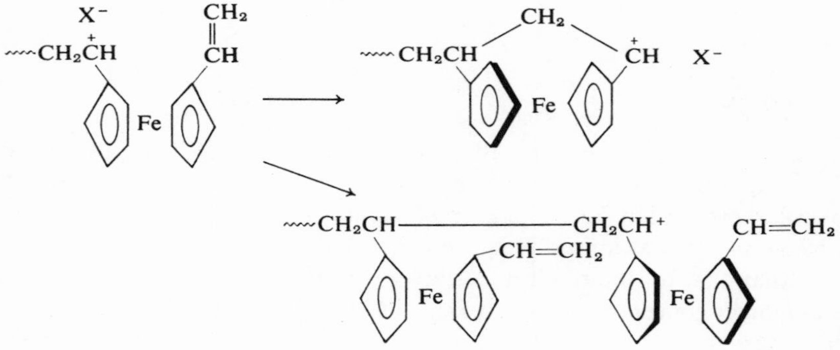

VI. OTHER MONOMERS

Organometallic acrylates and organometallic styrene derivatives have been made. For example, a wide variety of organotin acrylate homo- and copolymers have been prepared.[120-124] They are particularly promising as antifouling coatings.[125] However, these main-group metal polymers are outside the scope of this chapter. Similarly, a variety of vinyl and styryl lead derivatives have been prepared and incorporated into polymers. The prep- aration and polymerization of triphenyllead acrylate and methacrylate,[122,126] triethylvinyllead,[127] *p*-styrenyltriphenyllead[122,126,138,129], and *p*-styrenyltri-methyllead[130] have been described in great detail. Again, however these monomers of main-group metals will not be discussed here.

$$
\underset{\substack{| \\ \text{(vinyl-phenyl)} \\ PbPh_3}}{Bu_3SnOCC\!=\!CH_2} \overset{\displaystyle O}{\overset{\|}{}} \underset{CH_3}{\overset{\displaystyle |}{}}
\qquad
\underset{PbMe_3}{Ph_3PdOCCH\!=\!CH_2} \overset{\displaystyle O}{\overset{\|}{}}
\qquad
CH_2\!=\!CH\!-\!PbEt_3
$$

Fujita prepared five new palladium and platinum monomers containing a styryl group to include *trans*-Pd(PBu$_3$)$_2$-(C$_6$H$_4$CH$=$CH$_2$)X, where X = Cl, Br, CN, and Ph, and *trans*-Pt(PBu$_3$)$_2$-(C$_6$H$_4$CH$=$CH$_2$)Cl.[131] These monomers were readily homopolymerized in benzene using AIBN or BBu$_3$-O$_2$ as the initiator. Molecular weights from 3.6×10^3 to 11×10^3 resulted using AIBN while initiation with BBu$_3$-O$_2$ gave molecular weights as high as 20×10^{3}.[131]

Copolymerization studies with styrene have demonstrated that the Pd–C σ bond acts as a strongly electron-donating group to the phenyl ring[131]:

$$
M_1 \qquad + \qquad M_2 \qquad \xrightarrow[\substack{AIBN \\ r_1 = 0.45 \\ r_2 = 1.49}]{\text{benzene, 55°}} \qquad \left[\!\left(\!CH_2CH\!\right)\!-\!\left(\!CH_2CH\!\right)\!\right]
$$

Although only seven copolymerizations at different feed ratios were employed, and despite the fact that the differential form of the copolymer equation was used when yields of 6.5 to 21% were obtained, the values of $Q_1 = 0.41$ and $e_1 = -1.4$ did establish M$_1$ as an electron-rich monomer. If one applies the integral form of the copolymer in the correct nonlinear fashion, these same data lead to $e_1 = -1.62$. Thus, the Pd(PBu$_3$)$_2$Cl group is at least as strongly donating as a *p*-dimethylamino or a *p*-methoxy substituent.

Pd(PBu$_3$)$_2$-(C$_6$H$_4$CH$=$CH$_2$)Cl was synthesized by the reaction of two moles of *p*-styrylmagnesium bromide with bis(tributylphosphine)palladium dichloride to give Pd(PBu$_3$)$_2$-(C$_6$H$_4$CH$=$CH$_2$)$_2$, from which one mole of styrene was cleaved upon treatment with HCl at $-30°$ in ether. The platinum monomer was prepared by the same general route except that the HCl cleavage of styrene was performed at room temperature.

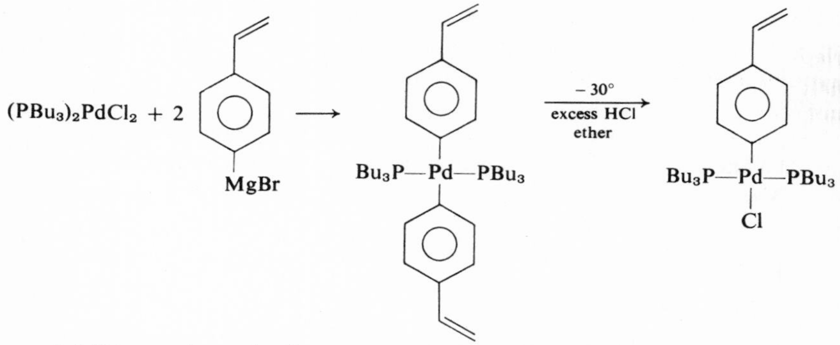

Addition polymerization or copolymerization of vinyl transition-metal compounds, where the vinyl group is directly attached to the metal, have not been reported. Such a study would be of great interest in assessing the effects of metals on the vinyl center. However, due to the great propensity for metal hydride elimination to occur, the resulting polymers would probably be quite unstable.

$$L_nM—CH=CH_2 \xrightarrow{\text{initiator}} \left(\begin{array}{c} -CH_2-CH- \\ | \\ ML_n \end{array}\right) \xrightarrow[\text{elimination}]{M—H} L_nMH + (CH=CH)$$

L = ligand

More studies of the electronic and steric effects on addition polymerization are needed for compounds with functional groups containing transition metals. Unfortunately, numerous examples do not exist. For comparison, vinyl monomers of the main group metals have been made with the vinyl group directly attached to the metal. Furthermore, they have been homopolymerized and copolymerized. From copolymerizations with styrene or methyl methacrylate the $Q-e$ scheme was applied for the series $CH_2=CH-R$, where $R = C(CH_3)_3$, $Si(CH_3)_3$, $Ge(CH_3)_3$, and $Sn(CH_3)_3$.[132-136] These results are summarized in Table VIII. The value of Q is low in every case,

TABLE VIII

$Q-e$ Vaules and Hammett Constants for the Monomer Series $CH_2=CHR$ (M_1)

R in CH=CHR	e_1	Q_1	σ_p	M_2	Ref.
$C(CH_3)_3$	-0.63	0.007	-0.197 ± 0.02	Styrene	132
$Si(CH_3)_3$	-0.14	0.035	-0.07 ± 0.01	Styrene	133, 134
	-0.10	0.031		Methyl methacrylate	133, 134
$Ge(CH_3)_3$	$+0.43$	0.005	0.0 ± 0.1	Styrene	135
	$+0.43$	0.037		Methyl methylacrylate	135
$Sn(CH_3)_3$	$+0.96$	0.005	0.0 ± 0.1	Styrene	136
	$+0.93$	0.036		Methyl methacrylate	136

indicating no great resonance stabilization is involved. Secondly, going from carbon to silicon, to germanium to tin there is a continuous and large change in e from negative to positive. This is in accord with a continuous increase in the electron deficiency at the vinyl group. It is curious, indeed, that the more electropositive metals appear to induce electron deficiency at the directly bonded vinyl group.

The radical-initiated polymerization of 1,1'-divinylferrocene gives mainly cyclopolymers containing crosslinks.[114] The overall activation energy was measured at 22.3 kcal mol^{-1}.[114] Macroradicals that were trapped within the crosslinked structure were claimed to account for the high observed rates.

The half-wave oxidation potential of poly(1,1'-divinylferrocene) has been measured in methylene chloride using $Bu_4N^+BF_4^-$ ($0.1M$) as the supporting electrolyte.[113] Bridged ferrocenes show lower oxidation potentials than ferrocene itself. Thus, it was not surprising that the oxidation potential of poly(1,1'-divinylferrocene), 0.43 V, was lower than that for polyvinylferrocene, 0.54 V, and ferrocene, 0.55 V. These results are summarized below.

VII. EXAMPLE MONOMER AND POLYMER SYNTHESES

A. Homopolymerization of Vinylferrocene

Radical Homopolymerization of Vinylferrocene.[16] Vinylferrocene (7.54 g), AIBN (0.074 g, recrystallized three times from absolute methanol, mp 102–103° dec), and benzene (21 ml, distilled from P_2O_5) were weighed into a glass tube. The reaction tube was then degassed at 10^{-3} mm by three alternate freeze–thaw cycles and placed in a constant-temperature bath at 70° for

96 hr. The tube was opened and the benzene solution was added dropwise to the petroleum ether with vigorous stirring. The precipitated polymer was reprecipitated two more times into methanol and the residual solvent was removed by heating at 70° (2 mm) for 24 hr. Polyvinylferrocene (5.24 g) with $\overline{M}_n = 9400$ was obtained in 69% yield. Using 8.20 g monomer, 0.083 g AIBN, and 7.10 ml benzene at 80° for 96 hr, 5.1 g polyvinylferrocene (68%) was obtained.

Cationic Homopolymerization of Vinylferrocene.[24] A Schlenk tube that had been thoroughly purged with nitrogen and cooled to 0° was charged with toluene (40 ml), a deoxygenated $2M$ toluene solution of vinylferrocene (50 ml), and 10 ml of a $0.5M$ toluene solution $BF_3 \cdot OEt_2$. The reaction solution was kept at 0° for 22 hr. Addition of methanol (10 ml) quenched the polymerization. The polymer solutions were then poured into methanol to give crude polymer which was collected by filtration. Purification was effected by two reprecipitations from benzene into methanol and drying *in vacuo*. A 52% conversion to polyvinylferrocene, $\overline{M}_n = 2200$, softening point 210° was obtained. When the same procedure was employed, using 10 ml of a 0.5 M solution of $Et_2AlCl \cdot t\text{-BuCl}$ in place of $BF_3 \cdot OEt_2$, a 28% conversion to polymer, $\overline{M}_n = 3500$, softening point 235°, resulted.

B. Copolymerization of Vinylferrocene

Copolymerization of Vinylferrocene with Styrene and Acrylonitrile.[12] Vinylferrocene was twice recrystallized from ethanol or triply sublimed (mp 50.5–51.5°). Styrene and acrylonitrile were washed with 10% NaOH, then with water, dried over anhydrous sodium sulfate, and then distilled. Only center cuts were used. Benzene was distilled from P_2O_5. Vinylferrocene (4.50 g), styrene (2.20 g), AIBN (0.007 g) and benzene (8 ml) were added to a glass tube. After three freeze–thaw cycles, it was held at constant temperature at 70° for 5 hr, removed and cooled, and the polymer solution was diluted with more benzene (25 ml). The polymer solution was added dropwise to excess (400 ml) petroleum ether to precipitate the copolymer. The polymer was then twice reprecipitated into petroleum ether and dried at 60° (2 mm) for 24 hr. The copolymer (0.45 g) contained 21.6 mol% vinylferrocene.

The same procedure was carried out using 5.00 g of vinylferrocene, 0.31 g of acrylonitrile, 0.005 g of AIBN, and 10 ml of benzene at 70° for 12 hr to give 0.37 g of copolymer containing 60 mol% vinylferrocene.

Copolymerization of Vinylferrocene with 1,3-Butadiene.[51] Vinylferrocene was purified by three successive sublimations at 30° and 0.1 mm (mp 53–53.5°).

Research-grade butadiene of 99.9% purity was distilled over potassium hydroxide and calcium hydroxide; 4,4'-azobis(4-cyanovaleric acid) was obtained commercially. 1,4-Dioxane was distilled at atmospheric pressure from sodium and then fractionally distilled from fresh sodium under reduced pressure.

Initiator, vinylferrocene, and solvent were weighed into heavy-walled glass vessels and degassed at $\approx 10^{-4}$ mbar by three freeze–thaw cycles. Butadiene was condensed into a second vessel, degassed in a similar manner and its volume measured at $-80°$. The amount of butadiene required to give a predetermined concentration (in mol) at the reaction temperature was distilled into the reaction vessel. The volumes of butadiene and vinylferrocene were found to be additive in solution and in good agreement with the volumes predicted from the densities of the two monomers. The density of vinyl-ferrocene as a function of temperature was determined. After separation from the manifold, the vessels were sheathed in glass cloth and placed in a water bath at $60 \pm 0.1°$ for 48 hr. The mole ratio of butadiene to vinylferrocene in the intial monomer feed was varied between 9:1 and 0.43:1. A total monomer concentration of 0.2 mol in 10 ml dioxane with a fixed initiator mole fraction of 0.4×10^{-2} was used in all reactions, based on the total monomer. The reactions were stopped by freezing in liquid nitrogen. Excess butadiene was allowed to evaporate and the contents of the vessels precipitated into methanol containing a trace of N-phenyl-1-naphthylamine. Dependent on the amount of vinylferrocene incorporated, the copolymers were either soft gums or tough elastomers and in all cases proved unsuitable for normal filtration techniques even at low temperature. Therefore, unreacted vinyl-ferrocene was removed by repeated washing in methanol, followed by decanting the supernatant liquid. The copolymers were then dissolved in benzene, reprecipitated into methanol and the washing process repeated. Final purification was achieved by Soxhlet extraction in methanol.

C. Preparation of Isopropenylferrocene[55]

Isopropenylferrocene was prepared from acetylferrocene by first reacting with methyl Grignard reagent followed by dehydration of the resulting 2-hydroxy-2-ferrocenylpropane.

2-Hydroxy-2-ferrocenylpropane. Acetylferrocene (0.44 mol) was placed in a predried glass reactor; 750 ml of THF and methyl Grignard reagent (0.75 mol) were added under a nitrogen atmosphere at $0–5°$ over a period of 1.15 hr. The mixture was stirred for approximately 16 hr. After adding 200 ml of ether, the mixture was treated with an ammonium hydroxide solution saturated with ammonium chloride until reflux stopped. After filtration, the filtrate was washed twice with water, dried over anhydrous magnesium sulfate, and the solvents removed by evaporation. A red oil which crystallized on standing was obtained in 95% yield.

Isopropenylferrocene. 2-Hydroxy-2-ferrocenylpropane (0.41 mol) was dissolved in 2.5 liters of dry methylene chloride containing 21 meq of *p*-toluenesulfonic acid and 9.1 meq of *N*-phenyl-2-naphthylamine. The mixture was heated to reflux under nitrogen for 2.5 hr with the water eliminated as the azeotrope. Upon completion, 10 ml of triethylamine was added to neutralize the acid and the solvent was completely removed by evaporation at low ambient temperature (0–10°). The oily residue was extracted several times with pentane, the extracts combined, methylene chloride added to give a 30:1 pentane–methylene chloride mixture, and this solution passed through a silica-gel column (5 cm i.d., 30 cm. high). One liter of the 30:1 solvent mixture was used to elucidate the column. Removal of the solvents gave a yellow solid (64.3%) which was recrystallized from methanol–water, mp 66.5–67°.

D. Synthesis and Homopolymerization of Ferrocenylmethyl Acrylate

Synthesis of Ferrocenylmethyl Acrylate.[22] Since this monomer is easily solvolyzed in methanol–water solvents, it should be used only in dry solvents. A 1-liter three-neck, round-bottom flask, equipped with a pressure-equalizing funnel, a condenser, and a mechanical stirrer, is placed in an ice bath. The flask is charged with 48.6 g (0.2 mol) of *N,N*-dimethylaminomethylferrocene in 40 ml of absolute methanol, and a solution of 45.2 g (0.3 mol) of iodomethane in 40 ml of absolute methanol is added dropwise through the pressure-equalizing funnel. The clear solution is heated under reflux for 5 min., and 500 ml of diethyl ether is then added. The quaternary salt immediately precipitates and is collected by filtration on a Buchner funnel, washed with ether until colorless, and dried to yield 73 g (95%) of the methiodide as yellow crystals which decompose slowly on heating to 220°.

A 2-liter, one-necked flask, equipped with a condenser, is charged with 200 g (0.52 mol) of the methiodide salt and 700 ml of 3-*N* NaOH. The mixture is heated under reflux for 6 hr with evolution of trimethylamine from the top of the condenser. After the mixture has been allowed to cool, the resulting oily material is extracted into 500 ml of diethyl ether. The ether solution is washed with distilled water until the washings are neutral to litmus and dried over anhydrous sodium sulfate; the mixture is then filtered. The solvent is removed from the filtrate by rotary evaporation and the residual powder is recrystallized from hexane to give 101 g (90%) of hydroxymethylferrocene as yellow plates, mp 81–82°.

Hydroxymethylferrocene is esterified either with acrylyl chloride and pyridine, or directly with acrylic acid (described below).

A 500-ml, three-necked flask, equipped with a condenser, a nitrogen inlet, and a mechanical stirrer, is charged with 10.8 g (0.05 mol) of hydroxy-methylferrocene, 18.0 g (0.25 mol) of unpurified commercial acrylic acid, 0.1 g of hydroquinone, 0.05 g of p-toluenesulfonic acid, and 300 ml of methylene chloride. The mixture is heated under reflux for 5.0 hr and then cooled. The solution is filtered and the filtrate is washed with 250 ml of a 20% aqueous sodium carbonate solution at 0°, then several times with distilled water, and dried over anhydrous sodium sulfate. The mixture is then filtered and the filtrate evaporated to dryness. The residual powder is recrystallized from hexane to give 11.4 g (85%) of ferrocenylmethyl acrylate (mp 42–43°).

Homopolymerization of Ferrocenylmethyl Acrylate.[57] Ferrocenylmethyl acrylate (15 g, 0.088 mol) prepared shortly before use and stored at $-15°$ in the dark, is weighed into a Fischer-Porter aerosol compatibility tube. Reagent-grade benzene (20 ml) distilled over phosphorus pentoxide shortly before use, and 0.1 g (6.1 × 10^{-4} mol) of recrystallized azobisisobutyronitrile are added. The tube is then equipped with its valve, and the solution is degassed at 10^{-3} mm by three freeze–thaw cycles. After the degassing is complete, the tube is placed in a constant-temperature water bath at 60° (% 0.1) for 120 hr. The tube is then removed, cooled, and the polymer precipitated on the addition of the benzene solution to 800 ml of 30–60° petroleum ether with strong stirring. The polymer is collected by filtration, redissolved in benzene, and reprecipitated two more times. The solvent is removed under vacuum to g'. *:* 7.0 to 9.3 g (47–62%) of poly(ferrocenylmethyl acrylate). The molecular weight was $\overline{M}_n = 17,300$, $\overline{M}_w = 36,800$. This method can also be used to produce polymers of ferrocenylmethyl methacrylate.

E. Synthesis and Homopolymerization of 1-Ferrocenyl-1,3-butadiene

Synthesis of 1-Ferrocenyl-1,3-butadiene.[66] To 600 ml of dry benzene in a 2-liter flask was added allyltriphenylphosphonium bromide (92.4 g, 0.24 mol). After fitting with a condenser, rubber septum, and a nitrogen inlet tube, the flask was flushed with dry, oxygen-free nitrogen for 15 min. Then a 2.2-M hexane solution of butyllithium (109 ml, 0.24 mol) was added and the mixture was stirred mechanically for 1.5 hr. A deep-red color appeared upon the addition of the n-butyllithium indicating -ylid formation. Formylferrocene (32.8 g, 0.152 mol) in dry benzene (100 ml) was then added dropwise over 30 min. Stirring was continued for 20 hr. The solution was filtered and the solvent of the filtrate was removed *in vacuo* at below room temperature. The resulting thick viscous liquid was eluted from a silica gel column (2m × 2.5 cm) using benzene. The first compound to be eluted was 1-ferrocenyl-1,3-butadiene which was recrystallized from hot methanol to give 23.8 g (65% yield), mp 80°.

Anionic Polymerization of 1-Ferrocenyl-1,3-Butadiene.[68] A magnetic stirring bar and 1-ferrocenyl-1,3-butadiene (9.3 g) was placed into a 250-ml round-bottom flask fitted with a side delivery tube (septum covered). A vacuum manifold was employed to vapor transfer anhydrous toluene (50 ml) into the polymerization flask. The required quantity of *n*-butyllithium was then injected into the rapidly stirring monomer solution. Mole ratios of 4.3:1 to 52:1 of monomer to initiator were employed. The polymerizations were allowed to continue for 16 hr, followed by quenching in methanol. The crude polymer was collected and subjected to analysis by gel permeation chromatography (gpc) to determine conversion. Finally the powdered polymer was washed with hexane to remove any residual monomer. Polymers were dried *in vacuo* at 60° for 24 hr. With a monomer:initiator ratio of 13 and 52, the values of $\overline{M}_n$ were 8600 and 40,000, respectively.

F. Synthesis and Polymerization of Ethynylferrocene

Synthesis of Ethynylferrocene.[69] A solution of 5 ml of phosphorus oxychloride (55 mmol) in 20 ml of dimethylformamide, prepared at 0°, was added dropwise to a stirred solution of 5.7 g of acetylferrocene (25 mmol) in 20 ml of dimethylformamide, kept in a nitrogen atmosphere, and cooled to 0°. After 15 min the deep-green solution turned purple. The reaction was allowed to proceed at 0° for 2 hr and continued at room temperature for 1 hr; then the solution was poured into 150 ml of 20% sodium acetate solution and stirred in a nitrogen atmosphere for 1.5 hr. This solution was extracted several times with methylene chloride, and the combined organic extract was, in turn, washed with water to remove dimethylformamide, dried over magnesium sulfate, and concentrated to a red oil. This was immediately placed onto a chromatographic column prepared from 500 g of alumina (treated with ethyl acetate). Elution with Skellysolve B gave two major bands. The first afforded 3.5 g of (1-chlorovinyl)ferrocene (59% yield). Rapid recrystallization from ether–pentane afforded the substance as yellow plates, mp 43–44.5°, which decomposed slowly at 0°, evolving HCl. The second band, on elution with 20% ether in Skellysolve B, and recrystallization of the product from the same solvent mixture gave 1.17 g of (2-formyl-1-chlorovinyl)ferrocene, mp 76–78°, as red prisms (17% yield).

To (1-chlorovinyl)ferrocene (2.55 g, 10.3 mmol) in 15 ml of dimethyl-formamide, cooled to 0° in a nitrogen atmosphere, was added 5.0 ml (55 mmol) of phosphorus oxychloride, dissolved in 15 ml of the same solvent. After addition was complete, the icebath was removed and the deep-purple solution was stirred for 1 hr. The reaction was then quenched by the addition of 50 ml of an aqueous 35% solution of sodium acetate, and stirring was continued in a nitrogen atmosphere for 12 hr. The aqueous solution was then extracted several times with methylene chloride and the combined organic extract was washed with water and dried over magnesium sulfate. The organic solution was then filtered twice through alumina and the solvent was allowed to evaporate leaving 2.05 g (73%) of (2-formyl-1-chlorovinyl)ferrocene.

(2-Formyl-1-chlorovinyl)ferrocene (476 mg, 1.73 mmol) was dissolved in 15 ml of dioxane and the solution was heated in a nitrogen atmosphere to reflux. To this solution was added rapidly 10 ml of 0.5 N NaOH solution. Reaction was allowed to continue for 5 min and the solution was then poured into 5 ml of cold water. After acidification with HCl, the solution was extracted exhaustively with ether, and the combined ether extract was washed to neutrality and dried over magnesium sulfate. Solvent was removed and the residue was chromatographed on 25 g of Fisher alumina (activity 2) washed with ethyl acetate. Elution with Skellysolve B led to the development of two bands. The first yellow band afforded 305 mg (88%) of ferrocenyl-acetylene, mp 51–53.5°.

Bulk Homopolymerization of Ethynylferrocene Initiated by Triisopropyl-borane.[25] Ethynylferrocene (5 g) and triisopropylborane were weighed into a glass ampoule which was sealed under an atmosphere of air. It was then heated at the desired temperature in an oven for 8 hr and cooled; the ampoule was opened, and the resulting polymer was extracted with excess boiling benzene (or some other desired solvent). Poly(ethynylferrocene) could be purified by reprecipitations with methanol or petroleum ether. Using 5.3 wt% triisopropylborane at 190° gave a 26% conversion to benzene-soluble polymer where $\overline{DP} = 7$. Using a 2.7 wt% triisopropylborane at 230° gave 30% conversion to benzene-soluble polymer, $\overline{DP} = 13$. Using 3 to 12 wt% initiator at 300° gave 70–80% conversion to an insoluble product.

Copolymerization of Ethynylferrocene with Isobutylene.[83] The copolymeri-zation was carried out in a glass reactor, using a methylene chloride solution with boron trifluoride as catalyst. A mixture of ethanol and liquid nitrogen was used for externally cooling the reaction flask. Into the reactor, cooled to −40°, was placed a solution of 5 g of ferrocenylacetylene (24 mmol) and 60 ml of methylene chloride and 15 g (268 mmol) of isobutylene. The mixture was cooled to the required temperature with constant stirring and gaseous anhydrous BF_3 was passed in for the prescribed time. After discontinuing the BF_3 supply, the reaction mixture was stirred for several hours; then 50 ml of cold absolute diethyl ether was added to bind the BF_3 and the temperature of the reaction mixture gradually rose to room temperature. The solvent was distilled from the reaction mixture at a pressure of 60 mm without heating. The residue was dissolved in xylene, washed repeatedly with a 2–3% aqueous solution of ascorbic acid (in order to reduce the ferricinium cation which was formed) and, then with water. The copolymer was isolated from the xylene solution by precipitation in methanol (1:5), purified by reprecipitation four times and then dried in air and in vacuum at room temperature. The copolymers of isobutylene with ferrocenylacetylene were resins or rubber-like products with an amber to a dark brown color.

G. Synthesis and Polymerization of Vinylcyclopentadienyltricarbonylmanganese

Synthesis of Vinylcyclopentadienylcarbonylmanganese.[50,84] Cyclopenta-dienyltricarbonylmanganese (Ethyl Corporation) (49 g, 0.24 mol) and acetyl chloride (27.5 g, 0.35 mol) in carbon disulfide were added slowly, from separate addition funnels, to a stirred solution of aluminum chloride (42.5 g) in carbon disulfide at room temperature. The solution was refluxed 1 hr, cooled to 0°, and quenched by addition of 500 g of ice and 100 ml of aqueous HCl. The organic layer was separated, washed three times with 10% HCl, then three times with distilled water, and dried over anhydrous sodium sulfate. The solution was concentrated and yellow crystals of acetylcyclopenta-dienyltricarbonylmanganese (50 g, 85%, mp 41–42°) formed as the solvent evaporated.

The acetyl derivative (78.4 g, 0.32 mol) was added to a solution of benzene (350 ml) and ethanol (250 ml). The solution was cooled to 0° and 24.2 g (0.64 mol) of sodium borohydride was slowly added with rapid stirring. After 12 hr at 20–25°, the solution was cooled to 0–5° and hydrolyzed with 400 ml of saturated aqueous NH_4Cl. After separating the organic layer, filtering, washing, and drying (over anhydrous sodium sulfate), solvent was removed *in vacuo* giving crude (α-hydroxyethyl)cyclopentadienyltricarbonyl-manganese (64.6 g, 81.4%).

The crude alcohol (25 g, 0.10 ml), $KHSO_4$ (1.0 g, 8 mmol), and a trace of hydroquinone were added to a vacuum distillation flask. Under a nitrogen purge the contents were heated at 190–200° for 10 min. The reaction mixture was cooled, the pressure reduced to 0.01 mm (or below) and vinylcyclopenta-dienyltricarbonylmanganese distilled. A center cut gave 16.1 g (70%) of this viscous yellow liquid (bp 64–65° at 0.01 mm). Yields vary between 65 and 80% and some dimer ether is also isolated.

Copolymerization of Vinylcyclopentadienyltricarbonylmanganese and Acrylonitrile.[50] Copolymerizations were carried out in closed glass tubes under nitrogen. The tube was charged with the manganese monomer (2.00 g), acrylonitrile (0.83 g, $M_1:M_2 = 30:70$), AIBN (0.0231 g) and ethyl acetate (20 ml). After three freeze–thaw deoxygenation cycles, the temperature was held at 70° for 53.5 hr cooled, diluted with ethyl acetate (100 ml), and filtered. The polymer was then precipitated in petroleum ether and purified by two additional reprecipitation cycles. The copolymer was dried for 40 hr *in vacuo* to give 1.29 g (41.8%) of polymer with 63.4 mol% acrylonitrile.

H. Synthesis and Copolymerization of Styrenetricarbonylchromium

Synthesis of Styrenetricarbonylchromium.[86] Triaminetricarbonylchromium was prepared from hexacarbonylchromium. It was then treated with styrene to give styrenetricarbonylchromium.

Hexacarbonylchromium (4.5 g, 20.4 mmol) and a solution of 7.75 g of potassium hydroxide in 120 ml of 95% ethanol were placed in a 200-ml pressure bottle. The bottle was filled with nitrogen and capped with a mechanical capper. The pressure bottle was wrapped in cloth and placed in a steam bath. After 6 hr, the bottle was cooled in ice and opened carefully. The orange solution was poured under nitrogen into a 1-liter flask. The bottle was washed with 30 ml of concentrated ammonium hydroxide and the washings were added to the flask. Subsequently, 120 ml of ammonium hydroxide was added to the flask and the mixture stirred for 1 hr under a nitrogen atmosphere. During this period the solution turned bright yellow. The yellow crystals were filtered under nitrogen using a glass frit (dec 100–105°, M_1) The yield of triaminetricarbonylchromium was 3.52 g (93%). Triaminetricarbonylchromium can be conveniently weighed in air, but is best stored for extended periods under nitrogen.

Into a 100-ml round-bottom flask, fitted with a nitrogen inlet tube, a magnetic stirring bar, and a condenser equipped with a mercury check valve, were placed 100 g (5.35 mmol) of triaminetricarbonylchromium, 0.50 ml (0.46 g, 4.4 mmol) of styrene, and 25 ml of dioxane. The reaction mixture was refluxed for 2.5 hr and then cooled in an ice bath. During the cooling period, a slight positive flow of nitrogen was maintained. The cooled reaction mixture was filtered under nitrogen through a glass frit (25–50 μm) to give an orange solution. The solvent was removed under reduced pressure and the remaining solid was sublimed (65° at 0.05 mm) in a water-cooled, cold finger sublimer to yield 0.655 g (62%) of orange crystals, mp 78–79°. An analytical sample of mp 80–81° was prepared by recrystallization of the product from hexane under nitrogen.

Copolymerization of Styrenetricarbonylchromium with Styrene.[87] Styrene-tricarbonylchromium (0.30 g, 1.25 mmol), styrene (0.13 g, 1.25 mmol), AIBN (0.10 g), and 10 ml of dry distilled benzene were placed in a glass tube. The solution was degassed by three freeze–thaw cycles and sealed. The polymerization was carried out at 70° for 24 hr. The reaction vessel was cooled, opened, and the solution was diluted with excess benzene. The resulting diluted polymer solution was precipitated into petroleum ether (1.6 liter). After washing with additional petroleum ether (2 liter), the polymer was dissolved through filter paper with dry benzene, evaporated to dryness, and dried *in vacuo* (100 μm) for 24 hr. The yield was 39%; mol% styrene = 79. When a reaction time of 96 hr was used, the yield was 49%, mol% styrene = 77.

I. Synthesis and Polymerization of Acrylic Monomers Containing η^6-Aryltricarbonylchromium

Synthesis of η^6-(2-Phenylethyl acrylate)tricarbonylchromium.[100] η^6-(2-Phenylethanol)tricarbonylchromium was prepared by the direct reaction of hexacarbonylchromium with 2-phenylethanol in a Strohmeier reactor and then conversion to the acrylate. A large Strohmeier reactor was flushed with dry nitrogen and charged with hexacarbonylchromium (50 g, 0.277 mol), 2-phenylethanol (278 g, 2.29 mol), and 1,2-dimethoxyethane (150 g, 1.67 mol). The 2-phenylethanol was dried over anhydrous calcium chloride and vacuum distilled before use. The 1,2-dimethoxyethane (DME) was purified by drying over sodium and then vacuum distilling before addition to the reactor. It serves as an inert solvent and lowers the boiling point of 2-phenylethanol.

A temperature of 140–150° was maintained in the reaction vessel for 144 hr or until there was no evidence of any white deposit of hexacarbonyl-chromium in the upper condenser portion of the Strohmeier apparatus. The greenish-yellow solution was allowed to cool and was then transferred to a 1-liter round-bottom flask. Unreacted 2-phenylethanol and DME were removed with a rotary evaporator at reduced pressure (100° at 0.1 mm) to give a viscous greenish-yellow oil. The viscous oil was dissolved in ethyl ether (500 ml) and filtered to remove the green colored insoluble chromium products. After removing the ether with a rotary evaporator, the remaining yellow viscous oil was recrystallized from a 75/25 v/v n-pentane–ethyl ether mixture. Upon filtering and drying in a vacuum oven (10^{-2} mm), a yield of 51.0 g (87.0%) of η^6-(2-phenylethanol)tricarbonylchromium was obtained as yellow needle-like crystals, mp 51–52°.

A three-necked flask, oven-dried for 24 hr and purged with nitrogen, was charged with 10.0 g (0.039 mol) of η^6-(2-phenylethanol)tricarbonylchromium,

550 ml of dry, distilled benzene, and 7.0 ml (0.082 mol) of pyridine. The reaction vessel was equipped with a mechanical stirrer, a nitrogen inlet, and cooled in an ice bath. Acetyl chloride (14.0 ml, 0.171 mol in 60 ml of benzene) was added dropwise to the rapidly stirred solution from the addition funnel. White pyridine hydrochloride precipitated immediately. The reaction was allowed to warm gradually to room temperature. After 8 hr the reaction was diluted with an additional 500 ml of benzene and filtered to remove the pyridine hydrochloride. Infrared spectroscopy showed that the precipitate contained some of the product. It was hydrolyzed with a 10% Na_2CO_3 solution saturated with NaCl and extracted with benzene until the extract was colorless. The extract was combined with the filtered reaction solution which was washed five times with 200-ml portions of 10% Na_2CO_3 solution, and then with 10% NaCl solution until neutral. If the wash became acidic, the basic wash was repeated. The benzene solution was finally washed five times with distilled water and dried over Na_2SO_4. The solvent was removed without heating *in vacuo*. The resulting brown, viscous oil was eluted down a silica gel column with a mixture of benzene and hexane (50:50) to effect separation from 2-phenylethyl acrylate and unreacted η^6-(2-phenylethanol)tricarbonyl-chromium. The 2-phenylethyl acrylate apparently originated from the esterification of uncomplexed 2-phenylethanol remaining in the crystals of η^6-(2-phenylethanol)tricarbonylchromium used as the starting material. The benzene was again stripped off to give 6.8 g (56.3%) of the very viscous yellow-brown (PEAC). Attempts to crystallize the liquid from numerous solvent mixtures were unsuccessful. However, it was observed to crystallize from a mixture of predominantly hexane and ethyl acetate of unknown ratio while being stored under refrigeration. The yellow powder had a melting point of 41–42°.

Copolymerizations of η^6-(2-Phenylethyl acrylate)tricarbonylchromium.[100] The monomer (PEAC) was copolymerized with styrene, methyl acrylate, acrylonitrile, and 2-phenylethyl acrylate. The first three comonomers were purified before use by washing with aqueous NaOH and distilled water, and then distilling. 2-Phenylethyl acrylate was synthesized by the esterification of 2-phenylethanol with acrylyl chloride and was vacuum distilled; AIBN was twice recrystallized from methanol before use (mp 102–103°).

Weighed batches (sufficient for three to five copolymerizations) of (PEAC) were dissolved in ethyl acetate, and measured amounts were charged into Fischer-Porter aerosol compatibility tubes with weighed amounts of comonomer and initiator. Additional solvent was added. All polymerizations were carried out in purified ethyl acetate. The tubes were triply degassed and immersed in a constant temperature bath for a predetermined time at 70 ± 0.01°. After polymerization, the tubes were cooled and the mixture diluted with ethyl acetate (ca. one-half the initial amount) and precipitated dropwise in 2-liter of petroleum ether, 30–60° (the styrene, methyl acrylate, and acrylonitrile copolymers) or in hexane (2-phenylethyl acrylate copolymers). After three reprecipitations the polymers were freeze-dried and then weighed.

Synthesis of η^6-(2-Phenylethyl methacrylate)tricarbonylchromium.[102] η^6-(2-Phenylethanol)tricarbonyl chromium (10.33 g, 0.04 mole) and dry pyridine (6.87 g, 0.086 mol) were added to a 1-liter flask containing dry, distilled benzene (400 ml) and equipped with a condenser, mechanical stirrer, and pressure-equalizing addition funnel. The flask and attachments had been dried in a hot oven (110°) and purged with dry nitrogen before addition of the solvent and reactants. After cooling in an ice bath (0°), a solution of methacrylyl chloride (9.0 g, 0.86 mol) in dry benzene (40 ml) was added dropwise to the rapidly stirred solution over a 30-min period. Pyridine hydrochloride precipitated immediately upon the addition of the acid chloride. After the addition was completed, the solution was allowed to warm-up to room temperature and stirring was continued for about 2 hr. The reaction product was diluted with additional benzene (200 ml) and filtered to remove the precipitated pyridine hydrochloride. Additional product was obtained by thoroughly washing the pyridine hydrochloride twice with benzene (150-ml portions) and combining the washings with the product solution. This benzene solution was then washed four times with saturated aqueous sodium bicarbonate solution (300-ml portions), three times with 10% sodium chloride solution (300-ml portions), and finally four times with distilled water (300-ml portions). These washings employed a mechanical stirrer to agitate the solutions to effect neutralization; then a separatory funnel was used to separate the organic phase from the aqueous phase. After drying over magnesium sulfate, the benzene solution was filtered and the solvent removed on a rotary evaporator (at room temperature in order to prevent possible homopolymerization of the monomer).

The crude product should be protected from light and air by keeping it dark and under a nitrogen atmosphere. Also during purification, the chromatographic column should be protected from light and a nitrogen atmosphere should be used during the separation of ester.

The resulting brownish-yellow viscous oil was purified by column chromatography on 70–325 mesh silica gel (30 g) using benzene as the eluant for the ester. After removing the benzene from the eluted product at room temperature, 7.5 g (57.6%) of viscous yellow η^6-(2-phenylethyl methacrylate)-tricarbonylchromium was obtained. The product was crystallized from n-hexane at −20° to yield a yellow powder, mp 52–53°.

J. Synthesis and Polymerization of
η^4-(2,4-Hexadien-1-yl acrylate)tricarbonyliron

Synthesis of η^4-(2,4-Hexadien-1-yl acrylate)tricarbonyliron.[103] First step: η^4-(2,4-hexadien-1-ol)tricarbonyliron was prepared by two methods (**A** and **B**).

A. A solution of 2,4-hexadien-1-ol (67.5 g, 0.69 mol) and pentacarbonyliron (160 g, 0.82 mol) was refluxed under nitrogen for 96 hr. The volatile products were removed under reduced pressure and crude η^4-(2,4-hexadien-1-ol)tricarbonyliron was distilled and purified by redistillation (96–97° at 0.3 mm) to give 35 g (0.15 mol). The complex solidified on refrigeration and was recrystallized from 30–60° petroleum ether to afford yellow needles, mp 37°.

B. A solution of freshly distilled 2,4-hexadien-1-al (55 g, 0.57 mol) and pentacarbonyliron (225 g, 1.15 mol) was refluxed 48 hr under nitrogen. The solution was cooled and the volatile products were removed under reduced pressure, followed by vacuum distillation of the residual oil which gave η^4-(2,4-hexadien-1-al)-tricarbonyliron (62.5 g, 2.66 mol), bp 102–104° at 1.0 mm. To a stirred methanol (250 ml) solution of η^4-(2,4-hexadien-1-al)tricarbonyliron (39.3 g, 0.166 mol), sodium borohydride (3.0 g, 0.079 mol) was added. The temperature was maintained at 25–30° during the addition and the reaction stirred an additional 30 min at this temperature. The reaction mixture was hydrolyzed with distilled water (200 ml) and the product was extracted with three 150-ml portions of ether. The combined extract was washed with water and dried over $MgSO_4$. The solvent was removed to give 37.5 g (0.15 mol) of crude alcohol which was purified by recrystallization from 30–60° petroleum ether to give the desired compound (mp 37°).

Second step: Preparation of η^4-(2,4-Hexadiene-1-yl acrylate)tricarbonyliron. Dry pyridine (14 ml) and η^4-(2,4-hexadien-1-ol)tricarbonyliron (19.04 g, 0.08 mol) were added to anhydrous ethyl ether (600 ml) and the solution was flushed with nitrogen. After cooling to 0°, a solution of acrylyl chloride (14.0 ml, 0.176 mol) in ether (50 ml) was added dropwise over a 30-min period with stirring. Reaction was immediate, as evidenced by the precipitation of pyridine hydrochloride. The mixture was warmed to 25° and stirring continued for 1 hr. The reaction was diluted with 200 ml of ether and filtered to remove the pyridine hydrochloride which was thoroughly washed with two 100-ml portions of ether. The combined ether layers were washed three times with 300-ml portions of saturated aqueous $NaHCO_3$ and twice with 300-ml portions of distilled water. After drying over $MgSO_4$, filtering, and evaporation of ether, the crude acrylate was isolated (17.0 g). It was purified by chromatography on 60–200-mesh silica gel with benzene as eluant. After removal of benzene under high vacuum, the purified acrylate (15 g, 0.05 mol) was obtainable as a brown liquid.

Copolymerization of η^4-*(2,4-Hexadien-1-yl acrylate)tricarbonyliron with Styrene and Vinyl Acetate.*[103] The initiator, AIBN, was twice recrystallized from methanol. Purified (HATI) was charged to a Fischer–Porter tube (1.49 g for vinyl acetate case and 2.66 g for styrene case), followed by AIBN (0.019 or 0.028 g, respectively) and comonomer. The amount of comonomer was sufficient in these two cases to give the following mole ratios: (HATI)/ styrene = 91/9, (HATI)/vinyl acetate = 51/49. Benzene (10 ml) was added in each case and the reaction solution was triply degassed using the freeze–thaw technique under vacuum. The tubes were heated to 80° for 2 hr in each case. The contents of the tubes were poured into excess petroleum ether and purified by three reprecipitations from benzene into petroleum ether followed by drying (50–60°) for 24 hr at 1–2 mm. The styrene copolymer was obtained in 25% yield. It contained 81.4 mol% of (HATI) and $\overline{M}_n = 9300$ and $\overline{M}_w = 28{,}000$. The vinyl acetate copolymer was isolated in 25% yield. It contained 86 mol% (HATI) and $\overline{M}_n = 11{,}200$ and $\overline{M}_w = 46{,}000$.

K. Synthesis and Polymerization of 1,1′-Divinylferrocene

Preparation of 1,1′-Divinylferrocene.[110] Ferrocene was diacylated using acetyl chloride and $AlCl_3$ in methylene chloride.[137] The resulting 1,1′-diacetylferrocene was reduced with excess $LiAlH_4$ in tetrahydrofuran to give 1,1′-di-(α-hydroxyethyl)ferrocene, mp 69.0–70.1°.[138] The diol was thoroughly ground with an excess of activated neutral alumina (Merck, Grade I) and heated in a sublimator to 155° under reduced pressure (13 mm). The 1,1′-divinylferrocene was obtained by dehydrative sublimation and purified by a second sublimation at 40° and 0.1 mm to give red needles, mp 33.5–34.5° (Ref. 139 , 40–41°).

Cationic Polymerization of 1,1′-Divinylferrocene.[110] The initiator $BF_3 \cdot OEt_2$ was purified by repeated distillations and methylene chloride was purified by filtering through alumina, distillation, and drying over a Linde molecular sieve, 3A. The polymerizations were carried out in glass-stoppered Schlenk tubes. To 50 ml of methylene chloride were added methylene chloride solutions of 1,1′-divinylferrocene and AIBN to give 75 ml of solution where the monomer concentration was 0.2 mol/liter and the initiator concentration was 0.01 mol/ liter. These solutions were charged to the polymerization tube under nitrogen. The reaction mixture was allowed to stand at room temperature for $8\frac{1}{2}$ hr and then poured into excess methanol. The polymer was recovered by filtration and extracted with benzene. The benzene-soluble fraction was then reprecipitated by methanol and dried *in vacuo* (73% yield, $\overline{M}_n = 10{,}000$). The benzene-insoluble fraction was washed with benzene and dried (1% yield).

Radical Polymerization of 1,1'-Divinylferrocene.[110] Radical polymerizations were performed in sealed ampoules. A measured amount of benzene, AIBN, and 1,1'-divinylferrocene were added to the ampoule. The mixture was degassed by three freeze–thaw cycles under nitrogen and the ampoule was sealed *in vacuo.* After heating to a predetermined temperature for the required time, the polymerization was terminated by pouring into methanol containing a small amount of 2,5-di-*t*-butyl-*p*-cresol. The products were worked up the same as in cationic polymerizations. Using [AIBN] = 2 × 10⁻³ mol/liter, [1,1'-divinylferrocene] = 1.0 mol/liter, and a reaction temperature of 70° for 22 hr resulted in 9% conversion to a benzene-soluble fraction and a 35% conversion to a benzene-insoluble fraction.

L. Synthesis and Polymerization of
trans-Chloro(4-vinylphenyl)bis(tributylphosphine)palladium(II)

Synthesis of trans-*Chloro(4-vinylphenyl)bis(tributylphosphine)palladium-(II).*[131] Magnesium turnings (0.73 g, 30 mg-atom) in THF (8 ml) were treated with 0.2 ml of ethyl bromide and the mixture was warmed to gentle reflux. A solution of freshly distilled *p*-chlorostyrene (2.1 g, 20 mmol) in THF (8 ml) was added slowly. The resulting mixture was refluxed until the magnesium disappeared. A solution consisting of *trans*-dichlorobis(tributylphosphine)-palladium(II) (6.01 g, 10 mmol) in THF (40 ml) was added dropwise to the Grignard reagent at room temperature, followed by stirring for 30 min at 40°. After removal of the reaction solvent, ether was added. The mixture was cooled with ice and a NH₄Cl solution added. Crude *trans*-di(4-vinylphenyl)-bis(tributylphosphine)palladium(II) was isolated from the ether layer and recrystallized from ether–ethanol as white needles (4.6 g, 71%), 114°(dec).

A mixture of this product (2.15 g, 3 mmol) in dry ether (10 ml) was treated with a solution of dry HCl in ether (10% excess). The ether and styrene (resulting from cleavage) were removed under reduced pressure to give white crystals (1.95 g, 100%), mp 38° .

Solution Polymerization Initiated by AIBN.[131] Homopolymerization was carried out in benzene or toluene when using AIBN. Benzene and toluene were freshly distilled from P₂O₅ before use and AIBN was recrystallized three times from ether. After degassing three times and sealing, an ampoule containing the monomer (1.95 g) and AIBN (0.29 g) in toluene (3 ml) was placed in a constant-temperature bath at 55 ± 0.5°. After 24 hr the ampoule was removed and the white polymer was precipitated in a rapidly stirring excess of methanol. The polymer was separated by centrifuging, washed with methanol, and reprecipitated from toluene into methanol two times. The product was dried to give 0.65 g (28%) of polymer with a $\overline{M}_n = 10,900$ (vapor pressure osmometry).

Solution Polymerization with n-Bu₃B-Oxygen.[131] A three-necked, 20-ml Schlenk tube was charged with monomer (0.98 g), tributylborane (0.27 g), 0.35 ml of air, and hexane (3 ml). The solution was degassed by melting under vacuum before introducing the tributylborane and air. A hexane solution of freshly distilled tributylborane, which had been dried over P_2O_5, was added by syringe technique. The reaction temperature was kept at 0° for 24 hr using an external bath. The polymer was isolated by precipitation into methanol and centrifugation. Purification was performed by reprecipitating in methanol three times, followed by drying to give 0.67 mg (6.8%) of the polymer with a $\overline{M}_n = 9000$.

Copolymerization with Styrene.[131] Styrene was washed, dried over anhydrous Na_2SO_4, and distilled as a center cut. *trans*-Pd(PBu₃)₂$+C_6H_4CH=CH_2$) Cl (1.6698 g), styrene (0.1671 g), and AIBN (0.0228 g) were weighed on an analytical balance and dissolved in benzene (1 ml) in an ampoule. After degassing three times and sealing, the ampoule was placed in a constant-temperature bath at 55 ± 0.5° for 3 hr. The polymer was isolated by adding the solution to rapidly stirring excess methanol and followed by centrifuging. After another reprecipitation and centrifugation, the polymer was dried to give a 21% yield of copolymer, containing 4.61% Cl, indicating that the mol% of styrene was 53.4.

REFERENCES

1. F. S. Arimoto and A. C. Haven, Jr., *J. Amer. Chem. Soc.*, **77**, 6295 (1955).
2. A. C. Haven, Jr., U. S. Patent 2,821,512.
3. E. W. Neuse and H. Rosenberg, *J. Macromol. Sci. Rev. Macromol. Chem.*, **4**, 1 (1970).
4. B. A. Bolto, in *Organic Semiconducting Polymers*, J. E. Eaton, Ed., Marcel Dekker, New York, 1968.
5. F. G. A. Stone and W. A. G. Graham, *Inorganic Polymers*, Academic Press, New York, 1962.
6. F. R. Mayo and F. M. Lewis, *J. Amer. Chem. Soc.*, **66**, 1594 (1944).
7. F. F. Mayo and C. Walling, *Chem. Rev.*, **46**, 191 (1950).
8. For a review, see T. Alfrey, Jr., J. J. Bohrer, and H. Mark, *Copolymerization* (High Polymers, Vol. III), Interscience, New York, 1952.
9. C. C. Price, *J. Polymer Sci.*, **3**, 772 (1948).
10. C. U. Pittman, Jr., *Chem. Tech.*, **1**, 416 (1971).
11. C. U. Pittman, Jr., *J. Paint Technology*, **43**, 29 (1971).
12. J. C. Lai, T. D. Rounsefell, and C. U. Pittman, Jr., *J. Polymer Sci.*, A-1, **9**, 651 (1971).

13. Y. H. Chen, M. Fernandez-Refojo, and H. G. Cassidy, *J. Polymer Sci.*, **40**, 433 (1959).
14. M. G. Baldwin and K. E. Johnson, *J. Polymer Sci.*, *A-1*, **5**, 2901 (1967).
15. M. H. George and G. F. Hayes, *J. Polymer Sci. Chem. Ed.*, **13**, 1049 (1975).
16. Y. Sasaki, L. L. Walker, E. L. Hurst, and C. U. Pittman, Jr., *J. Polymer Sci. Chem. Ed.*, **11**, 1213 (1973).
17. A. J. Tinker, M. H. George, and J. A. Barrie, *J. Polymer Sci. Chem. Ed.*, **13**, 2621 (1975).
18. A. J. Tinker, M. H. George, and J. A. Barrie, *J. Polymer Sci. Chem. Ed.*, **13**, 2133 (1975).
19. M. H. George and G. F. Hayes, *Polymer Letters*, **11**, 471 (1973).
20. C. Aso, T. Kunitake, and T. Nakashima, *Kogyo Kagaku Zasshi*, **72**, (6), 1411 (1969).
21. T. T. Ma, P. C. Yeh, C. C. Lu, and W. F. Wu, *KoFen. Tzu. Tung. Hsun*, **6**, 148 (1964).
22. C. U. Pittman, Jr., J. C. Lai, D. P. Vanderpool, M. Good, and R. Prados, *Macromolecules*, **3**, 746 (1970).
23. C. U. Pittman, Jr., J. C. Lai, D. P. Vanderpool, M. Good, and R. Prados in *Polymer Characterization, Interdisciplinary Approaches*, C. Craver, Ed., Plenum, New York, 1971, pp. 97–124.
24. C. Aso, T. Kunitake, and T. Nakashima, *Makromol. Chem.*, **124**, 232 (1969).
25. Cr. Simionescu, T. Lixandru, I. Negulescu, I. Mazilu, and L. Tataru, *Makromol. Chem.*, **163**, 59 (1973).
26. R. M. Golding and L. E. Orgel, *J. Chem. Soc.*, 363 (1962).
27. M. H. George and G. F. Hayes, *J. Polymer Sci. Chem. Ed.*, **14**, 475 (1976).
28. A. J. Tinker, M. H. George, and J. A. Barrie, *J. Polymer Sci. Chem. Ed.*, **13**, 2133 (1975).
29. A. J. Tinker, M. H. George, and J. A. Barrie, *J. Polymer Sci. Chem. Ed.*, **13**, 2005 (1975).
30. A. J. Tinker, M. H. George, and J. A. Barrie, *J. Polymer Sci. Chem. Ed.*, **13**, 2621 (1975).
31. M. H. George and G. F. Hayes, *Polymer*, **15**, 397 (1974).
32. C. U. Pittman, Jr. and P. L. Grube, *J. Polymer Sci.*, *A-1*, **9**, 3175 (1971).
33. C. U. Pittman, Jr., R. L. Voges, and J. Elder, *J. Polymer Sci.*, *B.*, **9**, 191 (1971).
34. C. U. Pittman, Jr. and P. L. Grube, *J. Applied Polymer Sci.*, **18**, 2269 (1974).
35. P. W. Tidwell and G. A. Mortimer, *J. Polymer Sci.*, *A-1*, **3**, 369 (1965).
36. P. W. Tidwell and G. A. Mortimer, *J. Macromol. Sci. Rev. Macromol. Chem.*, **C4**, 281 (1970).
37. M. Fineman and S. D. Ross, *J. Polymer Sci.*, **5**, 259 (1950).
38. T. Tsuruta and K. F. O'Driscoll, *Structure and Mechanism in Vinyl Polymerization*, Wiley, New York, 1957.
39. J. Brandrup and E. H. Immergut, Eds., *Polymer Handbook*, Interscience, New York, 1966.
40. E. M. Arnett and R. D. Bushick, *J. Org. Chem.*, **27**, 111 (1962).
41. C. U. Pittman, Jr., *Tetrahedron Lett.*, **37**, 3619 (1967).
42. T. G. Traylor and J. C. Ware, *J. Amer. Chem. Soc.*, **89**, 2304 (1967).
43. M. Cais, *Organometal. Chem. Rev.*, **1**, 435 (1966).
44. A. N. Nesmeyanov, E. G. Perevalova, S. P. Gubin, K. I. Grandberg, and A. G. Kozlovsky, *Tetrahedron Lett.*, **22**, 238 (1966).
45. C. Elschenborvich and M. Cais, *J. Organometal. Chem.*, **18**, 135 (1969).
46. A. A. Berlin, *Khim. Prom.*, 881 (1962).

47. E. G. Perevalova and Yu. A. Ustynayak, *Izvest. Akad. Nauk USSR*, **12**, 1776 (1963).
48. C. U. Pittman, Jr., unpublished studies.
49. K. Tada, H. Higuchi, M. Yoshimura, H. Mikawa, and Y. Shirota, *J. Polymer Sci. Chem. Ed.*, **13**, 1737 (1975).
50. C. U. Pittman, Jr., G. V. Marlin, and T. D. Rousefell, *Macromolecules*, **6**, 1 (1973).
51. M. H. George and G. F. Hayes, *Makromol. Chem.*, **177**, 399 (1976).
52. T. Nakashima, T. Kunitake, and C. Aso, *Makromol. Chem.*, **157**, 73 (1972)
53. C. U. Pittman, Jr., *J. Polymer Sci., B.*, **6**, 19 (1968).
54. C. U. Pittman, Jr. *J. Paint Technology*, **45** (582), 78 (1973).
55. M. Howard and S. F. Reed, Jr., *J. Polymer Sci.*, *A-1*, **9**, 2085 (1971).
56. H. M. Stanley, *Chem. Ind.*, **58**, 1080 (1939); A. B. Hershberger, J. C. Reid, and R. H. Heiligmann, *Ind. Eng. Chem.*, **37**, 1073 (1944).
57. C. U. Pittman, Jr., J. C. Lai, and D. P. Vanderpool, *Macromolecules*, **3**, 105 (1970).
58. J. C. Lai, T. D. Rousefell, and C. U. Pittman, Jr., *Macromolecules*, **4**, 155 (1971).
59. C. U. Pittman, Jr., R. L. Voges, and W. R. Jones, *Macromolecules*, **4**, 291 (1971).
60. C. U. Pittman, Jr., R. L. Voges, and W. R. Jones, *Macromolecules*, **4**, 298 (1971).
61. O. E. Ayers, S. P. McManus, and C. U. Pittman, Jr., *J. Polymer Sci., Chem. Ed.*, **11**, 1201 (1973).
62. C. U. Pittman, Jr. and B. Surynarayanan, *J. Amer. Chem. Soc.*, **96**, 7916 (1974).
63. V. V. Korshak, S. L. Sosin, T. M. Frunze, and I. I. Tverdokhlebova, *J. Polymer Sci., Part C*, **22**, 849 (1969).
64. S. L. Sosin, V. V. Korshak, T. M. Frunze, and I. I. Tverdokhlebova, *Dokl. Akad. Nauk SSSR*, **175** (5), 1076 (1967).
65. C. Hauser, R. Pruett, and A. Mashburn, *J. Org. Chem.*, **26**, 1800 (1961).
66. L. E. Coleman, Jr., and M. D. Rausch, *J. Polymer Sci.*, **28**, 207 (1958).
66a. D. H. Lewis, R. C. Kneisel, and B. W. Ponder, *Macromolecules*, **6**, 660 (1973).
67. D. C. Van Landuyt and S. F. Reed, Jr., *J. Polymer Sci.*, *A-1*, **9**, 523 (1971).
68. D. C. Landuyt, *J. Polymer Sci., B*, **10**, 125 (1972).
69. M. Rosenblum, N. Brown, J. Popemeier, and M. Applebaum *J. Organometal. Chem.*, **6**, 173 (1966).
70. V. V. Korshak, L. V. Dzhashi, and S. L. Sosin, *Nuova Chim.*, **49** (3), 31 (1973); *Chem. Abstr.*, **79**, 5644d (1973).
71. G. A. Yurlova, Yu. V. Chumakov, T. M. Ezhova, L. V. Dzashi, S. L. Sosin, and V. V. Korshak, *Vysokomol. Soedin.*, *Ser. A*, **13** (12), 276 (1971).
72. V. V. Korshak, L. V. Dzhashi, B. A. Antipova, and S. L. Sosin, *Dokl. Vses. Konf. Khim. Atsetilena 4th*, **3**, 217 (1972).
73. V. V. Korshak, T. M. Frunze, A. A. Izyneev, and V. G. Samsonova, *Vysokomol. Soedin.*, *Ser. A*, **15** (3), 521 (1973); *Chem. Abstr.*, **81**, 50069X (1974).
74. C. Simionescu, T. Lixandru, I. Maxilu, and L. Tatrau, *Makromol. Chem.*, **147**, 69 (1971).
75. C. U. Pittman, Y. Sasaki, and P. Grube, *J. Macromol. Sci. Chem.* **A8** (5), 923 (1974).
76. T. Nakashima, T. Kunitake, and C. Aso, *Makromol. Chem.*, **157**, 73 (1972).
77. B. F. Sokolov, *Neft Gaz Ikh Prod.*, 145 (1971); *Chem. Abstr.*, **78**, 44023z (1973).
78. C. U. Pittman, Jr. and Y. Sasaki, *Chem. Lett. Japan*, 383 (1975).
79. D. O. Cowan, J. Park, C. U. Pittman, Jr., Y. Sasaki, T. K. Mukherjee, and N. A. Diamond, *J. Amer. Chem. Soc.*, **94**, 5110 (1972).
80. T. P. Vishnyakova, L. I. Tolstykh, G. M. Ignat'eva, B. F. Sokolov, and Ya. M. Paushkin, *Dokl. Akad. Nauk SSSR*, **208** (4), 853 (1973).
81. V. V. Korshak and S. L. Sosin, *Vysokomolekul, Soedin.*, **B14** (3), 164 (1972).
82. W. J. Sparks and C. W. Muessig, U.S. Patent 2,255,396 (1941).

83. Ya. M. Paushkin, B. F. Sokolov, T. P. Vishnayakova, and G. Glazkova, *Dokl. Akad. Nauk SSSR*, **206**, 664 (1972).
84. J. Kozikowski and M. Cais, U. S. Patents 3,290,337 (1966) and 3,308,141 (1967).
85. C. U. Pittman, Jr., J. C. Lai, T. D. Rounsefell, G. V. Marlin, and J. Chapman, Abstracts of the Southeast-Southwest Regional Meeting of the American Chemical Society, New Orleans, La., Dec. 2–4, 1970, paper nos. 627 and 664–666.
86. M. D. Rausch, G. A. Moser, E. J. Zaiko, and A. L. Lipman, Jr., *J. Organometal. Chem.*, **23**, 185 (1970).
87. C. U. Pittman, Jr., P. L. Grube, O. E. Ayers, S. P. McManus, M. D. Rausch, and G. A. Moser, *J. Polymer Sci.*, *A-1*, **10**, 379 (1972).
88. G. R. Knox, D. G. Leppard, P. L. Pauson, and W. E. Watts, *J. Organometal. Chem.*, **34**, 347 (1972).
89. J. D. Holmes, A. K. Jones, and R. Pettit, *J. Organometal. Chem.*, **4**, 324 (1965).
90. D. A. Brown and J. R. Raju, *J. Chem. Soc.*, 40 (1966).
91. E. O. Fischer, K. Öfele, H. Essler, W. Fröhlich, J. P. Mortensen, and W. Semmlinger, *Chem. Ber.*, **91**, 2763 (1958).
92. V. S. Khandkarova, S. P. Gubin, and B. V. Kvasov, *J. Organometal. Chem.*, **23**, 509 (1970).
93. S. P. Gubin and V. S. Khandkarov, *J. Organometal. Chem.*, **22**, 449 (1970).
94. D. G. Carrol and S. P. McGlynn, *Inorg. Chem.*, **7**, 1285 (1968).
95. D. A. Brown and R. M. Rawlingon, *J. Chem. Soc.*, *Sect. A*, 1534 (1969).
96. R. S. Bly and R. C. Strickland, *J. Amer. Chem. Soc.*, **92**, 7459 (1970).
97. K. K. Wells and W. S. Trahanovsky, *J. Amer. Chem. Soc.*, **92**, 7461 (1970).
98. B. Nichols and M. C. Whiting, *J. Chem. Soc.*, 551 (1959).
99. C. U. Pittman, Jr., R. L. Voges, and J. Elder, *Macromolecules*, **4**, 302 (1971).
100. C. U. Pittman, Jr. and G. V. Marlin, *J. Polymer Sci. Chem. Ed.*, **11**, 2753 (1973).
101. G. V. Marlin, M.S. Thesis, University of Alabama, 1973.
102. C. U. Pittman, Jr., O. E. Ayers, and S. P. McManus, *Macromolecules*, **7**, 737, (1974).
103. C. U. Pittman, Jr., O. E. Ayers, and S. P. McManus, *J. Macromol. Sci. Chem.*, **A7** (8), 1563 (1973).
104. C. U. Pittman, Jr., *Macromolecules*, **7**, 396 (1974).
105. V. V. Korshak, A. M. Sladkov, L. K. Luneva, and A. S. Girshovich, *Vysokomol, Soedin.*, **5**, 1284 (1963).
106. R. Ralea, C. Ungureanu, and I. Maxim, *Rev. Roumaine Chim.*, **12**, 523 (1967).
107. S. L. Sosin, V. V. Korshak, and T. M. Frunze, *Dokl. Akad. Nauk SSSR*, **179**, 1124 (1968).
108. S. L. Sosin and L. V. Dzhashi, *Kinet. Mech. Polyreactions, Int. Symp. Macromol. Chem. Prepr.*, **1**, 327 (1969).
109. S. L. Sosin, L. V. Dzhashi, B. A. Antipova, and V. V. Korshak, *Vysokomol, Soedin., Ser B.*, **12** (9), 699 (1970).
110. T. Kunitake, T. Nakashima, and C. Aso, *J. Polymer Sci.*, *A-1*, **8**, 2853 (1970).
111. T. Kunitake, T. Nakashima, and C. Aso, *Makromol. Chem.*, **146**, 79 (1971).
112. G. A. Yurlova, Yu. V. Chumakov, and V. V. Korshak, *Vysokomol. Soedin., Ser A*, **13**, (12), 2761 (1971).
113. T. Nakashima and T. Kunitake, *Bull. Chem. Soc. Jap.*, **45**, 2892 (1972).
114. S. L. Sosin, L. V. Dzhashi, B. A. Antipova, and V. V. Korshak, *Vysokomol. Soedin., Ser. B* **16** (5), 347 (1974).
115. G. R. Knox and P. L. Pauson, *J. Chem. Soc.*, 4610 (1961).
116. R. Riemschneider and D. Helm, *Chem. Ber.*, **89**, 155 (1956).
117. K. Schlögl and M. Fried, *Monatsh. Chem.*, **95**, 558 (1964).

118. C. Aso, T. Kunitake, and R. Kita, *Makromol. Chem.*, **97**, 31 (1966).
119. C. Aso, T. Kunitake, Y. Matsuguma, and Y. Imaizumi, *J. Polymer Sci.*, *A-1*, **6**, 3049 (1968).
120. J. C. Montermoso, T. M. Andrews, and L. P. Marinelli, *J. Polymer Sci.*, **32**, 523 (1958).
121. M. M. Koton, T. M. Kiseleva, and F. S. Florinskiĭ, *Angew. Chem.*, **72**, 712 (1960).
122. M. M. Koton, T. M. Kiseleva, and F. S. Florinskiĭ, *Vysokomol. Soedin.*, **2**, 1639 (1960).
123. I. A. Shi Khiev, M. F. Shostakovskiĭ, and L. A. Kayutenko, *Dokl. Akad. Nauk Azerbaidzhan, SSR*, **14**, 687 (1958).
124. M. F. Shostakovskiĭ, S. P. Kalinina, V. N. Kotrelev, and D. A. Kochkin, *J. Polymer Sci.*, **52**, 223 (1961).
125. J. A. Montemarano and E. J. Dyckman, *J. Paint Technology*, **47**, 59 (1975).
126. M. M. Koton, T. M. Kiseleva, and F. S. Florinskiĭ, *J. Polymer Sci.*, **52**, 237 (1961).
127. V. V. Korshak, A. M. Polyakova, and M. D. Suchkova, *Vysokomol. Soedin.*, **2**, 13 (1960).
128. M. M. Koton, T. M. Kiseleva, and F. S. Florinskiĭ, *Izvest, Akad. Nauk. SSSR, Otdel. Khim. Nauk*, 948 (1959).
129. J. G. Noltes, H. A. Budding, and G. J. M. Van der Kerk, *Rec. Trav. Chim.*, **79**, 408 (1960).
130. J. G. Noltes, H. A. Budding, and G. J. M. Van der Kerk, *Rec. Trav. Chim.*, **79**, 1076 (1960).
131. N. Fujita and K. Sonogashira, *J. Polymer Sci. Chem. Ed.*, **12**, 2845 (1974).
132. B. G. Thompson, *J. Polymer Sci.*, **19**, 373 (1956).
133. C. E. Scott and C. C. Price, *J. Amer. Chem. Soc.*, **81**, 2670 (1959).
134. H. Ringsdorf and G. Greber, *Makromol. Chem.*, **31**, 27 (1959).
135. Y. Minoura and Y. Sakanaka, *J. Polymer Sci.*, *A-1*, **7**, 3287 (1969).
136. Y. Minoura, Y. Suzuki, and Y. Sakanaka, *J. Polymer Sci.*, *A-1*, **4**, 2757 (1966).
137. M. Rosenblum and R. B. Woodward, *J. Amer. Chem. Soc.*, **80**, 5443 (1958).
138. K. Yamakawa, H. Ochi, and K. Aracawa, *Chem. Pharm. Bull. (Japan)*, **11**, 905 (1963).
139. M. D. Rausch and A. Siegel, *J. Organometal. Chem.*, **11**, 317 (1968).

Chapter 2

Reactions of Metallocarboranes

RUSSELL N. GRIMES

I. INTRODUCTION AND SCOPE

The discovery in the 1950s of carboranes, which are boron cage compounds containing carbon in the skeletal framework, stimulated a considerable research effort in this area which has opened a vast new field of unlimited dimensions. Much of the research on carboranes[75,85,157,186,188,198,225] involves the derivative chemistry of polyhedral molecules such as $C_2B_{10}H_{12}$ and $C_2B_5H_7$ and has been largely oriented toward the synthesis of heat-stable polymers. However, an entirely different aspect of the field centers on the fact, first discovered by Hawthorne, Young, and Wegner[101] in 1965, that metal atoms can be inserted into carborane frameworks to produce metallocarboranes which maintain or even increase the stability of the original system. Hundreds of such compounds involving a wide variety of transition and main-group metals have been prepared, and this new field has in its short existence reached a surprisingly sophisticated level with rational synthetic routes and a qualitative theory of structure and bonding of considerable predictive power. The significance of the metallocarboranes in an ultimately practical sense has yet to be determined, although their exceptional stability, almost unlimited variety, and novel structures, and the possibility of unusual electronic or other properties, suggest that applications will one day be developed. However, the importance of metallocarboranes from a theoretical viewpoint is beyond doubt. This class of compounds occupies a strategic position overlapping several large areas of chemistry, namely the boranes,

RUSSELL N. GRIMES • Department of Chemistry, University of Virginia, Charlottesville, Virginia.

transition-metal complexes, metal clusters and other metal–metal bonded species, and organometallic compounds, particularly the metallocenes. The common features of geometry, electron distribution, and bonding which link the metallocarboranes with many of these chemically disparate groups of compounds (see following section) have only recently been dealt with explicitly and at some length in the literature, and the subject is certain to attract increasing attention as the general awareness of the polyhedral cage molecules grows among chemists in all fields.

The metallocarborane area has been reviewed in part several times,[13,75,77,78,86–88,92,188,198] although several of these articles have been essentially limited to their authors' own research. This review is concerned with the synthesis and reactions of carborane cage compounds containing one or more metal atoms *in the skeletal framework*, the term "metal" in this case signifying all transition elements and Groups II, III, and IV of the periodic table. Carboranes incorporating other heteroatoms are not included except when a metal atom, as defined above, is also present in the cage; thus, phospha-, arsa-, and stibacarboranes are excluded, but the species formed by insertion of a metal into these systems (e.g., metallophosphacarboranes) are included.

Derivatives of carboranes in which metal atoms are present only in ligands externally σ-bonded to the cage are not within the scope of this review, except for the few borderline cases in which a metal is linked to the cage via a three-center B–M–B bond as in μ-$(CH_3)_3SnC_2B_4H_7$. Finally, metalloborane cage species lacking carbon in the framework[154] are not discussed, except in conjunction with metallocarborane synthesis or structures. The scope thus delineated for this review is less arbitrary than it may at first appear, since the synthesis and the properties of the metallocarboranes *per se* are rather distinct from other types of metalloboron cage systems. In addition, the known metallocarborane species are far more numerous and have been examined in considerably more detail than the metalloboranes and other relatives, and thus warrant separate treatment.

This chapter was originally written in 1973–1974 and was subsequently updated to incorporate work reported through 1975, with the intent of providing reasonably complete coverage of the literature to that date.

II. STRUCTURES AND BONDING

A. General Observations

The metallocarboranes are normally regarded as analogs of carboranes in which one or more BH groups has been formally replaced by a metal atom and its associated ligand (e.g., C_5H_5Co) and like the carboranes, may be either closed polyhedra with triangulated faces, or open frameworks which can be viewed as polyhedra with one vertex missing. The two types are designated respectively as *closo* and *nido*, the latter term coined from the Greek word for nest. (In addition, the term *arachno* has been applied[224] to structures which are more open than the *nido* type and represent polyhedra with *two* missing vertices.) As examples, the species $1,2,4-(\eta-C_5H_5)CoC_2B_4H_6$ (Figure 1a) and $2,4-C_2B_5H_7$ (Figure 1b) are both pentagonal bipyramidal *closo* systems, with the 5-coordinate (apex) vertices 1 and 7 occupied in the former case by C_5H_5Co and BH groups, and in the latter molecule by two BH units. Similarly, $1,2,3-(CO)_3FeC_2B_3H_7$ (Figure 2a) is *formally* generated from the *nido*-carborane $2,3-C_2B_4H_8$ (Figure 2b) by replacement of the apex BH group in the latter with an $Fe(CO)_3$ moiety. A few metallocarboranes having no known carborane analogs have been prepared (e.g., the 13-atom $(\eta-C_5H_5)CoC_2B_{10}H_{12}$ system), but the theoretical implications of such cases are difficult to assess. Thus, while it is conceivable that a 13-vertex $C_2B_{11}H_{13}$ polyhedron will eventually be prepared, it is also possible that the cobalt atom in $(\eta-C_5H_5)CoC_2B_{10}H_{12}$ stabilizes the 13-atom cage in a way that an 11th boron atom cannot.

All carboranes, metallocarboranes, and other borane-cage molecules are formally "electron-deficient,"[125] an occasionally misinterpreted term which indicates only that the number of valence electrons is insufficient to describe the cage bonding entirely in terms of localized two-electron, two-center bonds. *Electron deficiency carries no implication whatsoever in regard to either molecular instability or electrophilicity*; indeed, most carboranes and metallocarboranes are extremely stable, and the degree of inductive electron-

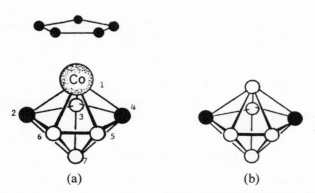

(a)   (b)

Figure 1. Structures of $1,2,4-(\eta-C_5H_5)CoC_2B_4H_6$ (a), and $2,4-C_2B_5H_7$ (b). ●, CH; ○, BH.

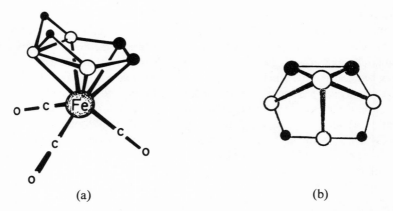

(a)   (b)

Figure 2. Structures of of 1,2,3-$(CO)_3FeC_2B_3H_7$ (a), and 2,3-$C_2B_4H_8$ (b). ●, CH; ○, BH; ●, H bridge.

withdrawing power exhibited by the cage varies considerably and is strongly dependent on the position of ligand attachment on the cage.[75] Many metallo-carboranes, indeed, have no discernible −I (negative inductive) effect at all.

The formal electron deficiency leads to delocalization of valence electrons over the cage framework, resulting in increased molecular stability in a fashion loosely analogous to resonance stabilization in aromatic hydrocarbon ring systems. The bonding in such systems is in general best described in terms of molecular orbital (MO) theory, particularly in the closed polyhedral cages. Although few quantitative studies of the electronic nature of metallo-carboranes have appeared, Wiersema and Hawthorne[222] examined the iso-tropic ^{11}B and ^{13}C nmr shifts in paramagnetic metallocarboranes of Cr(III), Fe(III), Ni(III), or Co(II) having 9–12 vertices, and concluded that electron delocalization is primarily of the σ-type and occurs via ligand-to-metal charge transfer, except in icosahedral cobalt(II); in the latter case, a metal-to-ligand shift is postulated. This particular study did not provide evidence for extensive transmission of electronic effects through cage molecular orbitals, in that only the nmr shifts of atoms directly linked to the metals were substantially affected, but such findings cannot be extrapolated to all types of metallo-carboranes; for example, species containing more than one metal atom in the cage were not examined.

B. Qualitative Bonding Descriptions

Although detailed MO treatments have been given for several borane anions and carboranes,[125] the introduction of metal atoms into the cage system greatly increases the complexity and cost of the calculations, and no quantitative metallocarborane MO study is available as yet. Fortunately, however, a simple electron-counting approach to the bonding, which correlates remarkably well with structural data, has been applied to the metallocarboranes and other borane-cage systems as well as the metal clusters, by several workers, particularly Wade,[205-208] Rudolph,[171] and Mingos.[133,153] The same basic ideas have been discussed by Grimes[74] and Hawthorne et al.[112] The essential points of interest here are as follows:

1. In polyhedral (*closo*) systems (defined above) all valence electrons in the cage framework are assumed to occupy delocalized MOs extending over the polyhedral surface.

2. Each framework atom (excluding hydrogen) in the polyhedron utilizes three of its atomic orbitals in bonding to the cage. For a boron, carbon, or other first-row atom, this leaves one remaining orbital that may be used for an exopolyhedral bond to H or other ligands. Transition-metal atoms, which have a total of nine valence orbitals, have six orbitals remaining after bonding to the cage framework; these six orbitals can be used for bonding to external ligands and for storage of nonbonding electron pairs.

3. The number of bonding molecular orbitals in a closed polyhedral cage of n atoms ($n \geq 5$) is $n + 1$, and therefore the number of valence electrons required for a filled-shell MO configuration *in the polyhedral cage* is $2n + 2$.[171,208] This postulate has been labeled the $(2n + 2)$ rule.[171] (If one includes, in addition, the valence electrons in metal nonbonding orbitals as well as those involved in metal–exopolyhedral ligand bonding, the rule becomes $2n + 14$[112]; this is, however, merely the "$(2n + 2)$ rule" in an arithmetically more cumbersome form, and is also inconsistent since it includes electrons in metal–ligand bonds but excludes electrons in boron–ligand and carbon–ligand bonds.)

4. When the number of valence electrons *in the cage skeleton* (exclusive of ligands external to the cage) exceeds $2n + 2$, the polyhedron undergoes distortion which commonly takes the form of the cage opening to form a *nido* structure. However, other forms of distortion are known, as in the slipped $(C_2B_9H_{11})_2M^q$-type complexes discussed in Section IV.

5. Metallocarborane *closo* or *nido* cages usually adopt the same gross geometry as the analogous isoelectronic n-atom carborane or borane anion framework. Thus, the ten-atom species $(\eta\text{-}C_5H_5)Co_2C_2B_6H_8$ and its analogs, $C_2B_8H_{10}$ and $B_{10}H_{10}^{2-}$ are all symmetrically bicapped square antiprisms.

In its simplest form, this electron-counting scheme invokes only the total numbers of framework valence electrons and orbitals, and requires no assumptions as to orbital hybridization or formal oxidation number (however, see Section II.C below). Thus, in the molecule $1,2,3\text{-}(CO)_3FeC_2B_3H_7$ (Figure 2a), the iron atom has eight valence electrons and a total of nine valence orbitals; since three of these orbitals accept three pairs of electrons from the CO groups, and three orbitals are utilized to bond the iron into the cage framework, there remain three "nonbonding" orbitals which can accommodate a total of six of the metal's valence electrons. This leaves two electrons to be donated to the cage system; i.e., a neutral $Fe(CO)_3$ group may be regarded as a formal two-electron donor. Since cage CH and BH groups are respectively three- and two-electron donors, and each of the two bridge hydrogens also contributes an electron, the total of valence electrons in the cage framework is 16, or two in excess of the 14 allowed by the $(2n + 2)$ rule for a six-vertex polyhedron. Hence the molecule is expected to have an open or *nido* geometry, and the structure thus predicted[76] (Figure 2a) has been confirmed by an X-ray diffraction study.[10] It has been noted earlier that $(CO)_3FeC_2B_3H_7$ is analogous to $C_2B_4H_8$.

Similar reasoning for the molecule $(CO)_3FeC_2B_3H_5$ yields the prediction[151] that it should be a closed polyhedral (octahedral) system, since its 14 cage valence electrons constitute the number required by the $(2n + 2)$ rule for a six-atom cage. The same is true of the analogous species $(\eta-C_5H_5)$-$CoC_2B_3H_5$,[151] since a neutral $(\eta-C_5H_5)Co$ moiety is, like $(CO)_3Fe$, a formal two-electron donor. Both of these *closo*-metallocarboranes are isoelectronic analogs (with respect to framework electrons) of *closo*-$C_2B_4H_6$, and one can simply regard the metal-containing group as replacing a BH unit in each case.

Further general discussion on these points, and additional examples including closely related metal cluster systems, are given in the references cited above, and the interested reader is referred to these. Specific problems of structure and bonding in individual species are dealt with throughout this chapter.

C. Metallocarboranes as Metal Complexes

A description of the bonding in $(CO)_3FeC_2B_3H_7$ which is qualitatively equivalent to that given above, yet schematically different, envisions a $(CO)_3Fe^{2+}$ group π-bonded to a formal $C_2B_3H_7^{2-}$ "cyclocarborane ring" ligand.[10,56,76,189] This approach is useful in comparing the molecule to its known metallocene analog, the $(\eta-C_5H_5)Fe(CO)_3^+$ cation, inasmuch as $C_2B_3H_7^{2-}$ and $C_5H_5^-$ are isoelectronic ligands (a free $C_2B_3H_7^{2-}$ ion has not, however, been found). Indeed, until recently it was fashionable to describe metallocarboranes in general as composed of metal ions π-bonded to the open faces of anionic carborane ligands; e.g., $(\eta-C_5H_5)FeC_2B_9H_{11}$ was pictured[100] as a sandwich complex of Fe^{3+} with a $C_5H_5^-$ and a $C_2B_9H_{11}^{2-}$ ligand. In this particular case this concept is quite appropriate, since the free 1,2- and 1,7-$C_2B_9H_{11}^{2-}$ ions are well characterized and it is reasonable (though not necessary) to emphasize their identities as ligands in their metalloborane derivatives. However, most carborane "ligands" which are implied by the metal-complex description are not, in fact, known as free species [e.g., $C_2B_8H_{10}^{4-}$ in $(\eta\text{-}C_5H_5)_2Co_2C_2B_8H_{10}$]. Largely for this reason, the more recent trend is to regard a metallocarborane cage as a single co-valently bonded molecule with no artificial separation between the metal and the remainder of the cage, and this approach will be emphasized in this review. Occasional exceptions seem warranted, for example, when analogies with metallocene systems are to be emphasized; a case in point is $1,7,2,3\text{-}(\eta\text{-}C_5H_5)_2\text{-}Co_2C_2B_3H_5$ (Section IV.F.2) which is a seven-vertex pentagonal bipyramid with $(\eta\text{-}C_5H_5)Co$ groups occupying the two five-coordinate vertices, but which can also be described[3,79] as a triple-decked complex in which two Co^{3+} ions are sandwiched between a central $C_2B_3H_5^{4-}$ ring and two $C_5H_5^-$ ligands. The formal $C_2B_3H_5^{4-}$ ring is isoelectronic with $C_5H_5^-$, and the "sandwich complex" description is therefore a useful means of affording comparisons with metallocene systems.

The same point can be made in reverse, in that a metallocene can be regarded as a cage system,[206] i.e., ferrocene consists of two electron-delocalized pentagonal pyramids fused at a common apex, the iron atom. None of these purely conceptual distinctions, of course, affects the electron–orbital balance or the correlation between electronic configuration and molecular structure as discussed above.

D. Nomenclature and Numbering

The problem of nomenclature in boron cage compounds has caused considerable consternation, owing to the inherent problems of dealing with three-dimensional cage structures of nearly unlimited variety, but also to some extent because of the divergent needs of different workers in the field. Thus, in comparing the icosahedral system $C_2B_{10}H_{12}$ to its derivative $C_2B_9H_{11}^{2-}$, which differ only by a BH^{2+} unit, some (including this author) have found it convenient to retain the same numbering in both cages, while other workers prefer to deal with the ion as a *nido*-borane, which entails a completely different numbering system. The most recent published effort at systematizing boron nomenclature is the IUPAC system,[111] followed in this review. The systematic naming of these cage compounds is exceedingly awkward and is generally avoided, heavy reliance being placed on the use of figures and line formulas to convey molecular structures. Numbering of cage systems herein follows the IUPAC rules to the extent that they are clearly set forth. However, these rules are ambiguous with respect to the order in which hetero (non-boron) atoms are to be numbered in a given polyhedral cage, when more than one type of heteroatom is present. The situation has understandably led to confusion and the use of different numbering conventions, often by the same

worker; thus, Hawthorne formerly assigned lowest framework numbers
to the metal atoms[42] but more recently has given priority to the carbon
atoms.[38] Others, including this author, have also been inconsistent in this
area; fortunately, the common use of diagrams has mitigated what would
otherwise be a chaotic situation.

In this review the metal atoms are given lowest heteroatom numbers
except for 12-vertex (icosahedral) cages; in the latter case, the framework
carbon atoms are given priority, as in $3,6,1,2\text{-}(\eta\text{-}C_5H_5)_2Co_2C_2B_8H_{10}$, in
which the carbon atoms are in vertices 1 and 2 and the metal atoms are in
vertices 3 and 6. This special treatment of icosahedra follows Hawthorne's
recent scheme[38] and has the advantage that the carbons are always in the
1,2 (ortho), 1,7 (meta), or 1,12 (para) positions (the only possible arrangements
in an icosahedron), and thus form a useful frame of reference for comparison
of icosahedral isomers having metal atoms in varying locations. For all other
polyhedra, lowest numbers are given to the metals.

In the interest of conserving space, ligand groups attached to the metal
atom(s) are not designated by number, and ligands not specified by a number
may be assumed to be bonded to the metal. Thus, in $1,2,4\text{-}(\eta\text{-}C_5H_5)CoC_2B_4H_6$
(Figure 1a), the cobalt atom occupies vertex 1, the carbons vertices 2 and
4, and the cyclopentadienyl group is attached to the cobalt.* Replacement
of the BH group in position 7 by a second $(\eta\text{-}C_5H_5)Co$ moiety would give
$1,7,2,4\text{-}(\eta\text{-}C_5H_5)_2Co_2C_2B_3H_5$. The application of these rules is generally
straightforward throughout the text.

III. SYNTHETIC ROUTES TO METALLOCARBORANES

A. General Observations

Nearly all known metallocarboranes have been prepared by the insertion
of metal atoms into a carborane framework, or by subsequent reactions of
metallocarboranes thus formed. The carborane starting material may be
either of the nido or closo type; in the former case, the metal is inserted into
the open face of the framework, while most syntheses utilizing closo-carbor-
anes have required a preliminary cage-opening step involving treatment with
strong bases or alkali metals prior to metal insertion. The distinction seems
to be largely a matter of convenience in the selection of carborane reagents
and has no particular mechanistic significance, since in either case the metal
atom effectively fills a hole in a nido-carborane substrate.

Recently, it has been found that polyhedral carboranes react directly
with certain metal reagents to effect insertion of the metal without an initial

* The formal designation of this compound would be $1\text{-}(\eta\text{-}C_5H_5)\text{-}1,2,4\text{-}CoC_2B_4H_6$.

cage-opening treatment. Such reactions can occur even in the gas phase, and may be assumed to be mechanistically quite distinct from the other routes mentioned.

The following discussion illustrates the various preparative methods as a background for the detailed treatment of metallocarborane synthesis and reactions in Sections IV–VII.

B. Synthesis from Open-Cage Carboranes

1. Small Nido-Carboranes

Open-cage (*nido*) carboranes in many cases react with metal reagents, losing the bridge hydrogen atoms and forming closed polyhedral metallo-carboranes.[77] Such reactions can occur directly in the vapor phase or in solution; in other cases, a bridge proton is first removed via base attack to create a *nido*-carborane anion which is reactive toward metal reagents. This method is limited in its general applicability, primarily because few stable and readily obtainable *nido*-carboranes are known. However, 2,3-dicarba-hexaborane(8), $C_2B_4H_8$ (Figure 2b) is conveniently prepared by the thermal reaction of B_5H_9 with acetylene and is an excellent precursor to polyhedral metallocarboranes, particularly those containing a C_2B_4M nucleus [3,76–81,189,190] (see Sections IV.D.1, IV.F.2, IV.H.1, VII.C.2, and VII.D.2, and Figures 12–14, 21, 25, and 33).

Metallocarboranes have also been prepared from the small *nido* species $2,3,4\text{-}CH_3C_3B_3H_6$ [106] and $1,2\text{-}C_2B_3H_7$ [56] (Sections IV.C.1 and IV.D.1), but unfortunately these starting materials are difficult to obtain in useful quantities. The structures of the parent $C_3B_3H_7$ and $C_2B_3H_7$ molecules are depicted in Figures 3a and 3b, respectively. Another member of this *nido*-carborane series, $2,3,4,5\text{-}C_4B_2H_6$, has a similar pentagonal pyramidal structure with all four carbons in the base. Due to its lack of bridge hydrogens it is expected

| (a) | (b) |

Figure 3. Structures of $2,3,4\text{-}C_3B_3H_7$ (a), and $1,2\text{-}C_2B_3H_7$ (b). ●, CH; ○, BH; •, H bridge.

to undergo direct metal insertion into the open face to form *closo*-metallo-carboranes having an MC_4B_2 cage system. This carborane, however, is also inconvenient to prepare,[6,145,158] and no such metallocarborane syntheses have been reported.

2. $C_2B_9H_{13}$ *and* $C_2B_7H_{13}$

The icosahedral-fragment species $1,2\text{-}C_2B_9H_{13}$ (Figure 4) which contains two B–H–B bridges on a C_2B_3 open face, has been employed occasionally in metallocarborane synthesis, primarily with nontransition-metal reagents such as dimethylberyllium etherate (Section VII.B).[166,167] Similar metal insertions with aluminum and gallium reagents have been reported, as discussed in Section VII.C.

The $C_2B_9H_{13}$ system is reversibly bridge-deprotonated to form the $1,2\text{-}C_2B_9H_{12}^-$ and $1,2\text{-}C_2B_9H_{11}^{2-}$ ions,[223] which have played a prominent role in metallocarborane chemistry. The most convenient synthesis of these ions, however, is via base degradation of $1,2\text{-}C_2B_{10}H_{12}$ (*ortho*-carborane) as described below.

The pyrolysis of $1,2\text{-}C_2B_9H_{13}$ or its isomer, $1,7\text{-}C_2B_9H_{13}$, produces an 11-vertex *closo*-carborane, $2,3\text{-}C_2B_9H_{11}$, which on treatment with oxidizing agents is degraded to a new open-cage species $1,3\text{-}C_2B_7H_{13}$ (Figure 5a).[195] An alternative, and better, synthesis of $C_2B_7H_{13}$ is the oxidation of the $1,7\text{-}C_2B_9H_{12}^-$ ion with chromic acid.[60]

The $C_2B_7H_{13}$ molecule contains two methylene groups and two bridging protons,[203] the latter being easily removed by treatment with sodium hydride, forming the *nido*-$C_2B_7H_{11}^{2-}$ ion[93] (Figure 5b). Scheme 1 summarizes these interconversions. The $C_2B_7H_{11}^{2-}$ species has been utilized sparingly in metallocarborane synthesis, reacting with manganese or cobalt reagents to

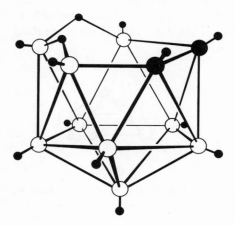

Figure 4. Structure of $1,2\text{-}C_2B_9H_{13}$. ●, C; ○, B; •, H.

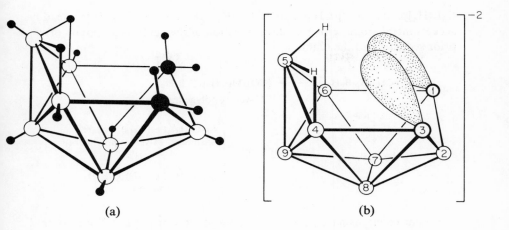

(a) (b)

Figure 5. Structure of 1,2-$C_2B_7H_{13}$ (a) (carbon atoms depicted as large solid circles), and probable structure of the $C_2B_7H_{11}^{2-}$ ion (b) (carbon atoms at positions 1 and 3) showing orbitals directed toward one of the vacant icosahedral vertices.[63] (From *J. Amer. Chem. Soc.*, **91**, 5475 (1969). Copyright by the American Chemical Society).

release H_2 and effect metal insertion into the open face.[61,63,93,96] The cobalt species formed are, as expected, C_2B_7M polyhedra, but in the manganese reaction a boron atom is lost from the cage and the main product is a C_2B_6M system.[61,96] These syntheses are described in Sections IV.C and IV.E.

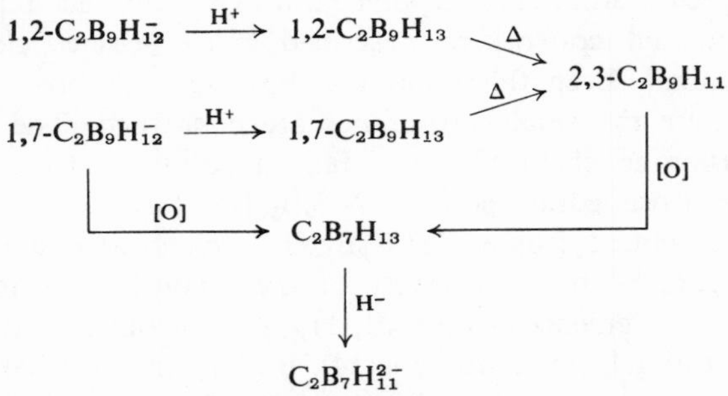

$$1,2\text{-}C_2B_9H_{12}^- \xrightarrow{H^+} 1,2\text{-}C_2B_9H_{13} \xrightarrow{\Delta}$$

$$\xrightarrow{\Delta} 2,3\text{-}C_2B_9H_{11}$$

$$1,7\text{-}C_2B_9H_{12}^- \xrightarrow{H^+} 1,7\text{-}C_2B_9H_{13}$$

$$\downarrow [O] \qquad\qquad\qquad\qquad \downarrow [O]$$

$$\xrightarrow{[O]} C_2B_7H_{13} \longleftarrow$$

$$\downarrow H^-$$

$$C_2B_7H_{11}^{2-}$$

SCHEME 1

3. $(CH_3)_4C_4B_8H_8$

The recently discovered[138,139] tetracarbon carborane $(CH_3)_4C_4B_8H_8$, a colorless air-stable solid which is obtained in the air oxidation of $[(CH_3)_2\text{-}$

$C_2B_4H_4]_2CoH$ or $[(CH_3)_2C_2B_4H_4]_2FeH_2$ (Section IV.F.2),[138,140] readily accepts metal atoms, either by direct reaction of the neutral compound or via prior reduction to the dianion.

$$(CH_3)_4C_4B_8H_8 \xrightarrow{\text{M(CO)}_6} (CO)_3M(CH_3)_4C_4B_8H_8$$

13-vertex polyhedra

$$\downarrow \text{Na}^0$$

$$[(CH_3)_4C_4B_8H_8]^{2-} \xrightarrow{\text{FeCl}_2,\ C_5H_5^-,\ O_2} \begin{cases} (\eta\text{-}C_5H_5)_2Fe_2(CH_3)_4C_4B_8H_8 \\ \text{4 isomers, 14-vertex polyhedra} \\ + \\ (\eta\text{-}C_5H_5)Fe(CH_3)_4C_4B_7H_8 \\ \text{12-vertex open cage} \end{cases}$$

Similar reactions have generated a series of tetracarbon metallocarboranes containing iron, cobalt, or nickel[136] (Sections IV.D.5, IV.F.4, IV.F.5a, IV.F.6, IV.H.3a, and IV.H.4).

C. Synthesis from Polyhedral Carboranes

1. Methods and Starting Materials

Each of the polyhedral carboranes of the $C_2B_{n-2}H_n$ series[75] has been successfully employed, in at least one isomeric form, in metallocarborane synthesis. At present, no single method has been shown to be applicable to all of these species, although the polyhedral expansion technique developed by Hawthorne and co-workers, described below, comes close to being completely general. Even this approach, however, has proved largely unpredictable where the small carboranes are concerned. The development of metallocarborane chemistry thus far has centered to a considerable degree on the icosahedral species $1,2\text{-}C_2B_{10}H_{12}$ (*ortho*-carborane) and its thermal-rearrangement isomer $1,7\text{-}C_2B_{10}H_{12}$ (*meta*-carborane) (Figure 6) as a consequence of the availability of these compounds in bench-scale quantities. The convenience of $1,2\text{-}C_2B_{10}H_{12}$ as a starting material is based on its facile preparation from decaborane(14), $B_{10}H_{14}$, an air-stable solid hydride which was prepared in ton quantities as part of a rocket-fuel development program in the 1950s. As a result, the majority of the known metallocarboranes (and nearly all of those reported before 1970) were obtained from $C_2B_{10}H_{12}$ or from intermediate-size carboranes derived from this compound by degradative processes.[75] More recent work[77] has produced syntheses of metallocarboranes from the lower polyhedral species such as $C_2B_3H_5$, $C_2B_4H_6$, and $C_2B_5H_7$, but the scale of these reactions is limited both by the relative scarcity

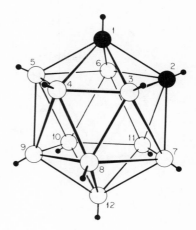

Figure 6. Structure of $1,2-C_2B_{10}H_{12}$, showing numbering system for icosahedral structures. ●, C; ○, B; •, H.

of these materials and by their high volatility, which necessitates the use of vacuum equipment. However, reactions of these lower *closo*-carboranes and of $2,3-C_2B_4H_8$ provide, at present, virtually the only routes to the smallest and structurally simplest metallocarboranes. Fortunately, most of these small-cage species are air-stable crystalline solids comparable to their larger homologs.

The most commonly employed methods of metal insertion into poly-hedral carboranes are the base-degradation and reductive cage-opening (polyhedral expansion) routes. Both procedures involve the initial formation of an open-cage anionic species, but they differ in that base degradation effects the removal of a cage BH group which is then replaced by a metal atom, while in a polyhedral expansion reaction all of the original atoms are retained and the cage is enlarged by the incorporation of the metal. A newer method, utilizing direct reaction of polyhedral carboranes with metal reagents (described in Section III.C.5) is an important development.

2. Base Degradation

The base-degradation technique has in general been successful only with 1,2- and $1,7-C_2B_{10}H_{12}$ and with certain metallocarboranes (see below). These icosahedral carboranes are exceedingly stable molecules, for example failing to react with hot concentrated H_2SO_4, but in the presence of strong Lewis bases such as trimethylamine or ethoxide ion, a boron atom is removed forming a $C_2B_9H_{12}^-$ ion which contains a bridge hydrogen on the open face:[87]

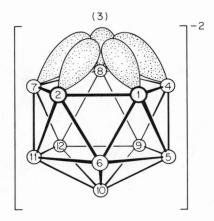

Figure 7. Schematic drawing of the 1,2- and $1,7\text{-}C_2B_9H_{11}^{2-}$ ions showing orbitals directed toward the vacant icosahedral vertex, with carbon atoms at either the 1 and 2 or 1 and 7 positions.[100] (From *J. Amer. Chem. Soc.*, **90**, 879 (1968). Copyright by the American Chemical Society.)

$$1,2\text{-}C_2B_{10}H_{12} + C_2H_5O^- + 2\ C_2H_5OH \xrightarrow[\text{fast}]{85°} B(OC_2H_5)_3 + H_2 + 1,2\text{-}C_2B_9H_{12}^-$$

$$\downarrow 450° \qquad\qquad\qquad\qquad\qquad\qquad\qquad\qquad\qquad\qquad\qquad\qquad \downarrow 350°$$

$$1,7\text{-}C_2B_{10}H_{12} + C_2H_5O^- + 2\ C_2H_5OH \xrightarrow[\text{slow}]{85°} B(OC_2H_5)_3 + H_2 + 1,7\text{-}C_2B_9H_{12}^-$$

The reaction is specific at positions B(6) or B(9) in $1,2\text{-}C_2B_{10}H_{12}$, and at B(2) or B(3) in $1,7\text{-}C_2B_{10}H_{12}$, in both cases corresponding to the most electropositive boron atom in the cage (Figure 6). The bridge proton in 1,2- or $1,7\text{-}C_2B_9H_{12}^-$ is readily removed by strong bases in aqueous or non-aqueous media to give the 1,2- or $1,7\text{-}C_2B_9H_{11}^{2-}$ ion (Figure 7), which corresponds structurally to an icosahedron with one vertex vacant[87]:

$$C_2B_9H_{12}^- + NaH \xrightarrow{THF} Na^+C_2B_9H_{11}^{2-} + H_2$$

$$C_2B_9H_{12}^- + OH^- \underset{}{\overset{H_2O}{\rightleftharpoons}} C_2B_9H_{11}^{2-} + H_2O$$

The complete icosahedron can be restored by the insertion of a metal (or nonmetal) atom into the open face of 1,2- or $1,7\text{-}C_2B_9H_{11}^{2-}$, and a large number of heteroatom-containing icosahedral carboranes have been prepared in this manner. In the preparation of metallocarboranes, it is not usually necessary to isolate the $C_2B_9H_{11}^{2-}$ salt; indeed, if the metal reagent and the product can tolerate aqueous basic media, the reaction may be conducted directly on $C_2B_9H_{12}^-$ in the presence of excess OH^- ion, since $C_2B_9H_{11}^{2-}$ is present in high concentration under such conditions.[75,87,92] The following equations are illustrative for the preparation of a bis (carboranyl)cobalt species:

$$\tfrac{3}{2}\ CoCl_2 + 2\ C_2B_9H_{11}^{2-} \xrightarrow{THF} (C_2B_9H_{11})_2Co^- + 3\ Cl^- + \tfrac{1}{2}\ Co^0$$

$$CoCl_2 + 2\ C_2B_9H_{12}^- \xrightarrow[\text{NaOH}]{\text{hot aq.}} (C_2B_9H_{11})_2Co^{2-} \xrightarrow{\text{air}} (C_2B_9H_{11})_2Co^- + Co^0$$

In the presence of $C_5H_5^-$ ion, mixed-ligand species such as $(\eta\text{-}C_5H_5)$-$CoC_2B_9H_{11}$ (Section IV.F.5) are formed. A recent improvement on this approach utilizes the addition of the metal salt and neutral cyclopentadiene directly into a $C_2B_9H_{12}^-$ solution which has been prepared *in situ* by alkaline methanolysis of 1,2-$C_2B_{10}H_{12}$.[116,164]

Attempts to extend the base-degradation technique to lower carboranes have been unsuccessful, owing to the tendency of these smaller cages to disintegrate under strongly basic conditions. For example, the pentagonal bipyramidal carborane 2,4-$C_2B_5H_7$ (Figure 1b), might be expected to lose an apex boron atom in reaction with ethanolic KOH and form an open cage $C_2B_4H_6^{2-}$ or $C_2B_4H_7^-$ species, but the carborane instead undergoes cage degradation[156]; however, the 2,3-$C_2B_4H_7^-$ ion is easily obtained from the *nido* species 2,3-$C_2B_4H_8$, as described above.

Although the base-degradation process has not been successfully applied to carboranes other than the $C_2B_{10}H_{12}$ isomers, it has been used to open icosahedral metallocarboranes and to effect the incorporation of a second metal atom.[54] The product contains a central $Co_2C_2B_8$ icosahedron, with each cobalt also facially bonded to a $C_2B_9H_{11}^{2-}$ ligand. Again, however, the method appears to be limited to icosahedral species.

$$(C_2B_9H_{11})_2Co^- + CoCl_2 \xrightarrow[\text{NaOH}]{\text{hot aq.}} [(C_2B_9H_{11})_2Co_2(C_2B_8H_{10})]^{2-}$$

3. Reductive Cage Opening (Polyhedral Expansion)

A far more general approach to metallocarborane synthesis was originated by Dunks and Hawthorne in 1970,[34] and is based on the fact that the polyhedral carboranes are filled-shell electronic systems that cannot accommodate additional electrons without significant distortion of the polyhedron as discussed in Section II. This distortion in most instances takes the form of cage opening, so that the addition of two electrons to a polyhedral carborane results in the formation of an open-cage dianion whose structure corresponds to the next larger polyhedron with one vertex vacant. It must be added, however, that few such anions have been well characterized, the evidence for them being primarily the structures of their derivative metallocarboranes.[42,44] Thus, the reduction of 1,6-$C_2B_6H_8$ with sodium in the presence of naphthalene yields a product assumed to be a *nido*-$C_2B_6H_8^{2-}$ ion; addition of $CoCl_2$ and NaC_5H_5 to this intermediate gives $(\eta\text{-}C_5H_5)CoC_2B_6H_8$, a nine-atom metallocarborane, and $(\eta\text{-}C_5H_5)_2Co_2C_2B_6H_8$, a ten-atom system (Section IV.F.3).[34,42]

Similar treatment of the polyhedral $C_2B_7H_9$, $C_2B_8H_{10}$, $C_2B_9H_{11}$,[42] and $C_2B_{10}H_{12}$[36] systems yield "expanded" metallocarborane products, the last being particularly interesting in that 13-atom cages are formed. (Detailed discussions of these reactions and the structures of the products will be found in Section IV.) The same procedure can be conducted with metallocarboranes themselves, to insert a second, or even a third, metal atom into the original cage.[43,46,47] Clearly, this type of reaction constitutes a powerful approach to metallocarborane synthesis which is not only broadly applicable but offers, up to a point, a rational means of preparing metallocarborane species of specific geometry and composition. However, the reactions occurring in many of the systems studied are extremely complex and frequently yield not only "expanded" metallocarboranes but also species having the same number of cage atoms, or even fewer, than the original carborane. In some instances, the "expanded" metallocarborane products are obtained only in low yield compared to the "nonexpanded" species; this is the case in the reaction of $2,4\text{-}C_2B_5H_7$ with sodium naphthalide followed by cobalt[146] or iron[189] insertion, as discussed in Section IV.

In this author's view, polyhedral expansion is properly regarded as an *effect* which is observed in the course of, or as a consequence of, certain reactions which may follow quite different pathways. For example, the direct insertion of metals in the vapor phase, discussed below, results in polyhedral expansion under conditions quite distinct from the sodium reduction experiments.

4. Polyhedral Contraction

All of the above synthetic routes to metallocarboranes feature the addition of metal atoms to a carborane cage, in some but not all cases retaining all of the original cage atoms. An entirely different approach is via "polyhedral contraction" in which a metallocarborane cage of n vertices is subjected to base degradation to yield an open-cage species of $(n - 1)$ atoms; removal of two electrons from this intermediate generates a closed polyhedron having $(n - 1)$ vertices. An example is the conversion of $(\eta\text{-}C_5H_5)\text{-}CoC_2B_9H_{11}$ to $(\eta\text{-}C_5H_5)CoC_2B_8H_{10}$ (Section IV.F.4).[113,115] While this is not

$$(\eta\text{-}C_5H_5)CoC_2B_9H_{11} \xrightarrow{\text{OH}^-} \text{red solution} \xrightarrow{\text{H}_2\text{O}_2}$$
$$\text{(intermediate)}$$

$$(\eta\text{-}C_5H_5)CoC_2B_8H_{10} + 2,6,7\text{-}(\eta\text{-}C_5H_5)CoC_2B_7H_9 + (\eta\text{-}C_5H_5)CoC_2B_7H_9$$
$$(63\%) \qquad\qquad (8\%) \qquad\qquad (\sim 2\%)$$

a primary synthetic route to metallocarboranes, in the sense that it does not involve the initial incorporation of metal atoms into a cage, it does seem likely to have considerable importance as a preparative tool in the interconversion of metallocarborane species.

5. Direct Metal Insertion

The reaction of $1,5-C_2B_3H_5$, $1,6-C_2B_4H_6$, or $2,4-C_2B_5H_7$ with various organometallics in the vapor phase at elevated temperatures produces metallocarboranes in fair to good yields.[77,151,189] Surprisingly, this reaction has been employed with some success even with the very stable $1,2-C_2B_{10}H_{12}$ system. In most cases, the products are "expanded," i.e., the metal insertion proceeds without loss of boron from the cage. However, $1,2-C_2B_{10}H_{12}$ reacts with $(\eta-C_5H_5)Co(CO)_2$ to give primarily $(\eta-C_5H_5)CoC_2B_9H_{11}$ species in which the metal replaces a boron atom. (It is likely that in this case a 13-atom $(\eta-C_5H_5)CoC_2B_{10}H_{12}$ species forms initially by simple metal insertion, but subsequently undergoes thermal expulsion of a BH group.[151]) This approach at present provides the only known method of metallo-carborane synthesis from $1,5-C_2B_3H_5$, since attempts to effect metal insertion via reduction with sodium have been unsuccessful.[79]

Direct insertions of metal atoms into carboranes of 7 to 11 vertices have also been conducted in ether or hydrocarbon solutions, yielding polyhedral metallocarboranes of nickel, platinum, and palladium (see Sections IV.H and IV.I).[2,68–71,191]

Effectively, most such reactions involve displacement of neutral ligands such as CO or olefins by the carborane, and thus do not *require* description in terms of a change in oxidation number of the metal. For example, while $(CO)_3FeC_2B_4H_6$ is conventionally viewed as a formal $C_2B_4H_6^{2-}$ ligand complexed to formal Fe(II), it is also valid (and equivalent) to regard a neutral $C_2B_4H_6$ ligand as replacing two neutral CO groups in $Fe(CO)_5$. (The $C_2B_4H_6$ ligand does, however, undergo a change in structure from octahedral to pentagonal pyramidal as a consequence of the introduction of the $:Fe(CO)_3$ moiety; see Section II).

The direct-insertion technique appears to have some advantages over synthetic routes requiring a prior cage-opening of the polyhedron, particularly the fact that it is a simple one-step operation of evidently broad applicability.

6. Metal Replacement in Metallocarboranes

The extraction of a metal atom from a metallocarborane and insertion of a different metal is a potentially useful method of metallocarborane synthesis, but few examples are known. A short communication[191] has appeared describing the reaction of thallium carborane anions, $TlR_1R_2C_2-B_9H_9^-$ (R_1, R_2 = H or CH_3), with metal halide compounds of iron, cobalt, platinum, or palladium in which thallium is replaced in the polyhedron by the transition metal. At this writing, the details of these syntheses have not been published.

D. Synthesis from Metalloboranes

A potential route to metallocarboranes, in principle, is the insertion of carbon into a metalloborane cage, but the only clear example of such a process is the reaction of acetylene with $2\text{-}(\eta\text{-}C_5H_5)CoB_4H_8$[147] to give the *nido*-metallocarborane species $1,2,3\text{-}(\eta\text{-}C_5H_5)CoC_2B_3H_7$.[77,148]

E. Synthesis from Boranes

The one reported case of metallocarborane synthesis directly from a boron hydride is the reaction of $Na^+B_{10}H_{13}^-$ with the hexacarbonyls of chromium, molybdenum, or tungsten to form a $(B_{10}H_{10}COH)M(CO)_4^-$ species in which the metal is incorporated into an icosahedral cage system[213] (Section V.C). Subsequent reactions of these products generate CO-bridged species and *nido*-metalloboranes. This reaction has not yet been described in detail, and its extension to other boranes or metal reagents has not been reported.

F. Other Potential Routes to Metallocarboranes

From a practical viewpoint, the variety of available carboranes (and preparative routes based on them) makes it likely that these materials will continue to be the major precursors to metallocarboranes. Nevertheless, it is worth considering other means by which stable metal–carbon–boron-cage systems might be synthesized. The metalloboranes form a potentially useful, if limited, class of starting materials, as was mentioned above. A completely unexplored possibility is the incorporation of boron into metallocarbon-cage compounds such as the $(C_6H_5)_4C_4Fe_3(CO)_8$ cluster[32] that contains a pentagonal bipyramidal C_4Fe_3 cage. The attractiveness of this idea is increased by the apparent electronic and structural analogy between such clusters and the boron cage compounds; for example, several authors[74,112,153,208] have recently stressed that all such compounds are formally electron-deficient systems stabilized by delocalization of electrons, and that the bonding and structural principles underlying both types of cage species are, to a degree, the same. Hence one might expect that the synthesis of metal–boron–carbon-cage systems could be approached from the metal–carbon clusters as well as from the carboranes, and the potential application of the reductive cage opening and direct metal-insertion techniques described above to metal cluster compounds seems rich with possibilities.

IV. DICARBON, TRICARBON, AND TETRACARBON TRANSITION-METAL METALLOCARBORANES

Most of the known and characterized metallocarborane species have two cage carbon atoms, primarily as a consequence of the fact that the most readily available carborane starting materials are dicarbon systems such as the

TABLE I

Dicarbon, Tricarbon, and Tetracarbon Transition-Metal Metallocarboranes

Compound[a]	Color	Mp, °C	Other data[b]	References
Titanium, zirconium, and vanadium				
13-Atom Cages				
$[(C_2H_5)_4N]_2[(C_2B_{10}H_{10})_2Ti]$	Red-orange		B, H, EC	179
$[(CH_3)_4N][(CH_3)_2C_2B_{10}H_{10})_2Ti]$	Bright red		B, H, EC, X	179
$[(C_2H_5)_4N]_2[(CH_3)_2C_2B_{10}H_{10})_2Zr]$	Purple		B, H, EC	179
$[(C_2H_5)_4N]_2[(C_2B_{10}H_{12})_2V]$	Red-brown		PR, EC	129, 179
Chromium				
12-Atom cages				
$Cs[(1,2-C_2B_9H_{11})_2Cr]$	Violet		IR, E, MAG	173
$[(CH_3)_4N][(1,2-(CH_3)C_2B_9H_{10})_2Cr]$	Violet		IR, E	173
$[(CH_3)_4N][(1,2-(CH_3)_2C_2B_9H_9)_2Cr]$	Blue-violet		IR, E	173
			B, MAG, PR	222
$Cs[(1,2-(CH_3)_2C_2B_9H_9)_2Cr] \cdot H_2O$	Dark red		X	175
$[(CH_3)_4N][(1,2-(C_6H_5)(C_2H_5)C_2B_9H_{10})_2Cr]$	Violet		IR, E	173
$3,1,2-(C_5H_5)CrC_2B_9H_{11}$	Dark red	248–249	IR, E	173
$1-CH_3-3,1,2-(C_5H_5)CrC_2B_9H_{10}$	Dark red	219–220	IR, E	173
$1,2-(CH_3)_2-3,1,2-(C_5H_5)CrC_2B_9H_9$	Dark red	261–262	IR, E	173
$1-C_6H_5-3,1,2-(C_5H_5)CrC_2B_9H_{10}$	Dark red	208–209	IR, E	173

(Continued)

[a] All compounds are closed polyhedral (*closo*) cages, except where the prefix *nido* is used to indicate open-cage structures. Isomers are numbered according to the IUPAC system (see also Section II) and indicate both established structures and proposed structures based on spectroscopic data. Formulas without prefix numbers represent compounds not structurally characterized. All C_5H_5 ligands listed are symmetrically π-bonded (pentahapto) to the metal atom.

[b] B = ^{11}B nmr data; H = ^{1}H nmr data; IR = infrared spectral data; MS = mass spectra or cutoff m/e values; X = X-ray diffraction data or structural information; MAG = magnetic susceptibility data; E = electronic spectral data; EC = electrochemical data; OR = optical rotation data; ORD = optical rotatory dispersion data; F = ^{19}F nmr data; C = ^{13}C nmr data; MB = Mössbauer data; NQR = nuclear quadrupole resonance data; PR = paramagnetic resonance data.

TABLE I —Continued

Compound[a]	Color	Mp, °C	Other data[b]	References
$[(CH_3)_4N]_2[3,1,2-(CO)_3CrC_2B_9H_{11}]$	Yellow		IR, B	97
$Cs[(1,7-C_2B_9H_{11})_2Cr]$	Brown		IR, E	173
$2,1,7-(C_5H_5)CrC_2B_9H_{11}$	Dark red	217–218	IR, E	173
Molybdenum				
12-Atom cages				
$[(CH_3)_4N]_2[3,1,2-(CO)_3MoC_2B_9H_{11}]$	Gray		IR	100
$[(CH_3)_4N]_2[1,2-(CH_3)_2-3,1,2-(CO)_3MoC_2B_9H_9]$	Pale yellow		IR	100
$[(CH_3)_4N][3,1,2-(CO)_3(CH_3)MoC_2B_9H_{11}]$	Tan		IR, H	100
$[(CH_3)_4N][3,1,2-(CO)_3HMoC_2B_9H_{11}]$	Red		IR	100
$[(CH_3)_4N]_2[3-Mo(CO)_5-1,2-(CH_3)_2-3,1,2-(CO)_3MoC_2B_9H_9]$	Yellow		IR	100
$[(CH_3)_4N]_2[3-W(CO)_5-3,1,2-(CO)_3MoC_2B_9H_{11}]$			IR, X	100
13-Atom cages, dicarbon				
$[(C_2H_5)_4N]_2[4,1,7-(CO)_3MoC_2B_{10}H_{12}]$	Bright yellow		H, B, IR	36
13-Atom cages, tetracarbon				
$(CO)_3Mo(CH_3)_4C_4B_8H_8$	Dark green		H, B, IR, MS	139
Tungsten				
12-Atom cages				
$[(CH_3)_4N]_2[3,1,2-(CO)_3WC_2B_9H_{11}]$	Yellow		IR	100
$[(CH_3)_4N][3,1,2-(CO)_3(CH_3)WC_2B_9H_{11}]$	Pale green		IR, H	100
$[(CH_3)_4N]_2[3-W(CO)_5-3,1,2-(CO)_3WC_2B_9H_{11}]$			IR	100
$[(CH_3)_4N]_2[3-Mo(CO)_5-3,1,2-(CO)_3WC_2B_9H_{11}]$			IR	100
13-Atom cages, dicarbon				
$[(C_2H_5)_4N]_2[4,1,7-(CO)_3WC_2B_{10}H_{12}]$	Orange		H, B, IR	36
13-Atom cages, tetracarbon				
$(CO)_3W(CH_3)_4C_4B_8H_8$	Blue-green		H, B, IR, MS	137, 139
Manganese				
7-Atom cages, tricarbon				

Compound	Color	M.p. (°C)	Methods	References
$2\text{-CH}_3\text{-}1,2,3,4\text{-(CO)}_3\text{MnC}_3\text{B}_3\text{H}_5$	Yellow	<20	B, H, IR, MS	106, 107
$2,3\text{-(CH}_3)_2\text{-}1,2,3,4\text{-(CO)}_3\text{MnC}_3\text{B}_3\text{H}_4$	Yellow-orange	<20	MS	106
9-Atom cages				
$[(\text{CH}_3)_4\text{N}][1,4,6\text{-(CO)}_3\text{MnC}_2\text{B}_6\text{H}_8]$	Amber		IR, B, H, E	61, 96
$[(\text{C}_6\text{H}_5)_3\text{PCH}_3][1,4,6\text{-(CO)}_3\text{MnC}_2\text{B}_6\text{H}_8]$			X	105
$[(\text{CH}_3)_4\text{N}][4\text{-C}_6\text{H}_5\text{-}1,4,6\text{-(CO)}_3\text{MnC}_2\text{B}_6\text{H}_7]$	Amber		B, H	61
$[(\text{CH}_3)_4\text{N}][4,6\text{-(CH}_3)_2\text{-}1,4,6\text{-(CO)}_3\text{MnC}_2\text{B}_6\text{H}_6]$	Orange		B, H	61
12-Atom cages				
$[(\text{CH}_3)_4\text{N}][3,1,2\text{-(CO)}_3\text{MnC}_2\text{B}_9\text{H}_{11}]$	Pale yellow		IR, H, E	100
$\text{Cs}[3,1,2\text{-(CO)}_3\text{MnC}_2\text{B}_9\text{H}_{11}]$	Yellow		IR, E	100
$\text{Cs}[3,1,2\text{-(CO)}_2[(\text{C}_6\text{H}_5)_3\text{P}]\text{MnC}_2\text{B}_9\text{H}_{11}]$		230–232 (dec)	IR	242
$\text{Cs}[1,2\text{-(C}_2\text{H}_3)_2\text{-}3,1,2\text{-(CO)}_3\text{MnC}_2\text{B}_9\text{H}_9]^c$	Bright yellow	285 (dec)	H, IR	231
$[(\text{CH}_3)_4\text{N}][(1,2\text{-C}_4\text{H}_4\text{-}1,2\text{-C}_2\text{B}_9\text{H}_9)\text{Mn(CO)}_3]$	Pale yellow	252 (dec)	H, E	134
$[(\text{CH}_3)_4\text{N}][(1,2\text{-C}_4\text{H}_6\text{-}1,2\text{-C}_2\text{B}_9\text{H}_9)\text{Mn(CO)}_3]$	Yellow	253 (dec)	H, IR, E	134
$[(\text{CH}_3)_4\text{N}]_2[(1,2\text{-C}_4\text{H}_6\text{-}1,2\text{-C}_2\text{B}_9\text{H}_9)\text{Mn(CO)}_2\text{Br}]_2$			H, IR, E	134
Rhenium				
12-Atom cages				
$[(\text{CH}_3)_4\text{N}][3,1,2\text{-(CO)}_3\text{ReC}_2\text{B}_9\text{H}_{11}]$	Pale yellow		IR, H, E	100
$\text{Cs}[3,1,2\text{-(CO)}_3\text{ReC}_2\text{B}_9\text{H}_{11}]$			IR, E, X	100, 243
Iron				
6-Atom cages				
$1,2,4\text{-(CO)}_3\text{FeC}_2\text{B}_3\text{H}_5$	Orange		B, H, IR, MS	151
$nido\text{-}1,2,3\text{-(CO)}_3\text{FeC}_2\text{B}_3\text{H}_7$	Pale yellow	~ −15	B, H, IR, MS	76, 189
			X	10
7-Atom cages, one metal atom				
$1,2,3\text{-(CO)}_3\text{FeC}_2\text{B}_4\text{H}_6$	Orange	<20	B, H, IR, MS	76, 189
$1,2,4\text{-(CO)}_3\text{FeC}_2\text{B}_4\text{H}_6$	Orange		B, H, IR, MS	189
$1,2,3\text{-(C}_5\text{H}_5)\text{FeC}_2\text{B}_4\text{H}_6$	Brown		B, H, IR, MS, MAG, PR	189, 190

(Continued)

^{c} C_2H_3 = vinyl.

TABLE I —*Continued*

Compound[a]	Color	Mp, °C	Other data[b]	References
1,2,4-$(C_5H_5)FeC_2B_4H_6$	Brown		MS	189
σ-$C_{10}H_7$-1,2,4-$(C_5H_5)FeC_2B_4H_5$	Green		B, H, MS	189
5-[2'-(2',4'-$C_2B_5H_6$)]-1,2,4-$(C_5H_5)FeC_2B_4H_5$	Lime green		B, H, IR, MS	189
1,2,3-$(C_5H_5)Fe(H)C_2B_4H_6$	Red-orange		B, H, IR, MS	189, 190
1,2,4-$(C_5H_5)Fe(H)C_2B_4H_6$	Red-orange		B, H, IR, MS	189
σ-(2',4'-$C_2B_5H_6$)-1,2,4-$(C_5H_5)Fe(H)C_2B_4H_5$			MS	189
[$(CH_3)_4N$]{[$(CH_3)_2C_2B_4H_4]_2Fe$}	Red-orange	243–246 (dec)	B, H, IR, MS	138
[2,3-$(CH_3)_2C_2B_4H_4]_2FeH_2$	Red		B, H, IR, MS	138
$nido$-μ-[$(C_5H_5)(CO)_2Fe$]-2,3-$C_2B_4H_7$	Light yellow		B, H, IR, MS	189, 190
7-Atom cages, two metal atoms				
1,2,3,5-$(CO)_6Fe_2C_2B_3H_5$	Orange		B, H, IR, MS	151
1,7-$(C_5H_5)Co(CO)_3Fe(CH_3)_2C_2B_3H_3$	Red-brown		B, H, IR, MS	138
8-Atom cages				
3,1,7-$(CO)_3FeC_2B_5H_7$	Orange		B, H, IR, MS	189
$(C_5H_5)CoFe(CH_3)_4C_4B_8H_8$[d]	Dark green		B, H, IR, MS, X	138, 141
9-Atom cages, one metal atom				
1,4,6-$(C_5H_5)FeC_2B_6H_8$	Brown		B, H, IR, MS, E, EC	42
			B, PR	222
4,1,8-$(C_5H_5)FeC_2B_6H_8$	Red	132–133	B, H, IR, MS, E, EC	42
			B, PR	222
9-Atom cages, three metal atoms				
$(CO)_6Fe_2(C_5H_5)Co(CH_3)_2C_2B_4H_4$	Yellow-brown		B, H, MS	135
10-Atom cages, one metal atom				
1,2,4-$(CO)_3Fe(CH_3)_2C_2B_7H_9$	Bright yellow		B, H, MS	137
10-Atom cages, two metal atoms				
1,6,2,3-$(C_5H_5)_2Fe_2C_2B_6H_8$	Green		B, MS, EC, MAG, X	12

11-Atom cages[e]				
$1,2,3\text{-}(C_5H_5)FeC_2B_8H_{10}$	Red-brown	275–277	B, H, IR, MS, E, EC	42
			B, PR, MAG	222
12-Atom cages, dicarbon				
$[(CH_3)_4N][3,1,2\text{-}(C_5H_5)Fe^{II}C_2B_9H_{11}]$	Orange	158	H, IR, E	100
$3,1,2\text{-}(C_5H_5)Fe^{III}C_2B_9H_{11}$	Purple	181–182	B, IR, E, EC	100
			MB, PR, X	7, 102, 132, 245
$[(CH_3)_4N][2,1,7\text{-}(C_5H_5)Fe^{II}C_2B_9H_{11}]$			MAG	65
			B, H, C, PR	222
			B, H	36 (p. 1114)
$2,1,7\text{-}(C_5H_5)Fe^{III}C_2B_9H_{11}$	Green		IR	36 (p. 1114)
				222
$[(CH_3)_4N]_2[(1,2\text{-}C_2B_9H_{11})_2Fe_2(CO)_4]$	Dark red		B, PR	97
			B, IR	72
$Cs_2[(1,2\text{-}C_2B_9H_{11})_2Fe^{II}]$			X	98
$[(CH_3)_4N]_2[(1,2\text{-}C_2B_9H_{11})_2Fe^{II}]$			B, E	98
$Cs[(1,2\text{-}C_2B_9H_{11})_2Fe^{II}H]$			H, E	100
$[(C_6H_5)_3PCH_3][(1,2\text{-}C_2B_9H_{11})_2Fe^{II}H]$	Orange	180–182 (dec)	B, EC	98
$Cs[(1,2\text{-}C_2B_9H_{11})_2Fe^{III}]$	Black	247–249	IR, E	98
			B, IR, E, EC, MAG	132
			PR	100
$[(CH_3)_4N][(1,2\text{-}C_2B_9H_{11})_2Fe^{III}]$	Red	>300	B, IR, E, EC, MAG	65, 100
			MB, PR	7, 102, 132
			B, PR	222
$[(CH_3)_4N]_2[(1,2\text{-}(CH_3)_2C_2B_9H_9)_2Fe^{II}]$	Lavender		H, E	100
$[(CH_3)_4N][(1,2\text{-}(CH_3)_2C_2B_9H_9)_2Fe^{III}]$	Red	247–249	B, IR, E, EC, MAG, PR	100, 132

(*Continued*)

[d] Two 8-atom cages fused on an edge.
[e] For 11-atom cages with Fe and CO atoms, see under cobalt compounds.

TABLE I —Continued

Compound[a]	Color	Mp, °C	Other data[b]	References
[(CH₃)₄N]₂[(1,2-(C₆H₅)C₂B₉H₁₀)₂Fe^II]	Blue		H, E	100
[(CH₃)₄N][(1,2-(C₆H₅)C₂B₉H₁₀)₂Fe^III]	Red	>300	B, IR, E, EC, MAG / PR	100 / 132
[(CH₃)₄N][(1-m-FC₆H₄-1,2-C₂B₉H₁₀)₂Fe^III]	Dark purple	160–200	F	1
[(CH₃)₄N][(1-p-FC₆H₄-1,2-C₂B₉H₁₀)₂Fe^III]		180–251	F	1
[8-[(C₂H₅)₂S]-1,2-C₂B₉H₁₀]₂Fe^II	Purple	215–217 (dec)	B, H, IR, E, EC	98
[(C₂H₅)₃PCH₃][(8-(C₂H₅)₂S-1,2-C₂B₉H₁₀)Fe^ii(1,2-C₂B₉H₁₁)]	Deep pink	227–228 (dec)	B, H, IR, E	98
[8-(C₂H₅)₂S-1,2-C₂B₉H₁₀Fe^III[1,2-C₂B₉H₁₁]	Dark red		IR, E, EC / B, PR	98 / 222
μ(8,8')-S₂CH-(1,2-C₂B₉H₁₀)₂Fe^III	Gray-green	d. 240	IR	52
[(CH₃)₄N][2-(1,12-C₂B₉H₁₁)₂Fe^III]			B, PR	222
12-Atom cages, tetracarbon				
nido-(C₅H₅)Fe(CH₃)₄C₄B₇H₈	Red-brown		B, H, IR, MS, X	11
13-Atom cages[f]				
[(C₂H₅)₄N]₂[(C₂B₁₀H₁₂)₂Fe^II]	Purple		B, H, IR, E, EC	36
[(CH₃)₄N][(C₅H₅)Fe^II C₂B₁₀H₁₂]	Orange		B, H	36
(C₅H₅)FeC₂B₁₀H₁₂	Violet		IR, E, EC, MAG	36
14-Atom cages, tetracarbon				
(C₅H₅)₂Fe₂(CH₃)₄C₄B₈H₈, isomer I	Brown		B, H, IR, MS, X	136, 142
isomer II	Gray-green		B, H, IR, MS, X	136, 142
isomer III	Gray-brown		B, H, IR, MS	136
isomer IV	Gray		MS	136
isomer V	Brown		B, H, IR, MS, X	136
Ruthenium				
3,1,2-[((C₆H₅)₃P)₂(CO)RuC₂B₉H₁₁	Yellow	315–316	H, IR	185
3,1,2-(CO)₃RuC₂B₉H₁₁·0.5C₆H₆	Yellow	<350 (dec)	B, H, IR, MS, E	185
Cobalt				
6-Atom cages				

Compound	Color		Methods	Ref.
$1,2,4\text{-}(C_5H_5)CoC_2B_3H_5$	Yellow-orange		B, H, IR, MS, E	151
$nido\text{-}1,2,3\text{-}(C_5H_5)CoC_2B_3H_7$	Pale yellow		B, H, IR, MS	79
$nido\text{-}2\text{-}CH_3\text{-}1,2,3\text{-}(C_5H_5)CoC_2B_3H_6$	Yellow		B, H, IR, MS	79
$[nido\text{-}2,3\text{-}(CH_3)_2C_2B_3H_5]CoH[closo\text{-}2,3\text{-}(CH_3)_2C_2B_4H_4]$	Yellow		B, H, MS	140
7-Atom cages, one metal atom				
$1,2,3\text{-}(C_5H_5)CoC_2B_4H_6$	Orange		B, H, IR, MS	79
$2\text{-}CH_3\text{-}1,2,3\text{-}(C_5H_5)CoC_2B_4H_5$	Orange		B, H, IR, MS	79
$2,3\text{-}(CH_3)_2\text{-}1,2,3\text{-}(C_5H_5)CoC_2B_4H_4$	Dark orange		X	215
$2\text{-}[(CH_3)_3Si]\text{-}1,2,3\text{-}(C_5H_5)CoC_2B_4H_5$	Orange		B, H, IR, MS	217
$B\text{-}[(CH_3)_3Si]\text{-}1,2,3\text{-}(C_5H_5)CoC_2B_4H_5$	Orange		B, H, IR, MS	217
$B\text{-}Br\text{-}1,2,3\text{-}(C_5H_5)CoC_2B_4H_5$	Orange		B, H, IR, MS	217
$1,2,4\text{-}(C_5H_5)CoC_2B_4H_6$	Yellow		B, H, IR, MS	146
$3\text{-}(2\text{-}C_{10}H_7)\text{-}1,2,4\text{-}(C_5H_5)CoC_2B_4H_5$	Yellow		B, H, IR, MS	146
$5\text{-}(1\text{-}C_{10}H_7)\text{-}1,2,4\text{-}(C_5H_5)CoC_2B_4H_5$	Yellow		B, H, IR, MS	79
$3\text{-}(2',4'\text{-}C_2B_5H_6)\text{-}1,2,4\text{-}(C_5H_5)CoC_2B_4H_5$	Yellow		B, H, IR, MS	146
$[(CH_3)_4N][(2,4\text{-}C_2B_4H_6)_2Co]$	Golden		B, H, IR, MS	79
$[2,3\text{-}(CH_3)_2\text{-}2,3\text{-}C_2B_4H_4]_2CoH$	Bright red		B, H, IR, MS	140
$[2,3\text{-}(CH_3)_2C_2B_4H_4]CoH[2,3\text{-}(CH_3)_2C_2B_4H_4]$	Yellow		B, H, IR, MS	140
$[(CH_3)_4N]\{[(CH_3)_2C_2B_4H_4]Co[(CH_3)_2C_2B_3H_5]\}$	Yellow	259–262	B, H, IR, MS	140
7-Atom cages, two metal atoms				
$1,7,2,3\text{-}(C_5H_5)_2Co_2C_2B_3H_5$	Red-brown		B, H, IR, MS	3, 79
$2\text{-}CH_3\text{-}1,7,2,3\text{-}(C_5H_5)_2Co_2C_2B_3H_4$	Red-brown		B, H, IR, MS, X	3, 79
$2,3\text{-}(CH_3)_2\text{-}1,7,2,3\text{-}(C_5H_5)_2Co_2C_2B_3H_3$	Red-brown		B, H, IR, MS	79
$2\text{-}(CH_3)_3Si\text{-}1,7,2,3\text{-}(C_5H_5)_2Co_2C_2B_3H_4$	Red		B, H, MS	216
$2\text{-}C_6H_5\text{-}1,7,2,3\text{-}(C_5H_5)_2Co_2C_2B_3H_4$	Red		B, H, MS	216
$5\text{-}Br\text{-}1,7,2,3\text{-}(C_5H_5)_2Co_2C_2B_3H_4$	Red		B, H, MS	216
$5\text{-}I\text{-}1,7,2,3\text{-}(C_5H_5)_2Co_2C_2B_3H_4$	Red		B, H, MS	216
$1,7,2,3\text{-}(CH_3C_5H_4)(C_5H_5)Co_2C_2B_3H_5$	Red		B, H, MS	216

(Continued)

^{f} For 13-atom cages with Fe and Co atoms, see under cobalt compounds.

TABLE I—*Continued*

Compound[a]	Color	Mp, °C	Other data[b]	References
$1,7,2,3\text{-}(C_2H_5C_5H_4)(C_5H_5)Co_2C_2B_3H_5$	Red		B, H, MS	216
$1,7,2,3\text{-}[(CH_3)_3Si\text{-}C_5H_4](C_5H_5)Co_2C_2B_3H_5$	Red		B, H, MS	216
$1,7,2,4\text{-}(C_5H_5)_2Co_2C_2B_3H_5$	Dark green		B, H, IR, MS	3, 77, 79
$2\text{-}CH_3\text{-}1,7,2,4\text{-}(C_5H_5)_2Co_2C_2B_3H_4$	Dark green		B, H, IR, MS	148
			X	169
$2\text{-}(CH_3)_3Si\text{-}1,7,2,4\text{-}(C_5H_5)_2Co_2C_2B_3H_4$	Green		B, H, MS	216
$2\text{-}C_6H_5\text{-}1,7,2,4\text{-}(C_5H_5)_2Co_2C_2B_3H_4$	Green		B, H, MS	216
$1,7,2,4\text{-}(CH_3C_5H_4)(C_5H_5)Co_2C_2B_3H_5$	Green		B, H, MS	216
$1,7,2,4\text{-}(C_2H_5C_5H_4)Co_2C_2B_3H_5$	Green		B, H, MS	216
$1,7,2,4\text{-}[(CH_3)_3SiC_5H_4]Co_2C_2B_3H_5$	Green		B, H, MS	216
$1,2,3,5\text{-}(C_5H_5)_2Co_2C_2B_3H_5$	Dark green		B, H, IR, MS, E	77, 151
$1,2,4,5\text{-}(C_5H_5)_2Co_2C_2B_3H_5$	Dark green		B, H, IR, MS, E	77, 151
$nido,closo\text{-}[(CH_3)_2C_2B_3H_5]CoH[(CH_3)_2C_2B_3H_3]Co(C_5H_5)$	Red-brown		B, H, IR, MS	140
$closo,closo\text{-}$ $(C_5H_5)Co[(CH_3)_2C_2B_3H_3]CoH[(CH_3)_2C_2B_3H_3]Co(C_5H_5)$	Black		B, H, IR, MS	140
8-Atom cages, one metal atom				
$3,1,7\text{-}(C_5H_5)CoC_2B_5H_7$	Orange		B, H, IR, MS, E	146, 151
8-Atom cages, two metal atoms[g]				
$3,5,1,7\text{-}(C_5H_5)_2Co_2C_2B_4H_6$	Green[h]		B, H, IR, MS, E	146, 151
9-Atom cages, one metal atom[i]				
$1,4,5\text{-}(C_5H_5)CoC_2B_6H_8$	Dark red	61–64	B, H, IR, E, EC	42
$1,2,6\text{-}(C_5H_5)CoC_2B_6H_8$ } identical compounds[j]	Red		B, H, IR, MS	146
$1,4,6\text{-}(C_5H_5)CoC_2B_6H_8$	Red-orange	120–122	B, H, IR, MS, E, EC	42
$[(C_6H_5)_4As][(4,5\text{-}C_2B_6H_8)_2Co]$	Dark brown		B, H, IR, E, EC	42
9-Atom cages, two metal atoms				
$1,8,5,6\text{-}(C_5H_5)_2Co_2C_2B_5H_7$	Red-brown		B, H, IR, MS	146, 151
			EC	43

Compound	Color	M.p.	Data	Ref.
$1,7,5,6\text{-}(C_5H_5)_2Co_2C_2B_5H_7$	Green		X	82
			B, H, IR, MS, E	146, 151
			X	82
10-Atom cages, one metal atom[i]				
$2,1,6\text{-}(C_5H_5)CoC_2B_7H_9$	Red	158–159	B, H, IR, MS, E	63, 93
$8\text{-}COCH_3\text{-}2,1,6\text{-}(C_5H_5)CoC_2B_7H_8$	Red	162	B, H, IR	66
$2,1,6\text{-}(C_5H_5)CoC_2B_7H_7Br_2$	Orange	129–130	H	66
$2,6,9\text{-}(C_5H_5)CoC_2B_7H_9$	Yellow		B, H, IR, MS, E, EC	113
$2,1,10\text{-}(C_5H_5)CoC_2B_7H_9$	Orange	113	B, H, IR, MS, E	62, 63
$[2,1,10\text{-}(C_5H_5)CoC_2B_7H_9]^-$			B	222
$2,3,10\text{-}(C_5H_5)CoC_2B_7H_9$	Yellow	57–58	B, H, IR, MS, E, EC	42
$(C_5H_5)CoC_2B_7H_9$ (isomer unknown)	Yellow		B, H, IR, MS, E, EC	113
$[(CH_3)_4N][(1,6\text{-}C_2B_7H_9)_2Co]$	Brown	240	B, H, IR, E	63, 93
$[(C_2H_5)_4N][(1,6\text{-}C_2B_7H_9)_2Co]$			X	127
$[(CH_3)_4N][(6,9\text{-}C_2B_7H_9)_2Co]$	Red	235	H, IR	63
$[(CH_3)_4N][(1,10\text{-}C_2B_7H_9)_2Co]$		192	B, H	63
$Cs[(1,10\text{-}C_2B_9H_9)_2Co]$	Orange	210	IR, E	62, 63
$[(CH_3)_4N][(1,2\text{-}C_2B_9H_{11})Co(1',6'\text{-}C_2B_7H_9)]$	Red		B, H, IR, E, EC	113
$[(CH_3)_4N][(1,2\text{-}C_2B_9H_{11})Co(1',10'\text{-}C_2B_7H_9)]$	Fawn		B, H, IR, E, EC	113
$nido\text{-}8,6,7\text{-}(C_5H_5)CoC_2B_7H_{11}$	Red	123	B, H, IR, MS, E, EC	113
			X	14
10-Atom cages, two metal atoms[i]				
$2,4,3,10\text{-}(C_5H_5)_2Co_2C_2B_6H_8$	Red-brown		B, H, IR, MS	146
$2,6,1,10\text{-}(C_5H_5)_2Co_2C_2B_6H_8$	Green	236–238	B, H, IR, MS, E, EC, X	34, 42
			X	104

(Continued)

[g] For an Fe–Co 8-atom system, see under iron compounds.

[h] Incorrectly reported as red in original paper.[146]

[i] See also monocarbon metallocarboranes in Table III.

[j] Thermal rearrangement data[38] suggest that the $1,4,5\text{-}(C_5H_5)CoC_2B_8H_8$ assignment is correct, although the ^{1}H and ^{11}B nmr data are consistent with both the 1,4,5-structure and the 1,2,6 geometry tentatively suggested earlier[146] (originally numbered 3,1,6).

TABLE I —*Continued*

Compound[a]	Color	Mp, °C	Other data[b]	References
2,7,1,10-$(C_5H_5)_2Co_2C_2B_6H_8$	Red		B, H, MS, E, EC	49,50
10-Atom cages, three metal atoms				
2,3,8,1,6-$(C_5H_5)_3Co_3C_2B_5H_7$	Green		B, H, IR, MS, E	151
2,3,4,1,10-$(C_5H_5)_3Co_3C_2B_5H_7$	Green		B, H, IR, MS, E	151
11-Atom cages, one metal atom, dicarbon				
1,2,3-$(C_5H_5)CoC_2B_8H_{10}$	Purple	297–300	B, H, IR, MS, E, EC	42
	Dark blue	294	MS, E	164
			B, PR	222
[1,2,3-$(C_5H_5)CoC_2B_8H_{10}$]$^-$				
2-[1'-(1',10'-$C_2B_8H_9$)]-1,2,3-$(C_5H_5)CoC_2B_8H_9$	Purple	179–180	B, H, IR, MS, E, EC	42
Cs[(2,3-$C_2B_8H_{10}$)$_2$Co]	Green	324 (dec)	B, H, IR, E, EC	42
[$(C_5H_5)_2$Co][(2,3-$C_2B_8H_{10}$)$_2$Co]	Green	310 (dec)	B, H	42
[$(CH_3)_4N$][(1,2-$C_2B_9H_{11}$)Co(2',3'-$C_2B_8H_{10}$)]			B, H, IR, E	114
1,2,4-$(C_5H_5)CoC_2B_8H_{10}$	Purple	168	B, H, IR, MS, E, EC	113
[7-C_5H_5N-1,2,4-$(C_5H_5)CoC_2B_8H_9$][PF$_6$]	Blue		B, H, IR, E	114
7-C_5H_{10}N-1,2,4-$(C_5H_5)CoC_2B_8H_9$	Green	151 (dec)	B, H, IR, MS, E	114
[$(CH_3)_4N$][(1,2-$C_2B_9H_{11}$)Co(2',4'-$C_2B_8H_{10}$)]	Blue		B, H, IR, E, EC	113
7'-C_5H_5N-(1,2-$C_2B_9H_{11}$)Co(2',4'-$C_2B_8H_9$)	Green	159	B, H, IR, E	114
nido-9,7,8-$(C_5H_5)CoC_2B_8H_{12}$	Mustard	128 (dec)	B, H, IR, MS, E	114
nido-[$(CH_3)_4N$][9,7,8-$(C_5H_5)CoC_2B_8H_{11}$]	Red		B, H, IR, E	114
nido-11-C_5H_5N-9,7,8-$(C_5H_5)CoC_2B_8H_{10}$	Red	224	B, H, IR, MS, E	114, 115
nido-11-C_5H_{10}NH-9,7,8-$(C_5H_5)CoC_2B_8H_{10}$	Red	> 200 (dec)	B, H, IR, MS, E	114, 115
[$(CH_3)_4N$]$_2$[*nido*-(1,2-$C_2B_9H_{11}$)Co(7',8'-$C_2B_8H_{11}$)]	Brown		B, H, IR, E	114
[$(CH_3)_4N$][*nido*-(1,2-$C_2B_9H_{11}$)Co(7',8'-$C_2B_8H_{12}$)]			B, H, IR, E	114
[$(CH_3)_4N$][*nido*-11'-C_5H_5N-(1,2-$C_2B_9H_{11}$)Co(7',8'-$C_2B_8H_{10}$)]	Brown		B, H, IR, E	114, 115
			X	22, 24
2,3-$(CH_3)_2$-10,2,3-$(C_5H_5)CoC_2B_6H_8$	Orange		B, H, IR, MS, E, EC	143

11-Atom cages, one metal atom, tetracarbon

Compound	Color	M.p.	Properties	Ref.
(C5H5)Co(CH3)4C4B6H6, isomer I	Yellow		B, H, IR, MS	136
isomer II	Golden yellow		B, H, IR, MS	136

11-Atom cages, two metal atoms

Compound	Color	M.p.	Properties	Ref.
1,4,2,3-(C5H5)2Co2C2B7H9	Green-brown	275–277	B, H, IR, E, EC	43
1,6,2,4-(C5H5)2Co2C2B7H9	Green-brown		B, H, E, EC	49
1,10,2,3-(C5H5)2Co2C2B7H9	Red-brown	219–220	B, H, IR, MS, E, EC	43, 50
1,7,2,4-(C5H5)2Co2C2B7H9				50
1,7,2,3-(C5H5)2Co2C2B7H9				50
1,8,2,3-(C5H5)2Co2C2B7H9	Red-brown	265–266	B, H	47
			E, IR, EC	43
8,9,2,3-(C5H5)2Co2C2B7H9	Orange		B, H, IR, MS, EC	49
1,8,2,3-(C5H5)2FeCoC2B7H9			B, MS, MAG	37

12-Atom cages, one metal atom, dicarbon

Compound	Color	M.p.	Properties	Ref.
3,1,2-(C5H5)CoC2B9H11	Yellow-orange	246–248	H, IR, E, EC	100, 118
[3,1,2-(C5H5)CoC2B9H11]⁻	Yellow-orange	229–230	B	222
1-(m-FC6H4)-3,1,2-(C5H5)CoC2B9H10	Yellow	250–251.8	F	1
1-(p-FC6H4)-3,1,2-(C5H5)CoC2B9H10		182–183	F	1
3,1,2-(C6H5C5H4)CoC2B9H11			H, IR	237
3,1,2-(m-FC6H4C5H4)CoC2B9H11			H, F	237
3,1,2-(p-FC6H4C5H4)CoC2B9H11			H, F	237
3,1,2-(C5H5)CoC2B9H10Br		276–278	H	235
3,1,2-(C5H5)CoC2B9H9Br2		269–272	H	235
3,1,2-(C5H5)CoC2B9H8Br3		360–363	H	235
1-CH3-3,1,2-(C5H5)CoC2B9H10	Orange	203–204	H	235
1-CH3-3,1,2-(C5H5)CoC2B9H9Br, isomer I		241–243	H	235
isomer II			H	235
1-CH3-3,1,2-(C5H5)CoC2B9H8Br2		205–208	H	235
1-COOH-3,1,2-(C5H5)CoC2B9H10	Yellow	290–291	H, IR	234, 236
1,2-(COOH)2-3,1,2-(C5H5)CoC2B9H9		298 (dec)	H, IR	234

(Continued)

TABLE I—*Continued*

Compound[a]	Color	Mp, °C	Other data[b]	References
1-(CH₃COO)-3,1,2-(C₅H₅)CoC₂B₉H₁₀			H, IR	234
1,2-(CH₃COO)₂-3,1,2-(C₅H₅)CoC₂B₉H₉		175–177	H, IR	234
3,1,2-(C₄H₉C₅H₄)CoC₂B₉H₁₁		61.5–62.5	H, IR	234
3,1,2-(C₆H₅C≡CC₅H₄)CoC₂B₉H₁₁		181–183	H, IR	234
1-(CH₂OH)-3,1,2-(C₅H₅)CoC₂B₉H₁₀	Yellow	208–210	H, IR	236
1-(CH₂=CH)-3,1,2-(C₅H₅)CoC₂B₉H₁₀	Yellow-orange	155.5–156.5	H, IR	236
1-(NCCH₂)-3,1,2-(C₅H₅)CoC₂B₉H₁₀	Yellow	233–235	H, IR	236
1-CH₃-2-(CH₃COO)-3,1,2-(C₅H₅)CoC₂B₉H₉	Yellow	219–220	H, IR	236
1-CH₃-1-(F₃C—CF=CF)-3,1,2-(C₅H₅)CoC₂B₉H₉	Yellow	146.5–148.5	H, IR	236
1-CHO-3,1,2-(C₅H₅)CoC₂B₉H₁₀	Yellow	186.5–188	H, IR	236
1-COCl-3,1,2-(C₅H₅)CoC₂B₉H₁₀	Yellow-orange	282–284	H, IR	236
1-C₆H₅CO-3,1,2-(C₅H₅)CoC₂B₉H₁₀	Yellow	232–233	H, IR	236
1,2-(CH₃)₂-3,1,2-(C₅H₅)CoC₂B₉H₉	Yellow-orange	>267	B, H, IR, E, EC	118
μ(1,2)-(CH₂)₃-3,1,2-(C₅H₅)CoC₂B₉H₉	Yellow-orange	302–303.5	B, H, IR, E, EC	118
2,1,7-(C₅H₅)CoC₂B₉H₁₁	Yellow-orange	239–239.5	B, H, IR, E, EC	118
1-(m-FC₆H₄)-2,1,7-(C₅H₅)CoC₂B₉H₁₁	Yellow	186–186.5	F	1
1-(p-FC₆H₄)-2,1,7-(C₅H₅)CoC₂B₉H₁₁	Yellow	209–210	F	1
1,7-(CH₃)₂-2,1,7-(C₅H₅)CoC₂B₉H₉	Yellow-orange	240–242	B, H, E, EC	118
4,1,7-(C₅H₅)CoC₂B₉H₁₁	Yellow-orange	152–153	B, H, IR, E, EC	118
1,7-(CH₃)₂-4,1,7-(C₅H₅)CoC₂B₉H₉	Yellow-orange	116–118	B, H, IR, E, EC	118
4,1,2-(C₅H₅)CoC₂B₉H₁₁	Yellow-orange	146–148	B, H, IR, E, EC	118
1,2-(CH₃)₂-4,1,2-(C₅H₅)CoC₂B₉H₉	Yellow-orange	114–116	B, H, E, EC	118
μ(1,2)-(CH₂)₃-4,1,2-(C₅H₅)CoC₂B₉H₉	Yellow-orange	197–198	B, H, IR, E, EC	118
9,1,7-(C₅H₅)CoC₂B₉H₁₁	Yellow-orange	116–116.5	B, H, IR, E, EC	118
1,7-(CH₃)₂-9,1,7-(C₅H₅)CoC₂B₉H₉	Yellow-orange	84–85	B, H, IR, E, EC	118
2,1,12-(C₅H₅)CoC₂B₉H₁₁	Yellow-orange	159–159.5	B, H, IR, E, EC	118
1,12-(CH₃)₂-2,1,12-(C₅H₅)CoC₂B₉H₉	Yellow-orange	94.5–96	B, H, IR, E, EC	118
5,1,7-(C₅H₅)CoC₂B₉H₁₁	Yellow-orange	128–130	B, H, IR, E, EC	118

Compound	Color	Temp	Methods	Ref.
1,7-(CH₃)₂-5,1,7-(C₅H₅)CoC₂B₉H₉	Yellow-orange	115–116.5	B, H, E, EC	118
μ(1,2)-(CH₂)₃-8,1,2-(C₅H₅)CoC₂B₉H₉	Yellow-orange	163–163.2	B, H, IR, E, EC	118
μ(1,2)-(CH₂)₃-9,1,2-(C₅H₅)CoC₂B₉H₉	Yellow-orange	130–132	B, H, E, EC	118
[(CH₃)₄N][3,1,2-(CO)₂CoC₂B₉H₁₁]	Yellow		B, IR	97
Cs[(1,2-C₂B₉H₁₁)₂Co]	Yellow	> 300	H, IR, E, EC	100
			NQR, X	84, 244
K[(1,2-C₂B₉H₁₁)₂Co]	Yellow			55
[(CH₃)₄N][(1,2-C₂B₉H₁₁)₂Co]			B	187
[(C₂B₉H₁₁)₂Co]⁻			B	222
[(CH₃)₄N][(1,2-C₂B₉H₆D₅)₂Co]			B	187
[C₆H₅N₂][(1,2-C₂B₉H₁₁)₂Co]	Yellow		IR	55
[CH₃C₆H₄N₂][(1,2-C₂B₉H₁₁)₂Co]				55
[(CH₃)₄N][(1-C₆H₅-1,2-C₂B₉H₁₀)₂Co]	Red	290–293	H, IR, E, EC	100
[(CH₃)₄N][(1-(m-FC₆H₄)-1,2-C₂B₉H₁₀)₂Co]	Red		F	1
[(CH₃)₄N][(1-(p-FC₆H₄)-1,2-C₂B₉H₁₀)₂Co]	Dark red		F	1
[(CH₃)₄N][(6-C₆H₅-1,2-C₂B₉H₁₀)₂Co]	Yellow	275–277	H, B	99, 187
[(CH₃)₄N][(8-OH-1,2-C₂B₉H₁₀)₂Co]			H, IR	52
[(CH₃)₄N][(8-SH-1,2-C₂B₉H₁₀)₂Co]			B	187
Cs[(8-SH-1,2-C₂B₉H₁₀)₂Co]			H, IR	52
[B-(C₂H₅)₂S-1,2-C₂B₉H₁₀]Co(1,2-C₂B₉H₁₁)			B, H, IR	98
[(CH₃)₄N][(1,2-(CH₃)₂-1,2-C₂B₉H₉)₂Co]	Red	273–275	IR, H, E, EC	100
[(CH₃)₄N][(1,2-(C₂H₃)₂-1,2-C₂B₉H₉)₂Co]		272 (dec)	H, IR	231
[(C₆H₅)₄As][(μ(1,2)-C₃H₄-1,2-C₂B₉H₉)₂Co]		203–204	H, IR	231
Cs[(μ(1,2)-C₃H₆-1,2-C₂B₉H₉)₂Co]	Red		IR	161
[(CH₃)₄N][(μ(1,2)-C₃H₆-1,2-C₂B₉H₉)₂Co]	Red		B, H, IR, EC	161
[(CH₃)₄N][(1,2-C₂B₉H₈Br₃)₂Co]	Orange		H, IR, E, EC	100
			X	30
Cs[μ(8,8')-S₂-(1,2-C₂B₉H₁₀)₂Co]	Deep purple		H, IR	52
μ(8,8')-S₂-CH-(1,2-C₂B₉H₁₀)₂Co	Brown-yellow	302–304	B, H, IR	52, 187
			X	21

(Continued)

TABLE I—*Continued*

Compound[a]	Color	Mp, °C	Other data[b]	References
μ(8,8')-SCH₃-(1,2-C₂B₉H₁₀)₂Co	Red	302–304	B, H, MS	163
Cs[μ(8,8')-S₂CH₂-(1,2-C₂B₉H₁₀)₂Co]			H, IR	52
[(CH₃)₄N][(1,2-C₂B₉H₁₀SH)Co(1,2-C₂B₉H₁₀SCHO)]			H, IR	52
μ(8,8')-O₂CCH₃-(1,2-C₂B₉H₁₀)₂Co	Golden		B, H, IR	52
[(CH₃)₄N][μ(8,8')-C₆H₄-(1,2-C₂B₉H₁₀)₂Co]	Red		B, H, C, IR	55
[(CH₃)₄N][μ(8,8')-CH₃C₆H₃-(1,2-C₂B₉H₁₀)₂Co]	Red		B, H, IR	55
[(CH₃)₄N][(1,2-C₂H₄-1,2-C₂B₉H₉)₂Co]	Red-brown	>300	H, IR, E	134
[(CH₃)₄N][(1,2-C₂H₄-1,2-C₂B₉H₉)Co(1,2-C₂B₉H₁₁)]		>300	IR, E	134
[(CH₃)₄N][(1,2-C₂H₄-1,2-C₂B₉H₉)₂Co]		125 (dec)	H, IR	134
[(CH₃)₄N][(1,2-C₂B₉H₁₁)Co(2,4-C₂B₈H₁₀)]	Blue		B, H, IR, E, EC	113
[(CH₃)₄N][(1,2-C₂B₉H₁₁)Co(1,6-C₂B₇H₉)]	Red		B, H, IR, E, EC	113
[(CH₃)₄N][(1,2-C₂B₉H₁₁)Co(1,10-C₂B₇H₉)]	Fawn		B, H, IR, E, EC	113
Cs[(1,7-C₂B₉H₁₁)₂Co]	Tan	>350	H, IR, E, EC	100
μ(8,8')-S₂CH-(1,7-C₂B₉H₁₀)₂Co	Orange	225–228	IR, MS	52
12-Atom cages, one metal atom, tetracarbon[k]				
(C₅H₅)Co(CH₃)₄C₄B₇H₇, isomer I	Red-brown		B, H, IR, MS	136
isomer II	Dark red		B, H, IR, MS	136
isomer III	Orange-brown		B, H, IR, MS	136
12-Atom cages, two metal atoms, dicarbon				
3,4,1,2-(C₅H₅)₂Co₂C₂B₈H₁₀		280–282	B, H, IR, MS, E	43
3,6,1,2-(C₅H₅)₂Co₂C₂B₈H₁₀	Red	>350	B, H, IR, MS, E, EC	116, 164
1-CH₂OH-3,6,1,2-(C₅H₅)₂Co₂C₂B₈H₉	Orange-brown	264–265	IR	240
1-CHO-3,6,1,2-(C₅H₅)₂Co₂C₂B₈H₉	Dark red	303–304	IR	240
2,3,1,7-(C₅H₅)₂Co₂C₂B₈H₁₀	Green	275–277	B, H, IR, MS, E, EC; X	42, 45, 46; 16
2,4,1,7-(C₅H₅)₂Co₂C₂B₈H₁₀	Red		B, H, IR, MS, E, EC	49
2,7,1,12-(C₅H₅)₂Co₂C₂B₈H₁₀	Green	334–335	B, H, IR, MS, E, EC	42

Compound	Color	M.p.	Methods	References
$2,9,1,12\text{-}(C_5H_5)_2Co_2C_2B_8H_{10}$			B, H, MS, IR, EC	45, 49
$6,9,1,7\text{-}(C_5H_5)_2Co_2C_2B_8H_{10}$			B, H, MS	45, 49
$3,9,1,7\text{-}(C_5H_5)_2Co_2C_2B_8H_{10}$			B, H, MS	45, 49
$4,10,1,7\text{-}(C_5H_5)_2Co_2C_2B_8H_{10}$	Orange		B, H, IR, MS, EC	49
$2,8,1,12\text{-}(C_5H_5)_2Co_2C_2B_8H_{10}$	Red-orange		B, H, IR, MS, EC	49
$4,2,1,7\text{-}(C_5H_5)_2Co_2C_2B_8H_{10}$	Orange		B, H, IR, MS, EC	49
$2,10,1,7\text{-}(C_5H_5)_2Co_2C_2B_8H_{10}$	Red		B, H, IR, E, EC	49
$2,9,1,12\text{-}(C_5H_5)_2Co_2C_2B_8H_{10}$	Red-purple		B, H, IR, MS, EC	49
$2,5,1,7\text{-}(C_5H_5)_2Co_2C_2B_8H_{10}$	Orange		B, H, IR, MS, EC	45, 49
$3,6,1,7\text{-}(C_5H_5)_2Co_2C_2B_8H_{10}$			B, H, MS	45
$Cs_2[3,6,1,2\text{-}(1',2'\text{-}C_2B_9H_{11})_2Co_2C_2B_8H_{10}]$	Red		B, H, IR, E, EC	54
$[(CH_3)_4N][3,11,1,7\text{-}(2',3'\text{-}C_2B_8H_{10})(C_5H_5)Co_2C_2B_8H_{10}]$	Red-brown		H, X	51
$[(CH_3)_4N]_2[3,6,1,2\text{-}(1',2'\text{-}C_2B_9H_{11})_2Co_2C_2B_8H_{10}]$	Brown		H, IR, E, EC	53
$Cs_2[3,6,1,2\text{-}(1',2'\text{-}C_2B_9H_{11})_2Co_2C_2B_8H_{10}]\cdot H_2O$			X	174
$[(CH_3)_4N][(3,1,2\text{-}[C_5H_5]CoC_2B_9H_{11})_2\text{-}6\text{-}Co]$	Brown		B, H, IR, E, EC	116
$[(C_2H_5)_4N]_3[[1,2\text{-}C_2B_9H_{11}]\text{-}3\text{-}Co[1,2\text{-}C_2B_8H_{10}])_2\text{-}6\text{-}Co]$	Deep red		X	28
$[(CH_3)_4N]_3[[[1,2\text{-}C_2B_9H_{11}]\text{-}3\text{-}Co[1,2\text{-}C_2B_8H_{10}])_2\text{-}3\text{-}Co]$	Brown		H, IR, E, EC	53
			B	46
12-Atom cages, two metal atoms, tetracarbon				
$4,7,1,2\text{-}(C_5H_5)Co[(C_2H_5)_3P]_2PtC_2B_8H_{10}$	Red	174	B, H, IR, MS, X	71
$(C_5H_5)CoC_2B_8H_{10}Pt(C_8H_{12})$	Dark brown		H	71
12-Atom cages, three metal atoms				
$(C_5H_5)_2Co_2(CH_3)_4C_4B_6H_6$	Red		B, H, MS	136
$2,3,5,1,7\text{-}(C_5H_5)_3Co_3C_2B_7H_9$	Green	237–240 (dec)	B, H, MS, E	47
			EC	43
13-Atom cages, one metal atom, dicarbon				
$4,1,7\text{-}(C_5H_5)CoC_2B_{10}H_{12}$	Red	250–251	B, H, IR, MS, E, EC	35, 36
			X	19, 20
$(C_5H_5)CoC_2B_{10}H_{12}$	Orange	250–251	B, H, IR, MS, E, EC	35, 36
	Red-orange	250–251	B, H, IR, MS, E, EC	35, 36

ᵏ See also 10- and 11-atom polyhedra (above) containing cobalt shared with a 12-atom $CoC_2B_9H_{11}$ system.

(Continued)

TABLE I—Continued

Compound[a]	Color	Mp, °C	Other data[b]	References
$[(C_6H_5)_4N][(C_2B_{10}H_{12})_2Co]$	Dark green		B, H, IR, E, EC	36
$[(C_6H_5)_4N][(C_2B_{10}H_{12})_2Co]$	Red		B, H, IR, E, EC	36
$[(C_2B_{10}H_{12})_2Co]^-$, 3 isomers derived from o,m- and p- $C_2B_{10}H_{12}$				8, 238
13-Atom cages, one metal atom, tetracarbon				
$(C_5H_5)Co(CH_3)_4C_4B_8H_8$	Brown-violet		B, H, IR, MS, X	11, 136
13-Atom cages, two metal atoms				
$(C_5H_5)_2Co_2C_2B_9H_{11}$	Green	293–295	B, H, IR, E	43
$4,5,1,8-(C_5H_5)_2Co_2C_2B_9H_{11}$	Dark green	295–296	B, H, IR, MS, E, EC	40
$1,8-(CH_3)_2-4,5,1,8-(C_5H_5)_2Co_2C_2B_9H_9$	Brown-green	234–235	B, H, IR, MS, E, EC	40
$4,5,1,13-(C_5H_5)_2Co_2C_2B_9H_{11}$	Green	277–278	B, H, IR, MS, E, EC	40
$1,13-(CH_3)_2-4,5,1,13-(C_5H_5)_2Co_2C_2B_9H_9$	Green	270–272	B, H, IR, MS, E, EC	40
$[(CH_3)_4N]\{[(C_5H_5)CoC_2B_9H_{11}]_2Co\}$	Black	> 300	B, H, IR, E, EC	40
$4,5,1,8-(C_5H_5)_2CoFeC_2B_9H_{11}$	Black	> 300	B, H, IR, MS, EC, E	37
$4,5-(CH_3)_2-4,5,1,8-(C_5H_5)_2CoFeC_2B_9H_9$	Black	> 300	B, H, IR, MS, E, EC	37, 40
14-Atom cages				
$1,4,2,10-(C_5H_5)_2Co_2C_2B_{10}H_{12}$			B, H, MS, EC	48
$1,4,2,9-(C_5H_5)_2Co_2C_2B_{10}H_{12}$			B, H, MS	48
Rhodium				
$nido$-μ-$[(C_6H_5)_3P]_3Rh$-$C_2B_4H_7$	Yellow		B, H, IR	131
$3,1,2-[(C_6H_5)_3P]_2Rh(H)C_2B_9H_{11}$	Yellow		B, H, ^{31}P, IR	160
$2,1,7-[(C_6H_5)_3P]_2Rh(H)C_2B_9H_{11}$	Yellow		B, H, ^{31}P, IR	160
$3,1,2-[(C_6H_5)_3P]_2ClRhC_2B_9H_{11}$	Orange	195–196 (dec)	B, H, IR, E	185
$3,1,2-[(C_6H_5)_4B][(C_6H_5)_3P]RhC_2B_9H_{11}$	Yellow	300 (dec)	B, H, IR, E, EC	185
$[(C_6H_5)_3P(C_6H_6)RhC_2B_9H_{11}]_2$	Purple	155	B, H, IR, E	185
$[(C_6H_5)_3P](H)RhC_2B_9H_{11}$	Orange		B, H, E	185
Nickel				
7-Atom cages				
$1,2,4-[(C_6H_5)_3P]_2NiC_2B_4H_6$	Red-orange		B, H, IR, E	151

Compound	Color	M.p. (°C)	Methods	References
$1,2,3\text{-}[(C_6H_5)_2PCH_2]_2NiC_2B_4H_6$	Brown		B, H, IR, MS	79
9-Atom cages				
$1,2,7,8\text{-}(C_5H_5)_2Ni_2C_2B_5H_7$	Brown		B, H, IR, MS	79
10-Atom cages[l]				
$nido\text{-}6,5,9\text{-}[(CH_3)_3P]_2NiC_2B_7H_{11}$	Red	152–154	H, X	67, 68
12-Atom cages[m]				
$3,1,2\text{-}(C_5H_5)NiC_2B_9H_{11}$	Green-brown		E, EC, MAG	226
$3,1,2\text{-}(1,5\text{-}C_8H_{12})Nii[(CH_3)_2C_2B_9H_9]$	Brown	207–210 (dec)	B, H, IR, MS	71, 191
$3,1,2\text{-}[(C_2H_5)_3P]_2Nii[(CH_3)_2C_2B_9H_9]$	Brown	225–227	IR	71, 191
$[(C_2H_5)_4N]_2[(1,2\text{-}C_2B_9H_{11})_2Ni^{II}]$	Brown		IR, MAG	100
			E	159
$Rb[(1,2\text{-}C_2B_9H_{11})_2Ni^{III}]$	Green-black		E	100
			X-ray photoelectron	165
$Rb[(1,2\text{-}D_2C_2B_9H_9)_2Ni^{III}]$			IR	212
$[(CH_3)_4N][(1,2\text{-}C_2B_9H_{11})_2Ni^{III}]$			EC, X	83, 212
	Yellow-brown	>300	IR, E, MAG	100
			B, PR	222
$(1,2\text{-}C_2B_9H_{11})_2Ni^{IV}$	Orange	265 (dec)	IR, E, EC	100, 212
			X	177
			X-ray photoelectron	165
$(1,2\text{-}D_2C_2B_9H_9)_2Ni^{IV}$	Orange		IR, MS	212
$Cs[(1,2\text{-}CH_3C_2B_9H_{10})_2Ni^{III}]$	Orange-brown	~250 (dec)	IR, E, EC	212
$(1,2\text{-}CH_3C_2B_9H_{10})_2Ni^{IV}$	Orange		H, MS, E, EC	212
$[(CH_3)_4N][(1,2\text{-}(CH_3)_2C_2B_9H_9)_2Ni^{III}]$	Deep red	200–210 (rearr)	IR, E, EC	212
			X-ray photoelectron	165
$[(CH_3)_4N][(1,2\text{-}(C_6H_5)C_2B_9H_{10})_2Ni^{III}]$	Orange-brown	~250 (dec)	IR, E, EC	212
$[1,2\text{-}(C_6H_5)C_2B_9H_{10}]_2Ni^{IV}$	Deep red		H, IR, MS, E, EC	212

(Continued)

[l] See Table III for 10-atom monocarbon systems.
[m] See also 1:1 charge-transfer adducts of $(1,2\text{-}C_2B_9H_{11})_2Ni^{IV}$ with Lewis-base donors (ref. 212). A mixed Ni–Co 12-vertex species is listed above under 12-vertex cobalt compounds (two metal atoms).

TABLE I—*Continued*

Compound[a]	Color	Mp, °C	Other data[b]	References
[1,2-(*m*-FC$_6$H$_4$)C$_2$B$_9$H$_{10}$]$_2$Ni^{IV}	Red	190 (dec)	F	1
[1,2-(*p*-FC$_6$H$_4$)C$_2$B$_9$H$_{10}$]$_2$Ni^{IV}	Red	190 (dec)	F	1
[(CH$_3$)$_4$N][(μ(1,2)-(CH$_2$)$_3$-1,2-C$_2$B$_9$H$_9$)$_2$Ni^{III}]	Red-brown		EC	212
[(C$_2$H$_5$)$_4$N][(μ(1,2)-(CH$_2$)$_3$-1,2-C$_2$B$_9$H$_9$)$_2$Ni^{III}]	Dark brown		IR, EC	161
1,2,1′,7′-(C$_2$B$_9$H$_{11}$)$_2$Ni^{IV}			EC	212
[(1,2-C$_4$H$_4$-1,2-C$_2$B$_9$H$_9$)$_2$Ni$_2$]			H, IR, E	134
[(CH$_3$)$_4$N][(1,2-C$_4$H$_4$-1,2-C$_2$B$_9$H$_9$)$_2$Ni^{III}]	Dark purple	300	H, IR, E	134
[(CH$_3$)$_4$N][1,2,1′,7′-((CH$_3$)$_2$C$_2$B$_9$H$_9$)$_2$Ni^{III}]	Yellow-brown		IR, E, EC, ORD, OR	212
1,2,1′,7′-[(CH$_3$)$_2$C$_2$B$_9$H$_9$]$_2$Ni^{IV}	Orange		B, H, IR, MS, E, EC	212
			X	23
			X-ray photoelectron	165
μ(1,2)(1′,7′)-[(CH$_2$)$_3$C$_2$B$_9$H$_9$]$_2$Ni^{IV}	Orange		B, IR, MS, EC	161, 212
			X	229
[CH$_3$N(C$_2$H$_4$)$_3$NCH$_3$][(1,7-C$_2$B$_9$H$_{11}$)$_2$Ni^{II}]			IR, E, MAG	100
[(CH$_3$)$_4$N][(1,7-C$_2$B$_9$H$_{11}$)$_2$Ni^{III}]	Olive green	>300	X-ray photoelectron	165
(1,7-C$_2$B$_9$H$_{11}$)$_2$Ni^{IV}	Orange		E, IR, EC	100, 212
[(CH$_3$)$_4$N][(1,7-(CH$_3$)$_2$C$_2$B$_9$H$_9$)$_2$Ni^{III}]	Deep golden		E, IR, EC, ORD, OR	212
[1,7-(CH$_3$)$_2$C$_2$B$_9$H$_9$]$_2$Ni^{IV}	Yellow		B, H, IR, MS, E, EC	212
[μ(1,7)-(CH$_2$)$_3$C$_2$B$_9$H$_9$]$_2$Ni^{IV}	Yellow		IR, EC	161, 212
12-Atom cages, tetracarbon				
[(C$_6$H$_5$)$_2$PCH$_2$]$_2$Ni[(CH$_3$)$_4$C$_4$B$_7$H$_7$, isomer I	Red-brown		B, H, IR, MS	136
isomer II	Yellow		B, H, IR, MS	136
13-Atom cages, dicarbon				
[(C$_2$H$_5$)$_4$N]$_2$[(1,2-C$_2$B$_{10}$H$_{12}$)$_2$Ni^{II}]	Orange-brown		H, IR, E, EC	36
(1,2-C$_2$B$_{10}$H$_{12}$)$_2$Ni^{IV} (not isolated)			B	36

	Color	M.p. (°C)	Characterization	Ref.
13-Atom cages, tetracarbon				
[(C$_6$H$_5$)$_2$PCH$_2$]$_2$Ni(CH$_3$)$_4$C$_4$B$_8$H$_8$, isomer I	Red		B, H, IR, MS	136
isomer II	Orange-brown		B, H, IR, MS	136
isomer III	Yellow		B, H, IR, MS	136
Palladium				
12-Atom cages				
3,1,2-[(C$_6$H$_5$)$_4$C$_4$]PdC$_2$B$_9$H$_{11}$	Red	310 (dec)	IR, E	100, 214
1,2-(CH$_3$)$_2$-3,1,2-[(C$_6$H$_5$)$_4$C$_4$]PdC$_2$B$_9$H$_9$	Red		IR, E	100, 214
3,1,2-[(C$_6$H$_5$)$_3$P]$_2$PdC$_2$B$_9$H$_{11}$				191
1,2-(CH$_3$)$_2$-3,1,2-[*t*-C$_4$H$_9$NC]$_2$PdC$_2$B$_9$H$_9$	Yellow	150 (dec)	B, H, IR	71, 191
[(C$_2$H$_5$)$_4$N]$_2$[(1,2-C$_2$B$_9$H$_{11}$)$_2$Pd^{II}]	Rust		B, IR, E	211
[(C$_2$H$_5$)$_4$N][(1,2-C$_2$B$_9$H$_{11}$)$_2$Pd^{III}]	Brown		IR, E	211
(1,2-C$_2$B$_9$H$_{11}$)$_2$Pd^{IV}	Yellow		IR, E, EC	211, 212
[(CH$_3$)$_4$N]$_2$[(1,2-(CH$_3$)$_2$C$_2$B$_9$H$_9$)$_2$Pd^{II}]	Deep red-brown		B, H, IR, MS, E, EC	212
(1,2-C$_2$B$_9$H$_{11}$)$_2$Pd^{IV} · I$^-$	Red		IR	212
(1,2-C$_2$B$_9$H$_{11}$)$_2$Pd^{IV} · pyrene	Brown		IR	212
[1,7-(CH$_3$)$_2$C$_2$B$_9$H$_9$]$_2$Pd^{IV}	Yellow		B, H, IR, MS, E, EC	212
Platinum				
9-Atom cages, one metal atom				
5,6-(CH$_3$)-4,5,6-[(CH$_3$)$_3$P]$_2$PtC$_2$B$_6$H$_6$	Yellow		X	70
nido-5,8-(CH$_3$)$_2$-6,5,8-[(C$_2$H$_5$)$_3$P]$_2$PtC$_2$B$_6$H$_6$	Orange		X	70
9-Atom cages, two metal atoms				
1,7,4,6-[(C$_2$H$_5$)$_3$P]$_4$Pt$_2$C$_2$B$_5$H$_7$	Red	170 (dec)	X	2
10-Atom cages				
2,7-(CH$_3$)$_2$-9,2,7-[(C$_2$H$_5$)$_3$P]$_2$PtC$_2$B$_7$H$_7$	Pale yellow	153	B, H, ^{31}P, ^{195}Pt	69
			X	219
11-Atom cages				
nido-μ(4,8)-{[(CH$_3$)$_3$P]$_2$Pt}-8,7,10-[(CH$_3$)$_3$P]$_2$PtC$_2$B$_8$H$_{10}$	Yellow	156 (dec)	H, X	2
nido-8,7,10-[(CH$_3$)$_3$P]$_2$PtC$_2$B$_8$H$_{10}$	Pale yellow	283	X	2

(Continued)

TABLE I—Continued

Compound[a]	Color	Mp, °C	Other data[b]	References
12-Atom cages				
3,1,2-[1,5-C_8H_{12}]$PtC_2B_9H_{11}$	Pale yellow		H IR, MS, E	191, 212
1,7-$(CH_3)_2$-2,1,7-[$(C_2H_5)_3P]_2PtC_2B_9H_9$	Yellow	235	IR	71, 191
1,7-$(CH_3)_2$-2,1,7-[$C_6H_5(CH_3)_2P]_2PtC_2B_9H_9$		187	H, IR	71, 191
			X	218
1,7-$(CH_3)_2$-2,1,7-[[$(CH_3)_3P]_2PtC_2B_9H_9$	Pale yellow	231	H, IR, MS	71
1-C_6H_5-3,1,2-[$(C_2H_5)_3P]_2PtC_2B_9H_{10}$	Orange	164–166	IR, H, E	130
1-C_6H_5-3,1,2-[$(n\text{-}C_3H_7)_3P]_2PtC_2B_9H_{10}$	Orange	155–156	IR, H, E	130
1,2-$(CH_3)_2$-3,1,2-[$(C_6H_5)_3P]_2PtC_2B_9H_9$	Yellow-orange	217–218	IR, H, E	130
Copper				
12-Atom cages				
[$(C_2H_5)_4N]_2$[(1,2-$C_2B_9H_{11})_2Cu^{II}$]	Deep blue		IR, E, EC	9, 100, 211
[$(CH_3)_4N$][(1,2-$C_2B_9H_{11})_2Cu^{III}$]			X	227
			E	211
[$(C_6H_5)_3PCH_3$][(1,2-$C_2B_9H_{11})_2Cu^{III}$]	Deep red		B, IR, EC	211
			X	228
Silver				
12-Atom cages				
[$(C_2H_5)_4N$][(1,2-$C_2B_9H_{11})_2Ag$]	Black			9
Gold				
7-Atom cages, bridged				
nido-μ-[$(C_6H_5)_3PAu$]-2,3-$C_2B_4H_7$	Red		B, H, IR, MS	131
12-Atom cages				
[$(C_2H_5)_4N]_2$[(1,2-$C_2B_9H_{11})_2Au^{II}$]	Blue		IR, E, EC	211
[$(C_2H_5)_4N$][(1,2-$C_2B_9H_{11})_2Au^{III}$]	Red		B, IR, E	211
[$(C_6H_5)_3PCH_3$][(1,2-$C_2B_9H_{11})_2Au^{III}$]	Red		EC	211
Mercury, bridged				
nido-μ-C_6H_5Hg-2,3-$C_2B_4H_7$	Light yellow		B, MS	131

$C_2B_{n-2}H_n$ polyhedral series[75] and nido-2,3-$C_2B_4H_8$. However, several tri- and tetracarbon species have been synthesized in our laboratory and will be included in this section, since their synthesis is closely related to the dicarbon metallocarboranes. The monocarbon metallocarboranes form a somewhat distinct class and are discussed separately in Section V.

The principal synthetic routes to the dicarbon compounds are outlined in the previous section. For purposes of convenience, the following detailed discussion of these materials is arranged by transition element, proceeding from left to right across the periodic table. All of the characterized dicarbon, tricarbon, and tetracarbon transition metal metallocarboranes are listed in Table I.

A. Titanium, Zirconium, and Vanadium Compounds

Until very recently there were no known examples of metallocarboranes of Groups IIIb, IVb, or Vb, but in 1975 complexes of Ti, Zr, and V were prepared from the $1,2\text{-}C_2B_{10}H_{12}^{2-}$ ion or its C,C'-dimethyl derivatives.[179]

$$1,2\text{-}R_2C_2B_{10}H_{10} \xrightarrow{\text{Na}^0} R_2C_2B_{10}H_{10}^{2-} \xrightarrow{\text{TiCl}_4,\ \text{ZrCl}_4,\ \text{or VCl}_3} (R_2C_2B_{10}H_{10})_2M$$

$R = H, CH_3$

$\qquad\qquad\qquad\qquad\qquad\qquad M = Ti, \quad R = H, CH_3;$

$\qquad\qquad\qquad\qquad\qquad\qquad M = Zr, \quad R = CH_3;$

$\qquad\qquad\qquad\qquad\qquad\qquad M = V, \quad R = H$

A crystallographic study of the titanium dimethyl derivative[129] established its geometry as that of two closed 13-vertex polyhedra fused at a common metal atom. Significantly, these systems are substantially more stable than their bis(cyclopentadienyl) analogs, in line with the general trend of superior stability for metallocarborane complexes in comparison with the respective metallocenes.

The fact that these complexes are "electron hyperdeficient,"[141] having fewer than the usual allotment of $(2n + 2)$ skeletal valence electrons per polyhedral cage, suggests[179] that some type of distortion from regular n-vertex polyhedral geometry is to be expected (Section II. B). The titanium species does exhibit unusually long metal-to-cage distances,[129] which may reflect the shortage of electrons. (It has been suggested elsewhere[74,79] that electronically induced cage distortions in metal–boron polyhedral systems may, under some circumstances, take the form of gross changes in bond length, as distinct from the more usual opening or closure of the polyhedron.)

B. Chromium, Molybdenum, and Tungsten Compounds

1. Twelve-Atom Cages

The 1,2- and $1,7\text{-}C_2B_9H_{11}^{2-}$ ions and their C-substituted derivatives (Section III) react with $CrCl_3$ in the absence of air and water to give the respective biscarboranyl cobalt compounds.[173]

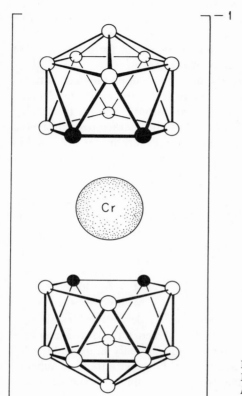

Figure 8. Structure of the 3,3'-Cr(1,2-C_2-B_9H_{11})$_2$ ion.[173] ●, CH; ○, BH. (From *Inorg. Chem.*, **7**, 2279 (1968). Copyright by the American Chemical Society.)

$$2\ C_2B_9H_{11}^{2-} + CrCl_3 \longrightarrow (C_2B_9H_{11})_2Cr^- + 3\ Cl^-$$

An X-ray study[175] of the $\{3,3'\text{-Cr}[1,2\text{-}(CH_3)_2\text{-}1,2\text{-}C_2B_9H_9)_2]\}^-$ ion established the structure illustrated in Figure 8, which consists of two icosahedral cages fused at the metal atom. These metallocarboranes containing formal Cr(III) are much more stable than their chromicinium analogs. For example, unlike the $Cr(\eta\text{-}C_5H_5)_2^+$ ion which is readily hydrolyzed, the $(C_2B_9H_{11})_2Cr^-$ salts are air stable and are unaffected even by hot concentrated sulfuric acid. Moreover, sodium amalgam fails to reduce these metallocarboranes to Cr(II) species.

The corresponding η-cyclopentadienyl chromium (II) compounds are obtained by introducing NaC_5H_5 into the reaction mixture.[173]

$$C_5H_5^- + C_2B_9H_{11}^{2-} + CrCl_3 \longrightarrow (\eta\text{-}C_5H_5)CrC_2B_9H_{11} + 3\ Cl^-$$

The photochemical reactions of chromium, molybdenum, and tungsten hexacarbonyls with the 1,2- and 1,7-$C_2B_9H_{11}^{2-}$ ions generate extremely air sensitive metallocarborane tricarbonyl anions.[97,100] The chemistry of the molybdenum and tungsten species is similar to that of the corresponding cyclopentadienyl metal tricarbonyls, for example, in their ease of protonation and alkylation.[100] Presumably, the air sensitivity reflects the fact that the metal in these species is in a formal oxidation state of zero, although the products of the reaction with air have not been described.

$$C_2B_9H_{11}^{2-} + M(CO)_6 \xrightarrow[M = Cr, Mo, W]{h\nu} [(CO)_3MC_2B_9H_{11}]^{2-} + 3\ CO$$

$$\xleftarrow[M = Mo\ or\ W]{CH_3I} \qquad\qquad \Big\downarrow {\text{anhydrous}\atop\text{HCl (M = Mo or W)}}$$

$$[CH_3(CO)_3MC_2B_9H_{11}]^- \qquad\qquad [H(CO)_3MC_2B_9H_{11}]^-$$

When excess $W(CO)_6$ or $Mo(CO)_6$ reacts with 1,2-$C_2B_9H_{11}^{2-}$ ion, or with the $(CO)_3MC_2B_9H_{11}^{2-}$ ions (M = W or Mo), metallocarboranes are obtained which contain an $M(CO)_5$ group linked to the cage metal atom via a metal–metal bond.[73] A series of such compounds containing both homo- and heteronuclear metal–metal bonds has been prepared, and the structure in Figure 9 has been proposed from [11]B and ^{1}H nmr spectra, supported by preliminary X-ray data (Ref. 100, footnote 34) which indicate the presence of metal–metal bonding and eight CO groups. These bimetallic derivatives are much more stable than the monometallic species described above, and the solids survive exposure to air for days although rapid decomposition occurs in solution.[100]

$$[(CO)_3MC_2B_9H_{11}]^{2-} + M'(CO)_6 \longrightarrow [(CO)_5M'(CO)_3MC_2B_9H_{11}]^{2-} + CO$$

$$M = Mo\ or\ W$$
$$M' = Mo\ or\ W$$

Several 12-atom monocarbon metallocarboranes and closely related 12-atom metallophospha- and metalloarsacarboranes of Cr, Mo, or W are discussed in Sections V and VI.

Experimental

Preparation of $Cs[(1,2-C_2B_9H_{11})_2Cr]$.[173] Chromium trichloride (2.06 g, 13.0 mmol) was added to a solution of $Na_2[1,2-C_2B_9H_{11}]$ (25.9 mmol from 5.0 g of $[(CH_3)_3NH][1,2-C_2B_9H_{11}]$) in 150 ml of THF. The cesium salt was precipitated by addition of aqueous cesium chloride, filtered, and washed with a small amount of ice water. Crystallization from water gave dark-violet crystals (3.98 g, 8.85 mmol, 68%).

Preparation of $3,1,2-(\eta-C_5H_5)CrC_2B_9H_{11}$.[173] Chromium trichloride (8.71 g, 55.0 mmol) was treated with the solution obtained by allowing

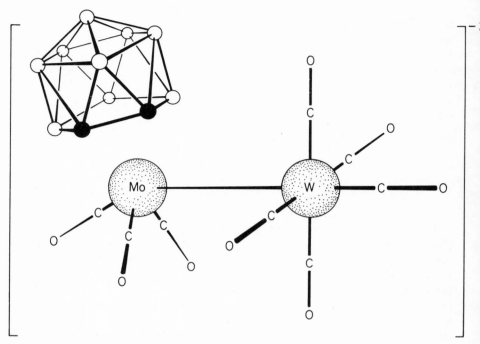

Figure 9. Structure of the $[(CO)_5W-(CO)_3MoC_2B_9H_{11}]^{2-}$ ion.[100] ●, CH; ○, BH.
(From *J. Amer. Chem. Soc.*, **90**, 879 (1968). Copyright by the American Chemical
Society.)

4.0 g (60.5 mmol) of C_5H_6 and 5.25 g (27.0 mmol) of $[(CH_3)_3NH][1,2-C_2B_9-H_{12}]$ to react with 3.9 g (162.5 mmol) of NaH (7.0 g of a 56% suspension in mineral oil) in 200 ml of THF. Dark-red crystals (1.55 g, 6.21 mmol, 23%) were obtained.

Preparation of $[(CH_3)_4N]_2$ $[(1,2-C_2B_9H_{11})W(CO)_3]$.[100] To a THF solution (100 ml) of 15.0 mmol of $Na_2[1,2-C_2B_9H_{11}]$ in a quartz tube was added 5.28 g (15.0 mmol) of tungsten hexacarbonyl; the tube was flushed with nitrogen and kept under a slight static pressure of the gas. The tungsten hexacarbonyl did not dissolve immediately and the mixture was stirred. The tube was irradiated with a Hanovia Type A 500-W mercury lamp for 37 hr. The solution immediately turned yellow with evolution of carbon monoxide upon irradiation, which was stopped when there was no more evolution of gas. The orange-yellow solution was poured into 1.5 liters of water and an excess of a 50% aqueous solution of tetramethylammonium chloride was added. The copious yellow precipitate was collected by filtration and dried *in vacuo* to afford 6.5 g (11.9 mmol, 79%) of crude product. It was recrystallized three times from acetone by slow evaporation of the solvent under reduced pressure. All recrystallization operations were performed in a drybox in an inert atmosphere.

2. Thirteen-Atom Cages

The recently discovered *nido*-1,2-$C_2B_{10}H_{12}^{2-}$ ion, obtained by reductive cage opening of 1,2-$C_2B_{10}H_{12}$ with sodium in THF, reacts with the hexacarbonyls of molybdenum and tungsten under uv radiation, displacing carbon monoxide and inserting the metal into the cage system[36]:

$$C_2B_{10}H_{12}^{2-} + M(CO)_6 \xrightarrow{h\nu} [(CO)_3MC_2B_{10}H_{12}]^{2-} + 3\ CO$$
$$M = \text{Mo or W}$$

The 13-atom metallocarborane ions are diamagnetic and highly air-sensitive, at least as tetraethylammonium salts, and no reactions have been reported. The structures are assumed to be analogous to the 13-atom cobalt species in Figure 32, whose geometry has been established in an X-ray-diffraction investigation (Ref. 36, footnote 10).

A prototype four-carbon metallocarborane has been synthesized from $(CH_3)_4C_4B_8H_8$ (itself a novel structure characterized as an opened icosahedron[139]) by thermal reaction with molybdenum hexacarbonyl[136,139]:

$$(CH_3)_4C_4B_8H_8 + Mo(CO)_6 \xrightarrow{\Delta} (CO)_3Mo(CH_3)_4C_4B_8H_8$$

The product is an asymmetrical 13-atom cage which is nonfluxional in solution on the nmr time scale, and is the first known neutral metallocarborane containing a formally zero-valent metal. The tungsten analog has been prepared in an analogous reaction,[139] and a variety of other 11- to 14-vertex metallocarboranes of Fe, Co, and Ni have been synthesized (see Section III. B.3).

Experimental

Preparation of $[(C_2H_5)_4N]_2$ $[(CO)_3WC_2B_{10}H_{12}]$.[36] To a solution of 12.5 mmol of $Na_2[C_2B_{10}H_{12}]$ in 100 ml of THF was added 4.4 g (12.5 mmol) of $W(CO)_6$. The mixture was stirred under strict exclusion of air for 48 hr while being irradiated with a Pen-ray uv lamp. The flask was then transferred to a nitrogen-filled dry box where the solvent was removed via rotary evaporation using a mechanical vacuum pump. The solid was redissolved in ethanol and added to an ethanol solution of tetraethylammonium bromide. The resulting orange precipitate was filtered, washed with ethanol, and recrystallized by the slow rotary evaporation of an acetone–ethanol solution. A pale-orange, microcrystalline product (1.1 g, 1.9 mmol, 15%) was isolated.

C. Manganese and Rhenium Compounds

1. Seven-Atom Cages

Metallocarboranes derived from the unique three-carbon *nido*-carborane $2,3,4\text{-}C_3B_3H_7$, of which only two examples have been prepared, are closely related structurally to the two-carbon, seven-vertex cages of iron, cobalt, and nickel. The $2,3,4\text{-}C_3B_3H_7$ system (Figure 3a) has been characterized in the form of the 2-methyl, 2,3-dimethyl, and 2,4-dimethyl derivatives, and is believed from nmr data to have pentagonal pyramidal geometry with an apical BH group and a planar $C_3B_2H_6$ basal ring, the "extra" hydrogen being present as a B–H–B bridge.[106] This bridging proton is reversibly removed by the action of sodium hydride in THF, giving a $C_3B_3H_6^-$ anion whose open face can π-bond to transition-metal atoms[106]:

$$CH_3C_3B_3H_6 + NaH \longrightarrow H_2 + Na^+CH_3C_3B_3H_5^- \xrightarrow[-NaBr]{BrMn(CO)_5,\ 25°}$$

$$[\sigma\text{-}(CO)_5MnCH_3C_3B_3H_5] \xrightarrow[\text{diglyme reflux}]{100°} 2\ CO + 2\text{-}CH_3\text{-}1,2,3,4\text{-}(CO)_3MnC_3B_3H_5$$
$$\text{red} \qquad\qquad\qquad\qquad\qquad\qquad\qquad\qquad\qquad\qquad \text{yellow}$$

$$175\text{--}200° \Big| -\tfrac{1}{2}\ H_2,\ 2\ CO$$

$$2\text{-}CH_3C_3B_3H_6 + \tfrac{1}{2}\ Mn_2(CO)_{10}$$

The red intermediate which is initially formed without evolution of carbon monoxide is assumed to have an $Mn(CO)_5$ group σ-bonded to the *nido*-carborane cage. On reflux at 100°, it converts in 20% yield, with loss of two moles of CO, to a seven-vertex metallocarborane whose nmr, ir, and mass spectra are consistent with a pentagonal bipyramid containing an $Mn(CO)_3$ unit in one apex, and three vicinal carbon atoms in the equator. The same compound is formed directly in 85–90% yield in the vapor-phase copyrolysis of $2\text{-}CH_3C_3B_3H_6$ and $Mn_2(CO)_{10}$ at 175–200°.[106,107] The analogous reaction of $2,3\text{-}(CH_3)_2C_3B_3H_5$ with $Mn_2(CO)_{10}$ at 200° generates a product whose mass spectrum corresponds to a $(CO)_3Mn(CH_3)_2C_3B_3H_4$ species, but which has not been further characterized.[106]

These compounds are the only known metallocarborane cage systems having three skeletal carbon atoms, and are isoelectronic analogs of the well-known, commercially important compound cymantrene $(\eta\text{-}C_5H_5)\text{-}Mn(CO)_3$; in this sense, the formal $C_3B_3H_6^-$ ligand (or its C-methylated derivatives) plays the role of the $\eta\text{-}C_5H_5^-$ ring and donates six delocalized electrons to the acceptor orbitals of the $Mn(CO)_3$ moiety. An equivalent description, in the language outlined in Section II, employs a three-electron donation from each cage CH, two electrons from each BH, and one electron from the $Mn(CO)_3$ group in the MnC_3B_3 polyhedron, yielding a total of 16 valence electrons in accordance with the $(2n + 2)$ rule.

Experimental

Reaction of $Na^+[2-CH_3-2,3,4-C_3B_3H_5]^-$ *with* $BrMn(CO)_5$.[106] A 3.0-ml diglyme solution containing 0.675 mmol of $Na^+CH_3C_3B_3H_5^-$ was added to 277.3 mg (1.01 mmol) of $BrMn(CO)_5$ in a previously evacuated flask equipped with a reflux condenser. After stirring at room temperature for ca. 1 hr, during which the color of the solution changed from yellow to red without evolution of CO, the mixture was refluxed at 100° for 12 hr, during which 0.95 mmol of CO was generated. Following the reaction, the solvent and volatile material were distilled from the flask to a vacuum line trap. The mixture was fractionated through traps cooled to -12, -45, and $-196°$. The material passing $-45°$ was purified by gas chromatography and gave a fraction (0.14 mmol) which was identified from its ir spectrum as $2-CH_3C_3B_3$-H_6. The material passing $-12°$ was diglyme. The fraction condensing at $-12°$ was further purified by passage through a trap cooled to 0° to give 23.5 mg (0.11 mmol, 21% yield) of a product identified from its ir and mass spectra as $2-CH_3-1,2,3,4-(CO)_3MnC_3B_3H_5$, identical with the compound obtained from $2-CH_3C_3B_3H_6$ and $Mn_2(CO)_{10}$ at 200°.[106] A small amount of $Mn_2(CO)_{10}$ was also recovered from the reaction mixture.

A red, nonsublimable solid which remained in the reaction flask was collected; its ir spectrum (Nujol mull) revealed three C–O stretching bands at 2010, 1975, and 1950 cm^{-1}.

2. Nine-Atom Cages

With the exception of the tricarbon manganese species just described, the only known subicosahedral metallocarboranes incorporating a manganese group atom are the $1,4,6-(CO)_3MnC_2B_6H_8^-$ ion and two C-substituted derivatives.[61,96] The originally proposed structure of this ion has been confirmed in an X-ray study[105] as a distorted tricapped trigonal prism containing manganese in a five-coordinate vertex (Figure 10). The syntheses of this species and its C-phenyl and C,C'-dimethyl derivatives are achieved by reaction of the $nido$-$C_2B_7H_{11}^{2-}$ ion (Section III.B.2), or its appropriate C-substituted derivatives, with bromomanganese pentacarbonyl in refluxing THF.[61,96] The parent metallocarborane has also been prepared in the reaction of $C_2B_7H_{11}^{2-}$ with $Mn_2(CO)_{10}$.[61]

$$C_2B_7H_{11}^{2-} + BrMn(CO)_5 \xrightarrow[THF]{\Delta} (CO)_3MnC_2B_6H_8^- + Br^- + H_2 + 2\ CO + solids$$

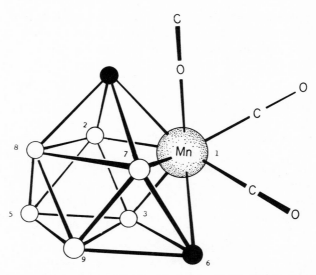

Figure 10. Structure of the 1,4,6-$(CO)_3MnC_2B_6H_8^-$ ion.[96,105] ●, CH; ○, BH. (From *J. Amer. Chem. Soc.*, **89**, 7115 (1967). Copyright by the American Chemical Society.)

Experimental

Preparation of [$(CO)_3MnC_2B_6H_8$]$^-$.[61] A solution of $BrMn(CO)_5$ (2.72 g, 9.9 mmol) in 100 ml of THF was added to an ethereal suspension of $Na_2B_7C_2H_{11}$ (10 mmol). The amber reaction mixture was heated vigorously for 6 hr, cooled, and filtered. Solvent was evaporated from the filtrate *in vacuo*. An ethereal solution of a major portion of the residue was shaken with a saturated aqueous solution of tetraethylammonium chloride, pentane was added, and the resulting precipitate was filtered and dried under vacuum. The dry salt was chromatographed on silica gel, eluting with dichloromethane. Recrystallization from dichloromethane–hexane yielded 1.53 g (40.7%) of amber [$(C_2H_5)_4N$] [$(CO)_3MnC_2B_6H_8$]. A minor portion of the crude sodium salt was chromatographed on silica gel, eluting with dichloromethane–acetone. An aqueous solution of the residue was shaken with an excess of cesium chloride and the resulting precipitate was recrystallized from aqueous acetone and from ether–hexane yielding 0.14 g (3.7%) of Cs[$(CO)_3MnC_2B_6H_8$]. The bis(pyridine)boronium salt was similarly prepared.

3. Twelve-Atom Cages

The insertion of manganese or rhenium into the *nido*-$C_2B_9H_{11}^{2-}$ ion (Section III.C) in reactions with bromometal pentacarbonyl takes place readily in refluxing THF.[89,100] On mixing the reagents at room temperature, a presumed (but not isolated) σ-bonded complex is formed without evolution of CO, but at reflux two moles of CO are released and the *closo*-metallocarborane is obtained:

$$C_2B_9H_{11}^{2-} + BrM(CO)_5 \longrightarrow \sigma\text{-}[(CO)_5M]C_2B_9H_{11}^- + Br^-$$

$$2\ CO + [3,1,2\text{-}(CO)_3MC_2B_9H_{11}]^- \xleftarrow[\text{THF}]{\text{reflux}}$$

$$M = Mn\ or\ Re$$

The structure[243] determined by X-ray of the rhenium ion is that of an icosahedron (Fig. 6) with the metal in the 3 position. The C,C'-divinyl derivative of the manganese ion has been prepared in the same way from $1,2\text{-}(C_2H_3)_2\text{-}1,2\text{-}C_2B_9H_9^{2-}$.[231]

"Benzodicarbollide" and "dihydrobenzodicarbollide" complexes of manganese, in which the adjacent cage carbons in the C_2B_9 ligand are linked by an external $-C_4H_4-$ or $-C_4H_6-$ chain, have also been reported (Table 1).[134]

The chemistry of these metallocarboranes has been little investigated. Degradation of the $3,1,2\text{-}(CO)_3MnC_2B_9H_{11}^-$ ion to MnO_2 and $C_2B_9H_{12}^-$ occurs on treatment with concentrated aqueous hydroxide ion at 100° in the presence of air.[89] The same metallocarborane ion undergoes carbonyl replacement on treatment with triphenylphosphine under uv radiation, forming the yellow $(C_6H_5)_3P(CO)_2MnC_2B_9H_{11}^-$ ion, but the reaction takes place slowly and is accompanied by considerable decomposition.[242] On the other hand, photolysis of the C,C'-divinyl derivative, mentioned above, in acetonitrile solution failed to displace CO with a vinyl group, possibly for steric reasons.[231]

It will be noted that all of these manganese and rhenium metallocarboranes contain a metal atom in a formal $+1$ oxidation state, so that species derived from $C_2B_9H_{11}^{2-}$ or other carborane dianions exist as negative ions. The neutral analogs of these ions which would contain formal Mn(II) or Re(II) are unknown, in contrast to the large number of metallocarboranes of iron, cobalt, and nickel in which the formal metal oxidation state is $+2$, $+3$, or even $+4$. Many of these latter species are electrically neutral [e.g., $(\eta\text{-}C_5H_5)M^{III}C_2B_9H_{11}$], and have been isolated as stable crystalline solids that are frequently sublimable and rarely present solubility problems. Largely for this reason, studies of metallocarboranes have centered primarily on the Group-VIII metals and other transition elements capable of achieving a $+2$ or higher oxidation state in a boron-cage environment. In the special case of cobalt, the extraordinary ability of the $(\eta\text{-}C_5H_5)Co^{2+}$ moiety to replace BH^{2+} in a variety of situations (see below) is an additional factor making this metal particularly attractive in metallocarborane and metalloborane work.

Experimental

Preparation of $3,1,2\text{-}(CO)_3MnC_2B_9H_{11}^-$.[100] A THF solution containing 5.18 mmol of $Na_2[1,2\text{-}C_2B_9H_{11}]$ was added under nitrogen to a THF solution of 1.42 g (5.18 mmol) of $BrMn(CO)_5$. The mixture was refluxed under nitrogen for 4 hr and filtered, the solvent removed *in vacuo*, and the residue taken up in 50 ml of water. The aqueous solution was treated with cesium chloride, and the resulting precipitate crystallized from warm water to give 1.16 g (2.87 mmol, 55.5%) of pale-yellow product.

D. Iron Compounds

1. Six-, Seven-, and Eight-Atom Cages

Metallocarboranes smaller than six-atom systems are unknown, although the five-atom *nido*-metalloboranes 1- and 2-$(\eta\text{-}C_5H_5)CoB_4H_8$ and 1-$(CO)_3$-FeB_4H_8, with no framework carbons, have been characterized as square pyramidal molecules analogous to B_5H_9.[73,147] The only known six-atom polyhedral metallocarboranes[151] are $1,2,4\text{-}(\eta\text{-}C_5H_5)CoC_2B_3H_5$, discussed below under cobalt compounds, and $1,2,4\text{-}(CO)_3FeC_2B_3H_5$. The iron species is prepared by the direct gas-phase reaction of iron pentacarbonyl with *closo*-$1,5\text{-}C_2B_3H_5$, a trigonal bipyramidal carborane which is the smallest homolog of the polyhedral $C_2B_{n-2}H_n$ series.[75] The synthesis is conducted in a hot–cold reactor, essentially an evacuated vertical Pyrex tube whose center section is heated to 230° and lower end is maintained at 25° for several hours.[151] A second product, characterized as a seven-vertex diiron species, $1,2,3,5\text{-}(CO)_6Fe_2C_2B_3H_5$, is obtained in small yield; the same compound is formed by the thermal reaction of $1,2,4\text{-}(CO)_3FeC_2B_3H_5$ with excess

$$Fe(CO)_5 + 1,5\text{-}C_2B_3H_5 \xrightarrow{\Delta} (CO)_3FeC_2B_3H_5 \xrightarrow[\Delta]{Fe(CO)_5} (CO)_6Fe_2C_2B_3H_5$$

$$13\% \qquad\qquad\qquad\qquad 1.5\%$$

$Fe(CO)_5$. Both products are volatile orange liquids, the monoiron species having sufficient vapor pressure to pass a −23° trap in an evacuated fractionation train, and have been structurally characterized as shown in Figure 11 from ^{11}B and ^1H nmr, infrared, and mass spectroscopic evidence.[151]

Since an $Fe(CO)_3$ group is a formal two-electron donor to the cage system (see Section II), the monoiron compound is regarded as an iso-electronic analog of *closo*-$C_2B_4H_6$ and the $B_6H_6^{2-}$ ion, both octahedral structures having been firmly established in X-ray studies. Together with the $(\eta\text{-}C_5H_5)CoC_2B_3H_5$ mentioned above, this compound is the smallest

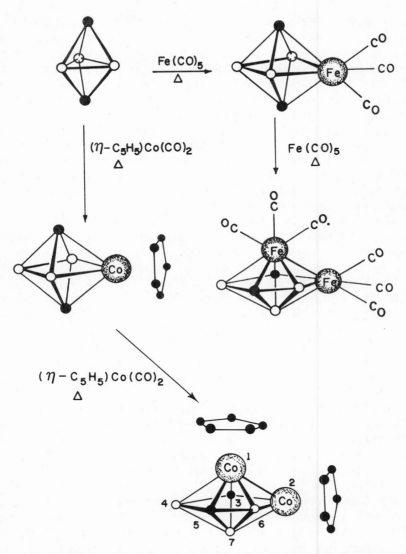

Figure 11. Gas-phase polyhedral expansion of 1,5-C$_2$B$_3$H$_5$,[151] showing proposed struc-
tures of 1,2,4-(CO)$_3$FeC$_2$B$_3$H$_5$, 1,2,3,5-(CO)$_6$Fe$_2$C$_2$B$_3$H$_5$, 1,2,4-(η-C$_5$H$_5$)CoC$_2$B$_3$H$_5$,
and 1,2,3,5-(η-C$_5$H$_5$)$_2$Co$_2$C$_2$B$_3$H$_5$. The compound 1,7,2,4-(η-C$_5$H$_5$)$_2$Co$_2$C$_2$B$_3$H$_5$ (Figure
20) is also obtained in the cobalt reaction sequence.

known metallocarborane polyhedron. The diiron metallocarborane is one of many known seven-atom cage systems (see Table 1) that are known or assumed to have a gross pentagonal bipyramidal structure analogous to *closo*-$C_2B_5H_7$. Its proposed[151] structure (Figure 11) is perhaps surprising in that the iron atoms assume adjacent four- and five-coordinate vertices rather than the more symmetrical 1 and 7 (apex) positions. The latter arrangement is found in the isoelectronically analogous "triple-decked" cobalt species, 1,7,2,4-(η-C_5H_5)$_2$Co$_2$C$_2$B$_3$H$_5$, whose structure has been crystallographically established (Section IV.F.2). However, at least three other isomers of (η-C_5H_5)$_2$Co$_2$C$_2$B$_3$H$_5$ have been found, one of which has geometry apparently analogous to 1,2,3,5-(CO)$_6$Fe$_2$C$_2$B$_3$H$_5$ (Section IV.F.2). Since both the iron and the cobalt 1,2-dimetallic isomers are formed at elevated temperature, at least reasonable thermal stability may be inferred for these species. Conceivably, the metal–metal bond may be a significant stabilizing factor in these molecules, which have no symmetry whatsoever and presumably have a highly asymmetric charge distribution.

The *nido*-carborane 2,3-$C_2B_4H_8$ (Figure 2b) is a highly useful reagent for the preparation of small iron metallocarboranes via either direct reaction in the vapor phase or deprotonation followed by metal insertion in solution. The gas-phase interaction with iron pentacarbonyl in a hot–cold reactor at 240/25° yields a 6-atom *nido* species, 1,2,3-(CO)$_3$FeC$_2$B$_3$H$_7$ and a seven-atom *closo* compound, 1,2,3-(CO)$_3$FeC$_2$B$_4$H$_6$.[76,189] The latter compound is thermally less stable than the former, to which it is converted on heating:

$$C_2B_4H_8 + Fe(CO)_5 \xrightarrow{\Delta} (CO)_3FeC_2B_3H_7 + (CO)_3FeC_2B_4H_6$$
$$5\text{–}10\%\qquad\qquad 5\text{–}10\%$$

The originally proposed[76] structure of pale-yellow, volatile (CO)$_3$FeC$_2$B$_3$H$_7$ (Figure 2a) has been confirmed in an X-ray study,[10] and is isoelectronically analogous to the (η-C_5H_5)Fe(CO)$_3^+$ ion. The chemistry of (CO)$_3$FeC$_2$B$_3$H$_7$ has not yet been explored, but some reactions of an analogous cobalt species, (η-C_5H_5)CoC$_2$B$_3$H$_7$, have been examined (Section IV.F.2).

The proposed[76,189] structure of 1,2,3-(CO)$_3$FeC$_2$B$_4$H$_6$ (Figure 12), a volatile orange compound, is based on spectroscopic data and on analogy with other seven-vertex metallocarborane systems. The isomeric species 1,2,4-(CO)$_3$FeC$_2$B$_4$H$_6$, in which the skeletal carbons are equatorial but nonadjacent, is formed in the thermal reaction of the octahedral *closo*-carborane 1,6-$C_2B_4H_6$ with Fe(CO)$_5$[151]; the same isomer is a product of the copyrolysis of 2,4-$C_2B_5H_7$ with Fe(CO)$_5$ (see below).

The 2,3-$C_2B_4H_7^-$ anion (obtained by bridge-deprotonation of 2,3-$C_2B_4H_8$ by the action of sodium hydride in THF) undergoes metal insertion with cyclopentadienyliron dicarbonyl iodide to give a metal-bridged species,

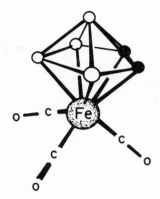

Figure 12. Probable structure of 1,2,3-
$(CO)_3FeC_2B_4H_6$.[189] ●, CH; ○, BH.

μ-[(η-C_5H_5)Fe(CO)$_2$]-$C_2B_4H_7$ (Figure 13).[189,190] This compound, whose proposed structure[189,190] is based primarily on [11]B and ^{1}H nmr data, undergoes reaction in THF under uv radiation to give two new compounds characterized as *closo*-metallocarboranes, 1,2,3-(η-C_5H_5)Fe[II]$C_2B_4H_7$ and 1,2,3-(η-C_5H_5)Fe[III]$C_2B_4H_6$, in a total yield of 90%. The first of these, a diamagnetic, red-orange solid, contains an "extra" hydrogen atom which has been tentatively assigned[189,190] to an Fe–H group on the basis of a high-field resonance in the ^{1}H nmr spectrum. This hydrogen is probably not bonded solely to the metal, but retains some association with the carborane ligand as well. This hypothesis, suggested schematically in Fig. 13, can be expressed in a slightly different manner in terms of a face-bonded bridging proton which tautomerizes rapidly between adjacent FeB$_2$ triangles on the polyhedral surface.[189] The suggestion of direct metal–proton interaction in this compound and related iron species (discussed below) is strengthened by other recent findings in metalloboron chemistry, particularly the discovery of manganaboranes,[59,129] cobaltaboranes,[147,149] and cobaltacarboranes[114] containing metal–hydrogen bonds. Also relevant is the reported protonation[98] of the $(C_2B_9H_{11})_2Fe^{2-}$ ion to $(C_2B_9H_{11})_2FeH^-$, although the location of the added proton was not established. Very recent work in our laboratory has indicated that the compounds $[(CH_3)_2C_2B_4H_4]_2FeH_2$ and $[(CH_3)_2C_2B_4H_4]_2CoH$ have protons linked to boron and the metal atom, in which tautomerism of the type described above is strongly supported.[138,140]

The second product, 1,2,3-(η-C_5H_5)Fe[III]$C_2B_4H_6$ (Figure 13) is a brown paramagnetic crystalline solid whose temperature-dependent magnetic susceptibility gives a magnetic moment of 2.1, corresponding to the expected low-spin d^5 configuration on the metal. These iron(II) and (III) species are easily and reversibly interconverted by acid–base and redox processes as

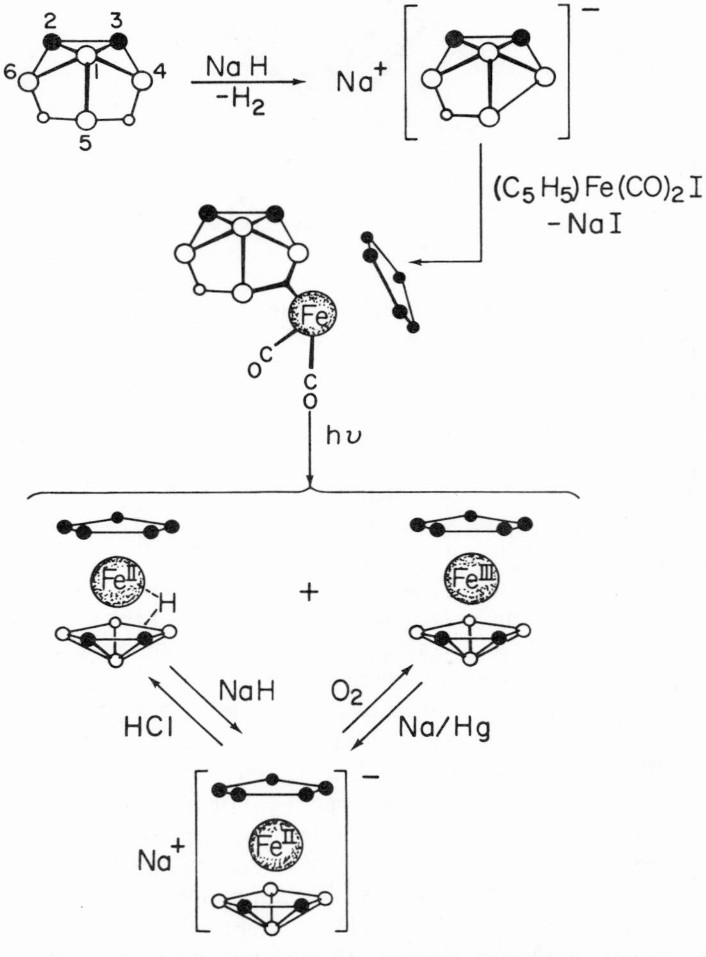

Figure 13. Reaction scheme for the formation of cyclopentadienyliron metallocarboranes from the 2,3-$C_2B_4H_7^-$ ion.[189] The metal-bound proton in 1,2,3-(η-C_5H_5)Fe(H)$C_2B_4H_6$, shown schematically, probably undergoes rapid tautometric migration through face-bonded positions on the polyhedral surface. (From *J. Amer. Chem. Soc.*, **95**, 6623 (1973). Copyright by the American Chemical Society.)

shown in Figure 13.[189] The light-orange $(\eta\text{-}C_5H_5)FeC_2B_4H_6^-$ intermediate has not been characterized.

The reaction of $C_2B_4H_7^-$ ion with $FeCl_2$ in the *absence* of $C_5H_5^-$ was expected to generate *commo*-metallocarborane* species of the type $[(C_2B_4H_6)_2Fe^{II}]^{2-}$, but in fact no isolable products were obtained.[140] However, when the C,C'-dimethyl derivative, $(CH_3)_2C_2B_4H_5^-$, was used instead, the neutral red solid compound $[(CH_3)_2C_2B_4H_4]_2FeH_2$ was isolated in good yield (Figure 14).[138] The two "extra" hydrogen atoms in this molecule are indicated from NMR evidence to be associated with the metal atom; rapid tautomerization through face-bonded locations above the centers of FeB_2 triangles on the polyhedral surface has been proposed.[138] The neutral compound is the focus of some interesting chemistry as shown in Figure 14. The species labeled V, $(\eta\text{-}C_5H_5)CoFe(CH_3)_4C_4B_8H_8$, was shown by X-ray examination[141] to have the unique structure depicted here. This molecule can be viewed as a pair of seven-vertex polyhedra fused at a common iron atom, with an extra BH unit wedged between them and capping both polyhedra; the "capping" effect is probably a consequence of the presence of fewer than $(2n + 2)$ valence electrons in each polyhedral cage, a situation we have labeled "electron hyperdeficiency."[141] Compound IV in Figure 14, characterized from nmr data as $1,7,2,3\text{-}(\eta\text{-}C_5H_5)Co(CH_3)_2C_2B_3H_3Fe(CO)_3$ may be an analog of the "triple-decked sandwich" dicobalt complexes discussed in Section IV.F.2.

The reductive cage opening of $closo\text{-}2,4\text{-}C_2B_5H_7$ with sodium naphthalide, followed by addition of ferrous chloride and sodium cyclopentadienide and exposure to air, yields a mixture of ferracarboranes incorporating the pentagonal bipyramidal $1,2,4\text{-}FeC_2B_4$ cage[189]:

$$C_2B_5H_7 \xrightarrow[C_{10}H_8]{Na} \xrightarrow[NaC_5H_5]{FeCl_2} \xrightarrow{O_2} \begin{array}{l} 1,2,4\text{-}(\eta\text{-}C_5H_5)Fe^{II}HC_2B_4H_6 \\ + \ 5\text{-}R\text{-}1,2,4\text{-}(\eta\text{-}C_5H_5)Fe^{III}C_2B_4H_5 \end{array}$$

$$R = 2\text{-}(2,4\text{-}C_2B_5H_6),\ C_{10}H_7,\ \text{or}\ H$$

In contrast to the corresponding reaction with cobaltous chloride, polyhedral expansion is not observed and no evidence of an FeC_2B_5 or larger cage system has been found in this reaction. The green linked-cage species containing a $C_2B_5H_6$ ligand σ-bonded to a ferracarborane cage is the major product and the parent compound (R = H) is isolated only in trace quantity. The Fe^{II} species is an isomer of $1,2,3\text{-}(\eta\text{-}C_5H_5)Fe^{III}C_2B_4H_7$ (Figure 13) having nonadjacent framework carbon atoms, and like the latter compound is assumed to have an Fe–H bonding interaction.[189]

Although the sodium/ferrous ion treatment of $2,4\text{-}C_2B_5H_7$ generates

---

* The prefix *commo*- indicates coordination of two carborane ligands to the same metal atom.

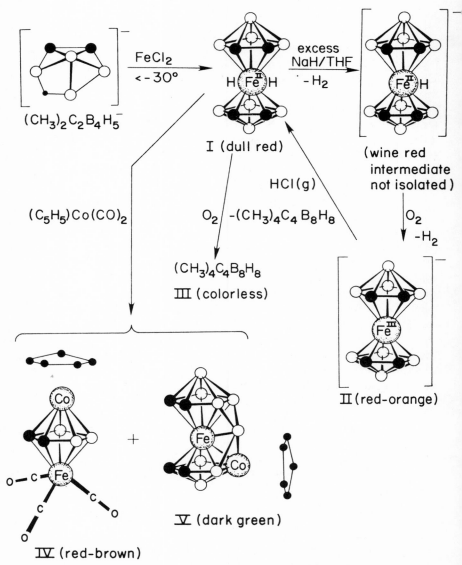

Figure 14. Preparation and chemistry of $[(CH_3)_2C_2B_4H_4]_2FeH_2$.[138] Metal-bonded hydrogens are shown schematically. The structure of compound V has been crystallographically established[141]; the others are postulated from ^{11}B and ^1H nmr data. ○, BH; ●, CCH₃; •, CH; •, H.

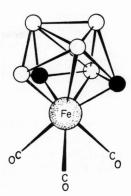

Figure 15. Proposed structure of $3,1,7\text{-}(CO)_3FeC_2B_5H_7$.[189]
●, CH; ○, BH.

only FeC_2B_4-type species and fails to give "expanded" metallocarboranes (in which iron is inserted without loss of boron), polyhedral expansion does occur in the gas-phase reaction of $2,4\text{-}C_2B_5H_7$ with iron pentacarbonyl in a hot–cold reactor at 280/25°. The major product is a seven-vertex species, $1,2,4\text{-}(CO)_3FeC_2B_4H_6$, mentioned earlier, but an eight-vertex polyhedron, characterized as $3,1,7\text{-}(CO)_3FeC_2B_5H_7$ (Figure 15) has been isolated in low yield[189]:

$$C_2B_5H_7 + Fe(CO)_5 \xrightarrow{\Delta} 1,2,4\text{-}(CO)_3FeC_2B_4H_6 + 3,1,7\text{-}(CO)_3FeC_2B_5H_7$$

The structure indicated in Figure 15 (formerly[189] numbered 5,1,8) has been proposed[189] on the basis of ^{11}B and ^1H nmr data, but is not unique and other possibilities exist. However, this compound does appear to be structurally analogous to the cobalt species $3,1,7\text{-}(\eta\text{-}C_5H_5)CoC_2B_5H_7$, described in Section IV.F.3.

Experimental

Preparation of $(CO)_3FeC_2B_3H_5$ *and* $(CO)_6Fe_2C_2B_3H_5$.[151] An evacuated cylindrical Pyrex tube, 24 mm in diameter, connected to a vacuum line with a greaseless Teflon stopcock containing a Viton O-ring and maintained in a vertical position with its central portion heated by a Variac-controlled heating tape, was charged with 2.78 mmol of $C_2B_3H_5$ and 6.0 mmol of $Fe(CO)_5$ *in vacuo*. The hot zone was heated to 230° while the lower end was maintained at 25–30° for 11 hr. At this time the tube was opened to the vacuum line and the volatiles removed. The pure orange liquids $1,2,4\text{-}(CO)_3FeC_2B_3H_5$ (64.7 mg, 0.320 mmol, 12.8%) and $1,2,3,5\text{-}(CO)_6Fe_2C_2B_3H_5$ (11.5 mg, 0.034 mmol, 1.4%) were obtained by repeated fractionation through $-23°$ and 0° traps, respectively. A 0.286-mmol quantity of $C_2B_3H_5$ was recovered.

Preparation of $(CO)_3FeC_2B_3H_7$ *and* $1,2,3\text{-}(CO)_3FeC_2B_4H_6$.[189] Into the cylindrical Pyrex reactor described above was distilled 3.24 mmol of $Fe(CO)_5$. While a liquid nitrogen trap was maintained on the reactor, 3.10 mmol of $C_2B_4H_8$ was added by distillation and the stopcock was closed. The reactor was warmed to room temperature, a heating tape controlled by a Variac was wrapped around the central portion of the reactor, and the reactor was clamped in a vertical position such that the bottom section containing a pool of liquid $Fe(CO)_5$ was immersed in a Dewar flask of water at 25°. The heating tape was maintained at 215–245° for 24 hr during which substantial dark nonvolatile deposits, including metallic iron, were formed on the reactor walls. The reactor was then cooled to $-196°$; noncondensables (ca.3 mmol) were pumped off and the remaining volatiles were distilled through a $-45°$ trap which allowed $Fe(CO)_5$, $C_2B_5H_7$, and unreacted $C_2B_4H_8$ to pass through. The $-45°$ condensate consisted of two components which were separated by repeated fractionation through a $-23°$ trap. The condensate at $-23°$ consisted of $1,2,3\text{-}(CO)_3FeC_2B_4H_6$, a yellow-orange liquid; the more volatile fraction was $(CO)_3FeC_2B_3H_7$, a pale-yellow liquid. Yields of the pure compounds were estimated to be 5–10% each.

Preparation of $\mu\text{-}[(\eta\text{-}C_5H_5)Fe(CO)_2]\text{-}C_2B_4H_7$.[189] A solution of 2.6 mmol of $NaC_2B_4H_7$ in 10 ml of THF was added dropwise under nitrogen to a solution of 2.6 mmol of $(\eta\text{-}C_5H_5)Fe(CO)_2I$ in 10 ml of THF. The mixture was allowed to react for 4 hr at 25°. The reaction was then exposed to the air and filtered, and the solvent was removed. The residue was chromatographed on a silica gel column. Hexane eluted $\mu\text{-}[(\eta\text{-}C_5H_5)Fe(CO)_2]\text{-}C_2B_4H_7$, a bright-yellow moderately air-sensitive solid. Due to decomposition on the column, precise yield data could not be obtained, but the estimated yield was at least 50%.

Preparation of $1,2,3\text{-}(\eta\text{-}C_5H_5)Fe^{II}HC_2B_4H_6$ *and* $1,2,3\text{-}(\eta\text{-}C_5H_5)Fe^{III}\text{-}C_2B_4H_6$.[189] A solution of ca. 0.7 mmol of $\mu\text{-}[(\eta\text{-}C_5H_5)Fe(CO)_2]C_2B_4H_7$ in 10 ml of THF was added to a quartz reaction flask under vacuum and irradiated for 12 hr with stirring at 30°. The flask was opened, the solvent removed, and the residue chromatographed on a silica gel column. Carbon tetrachloride eluted two bands: red-orange $(\eta\text{-}C_5H_5)FeHC_2B_4H_6$ (45.3 mg, 0.23 mmol), and brown $(\eta\text{-}C_5H_5)FeC_2B_4H_6$ (82.1 mg, 0.42 mmol). No other products were detected.

Conversion of $(\eta\text{-}C_5H_5)Fe^{II}HC_2B_4H_6$ *to* $(\eta\text{-}C_5H_5)Fe^{III}C_2B_4H_6$.[189] A solution of $1,2,3\text{-}(\eta\text{-}C_5H_5)FeHC_2B_4H_6$ (45.3 mg, 0.23 mmol) in THF was added dropwise to a suspension of sodium hydride (0.50 mmol, 56% mineral oil dispersion) in THF (5 ml) under nitrogen. Vigorous bubbling occurred at once and the solution changed from red-orange to light orange. After 10 min this solution was filtered under nitrogen. Upon exposure to air the solution immediately turned a dark brown color, characteristic of 1,2,3-$(\eta$-$C_5H_5)$-$FeC_2B_4H_6$. Purification of the product on a silica gel column yielded 33.1 mg (0.17 mmol, 73% yield) of 1,2,3-$(\eta$-$C_5H_5)FeC_2B_4H_6$, identified from its ir spectrum and tlc (thin-layer chromatography) R_f value.

Conversion of $(\eta$-$C_5H_5)Fe^{III}C_2B_4H_6$ *to* $(\eta$-$C_5H_5)Fe^{II}HC_2B_4H_6$.[189] A solution of 1,2,3-$(\eta$-$C_5H_5)FeC_2B_4H_6$ (72.6 mg, 0.37 mmol) in 10 ml of THF was added to 2 ml of sodium amalgam containing 1.70 mmol of Na, under an atmosphere of nitrogen. Within 20 min the solution had changed from dark brown to light orange. The solution was then decanted and filtered under nitrogen. The solvent was removed under vacuum and 0.60 mmol of HCl was added. After a reaction period of 10 min, fractionation of the volatiles showed that 0.34 mmol of HCl remained. The residue was chromatographed on a silica gel column to yield 23.3 mg (0.12 mmol, 31%) of 1,2,3-$(\eta$-$C_5H_5)$-$FeHC_2B_4H_6$, which was identified from its ir and boron-11 nmr spectra and its tlc R_f value.

Preparation of 3,1,7-$(CO)_3FeC_2B_5H_7$.[189] A cylindrical reactor of the type described above was charged with 3.2 mmol of $C_2B_5H_7$ and 5.2 mmol of $Fe(CO)_5$, and the central portion heated to 280° for 15 hr during which time the lower end was maintained at 25°. Following removal of noncondensables and unreacted starting materials, the two principal products, which were retained in a $-35°$ trap, were separated by repeated fractionation through a trap at $-23°$ to give 20 mg (0.10 mmol) of orange 1,2,4-$(CO)_3FeC_2B_4H_6$ and 14 mg (0.063 mmol) of orange 3,1,7-$(CO)_3FeC_2B_5H_7$. A total of 1.0 mmol of $C_2B_5H_7$ was recovered.

2. Nine-, Ten-, and Eleven-Atom Cages

The reduction of *closo*-1,7-$C_2B_6H_8$ with sodium naphthalide in THF at $-20°$, followed by reaction with $FeCl_2$ results in polyhedral expansion, giving two isomers of a $(\eta$-$C_5H_5)FeC_2B_6H_8$ nine-vertex system.[42] Assuming that the gross geometry is that of a tricapped trigonal prism, it has been proposed from [11]B and ^{1}H nmr data[42] that these isomers are 1,4,6- and 4,1,8-$(\eta$-$C_5H_5)FeC_2B_6H_8$, respectively. The latter species has the metal in a four-coordinate vertex (Figure 16), an unusual situation in that the metal atoms in metallocarboranes exhibit a distinct tendency to occupy five-coordinate vertices [see, however, the previously mentioned 1,2,3,5-$(CO)_6$-$Fe_2C_2B_3H_5$ and the analogous $(\eta$-$C_5H_5)_2Co_2C_2B_3H_5$ isomer, discussed below]. The treatment of the seven-vertex cobaltacarborane 1,2,3-$(\eta$-$C_5H_5)Co$-$(CH_3)_2$-

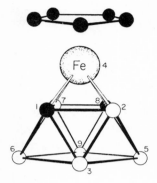

Figure 16. Proposed structure of 4,1,8-(η-C_5H_5)-$FeC_2B_6H_8$.[42] ●, CH; ○, BH. (From *J. Amer. Chem. Soc.*, **95**, 4565 (1973). Copyright by the American Chemical Society.)

$C_2B_4H_6$ with excess $Fe(CO)_5$ at 125° in nonane effects the addition of two $Fe(CO)_3$ units to the cage, producing the nine-vertex mixed-metal species $(CO)_6Fe_2(\eta$-$C_5H_5)Co(CH_3)_2C_2B_4H_4$.[135] This compound represents the first known example of a trimetallocarborane of fewer than 10 vertices (although several polyhedral cobaltaboranes containing up to four metal atoms in six- to eight-vertex cages have been reported[149]) and is the only reported mixed-metal trimetallocarborane of any size.

At present, only two 10-vertex ferracarborane systems have been reported, but both are of unusual interest, in one instance because of the structure and in the other because of the method of synthesis. The treatment of *closo*-4,5-$C_2B_7H_9$ with sodium followed by addition of ferrous chloride and sodium cyclopentadienide generated several species, including green $(\eta$-$C_5H_5)_2Fe_2C_2B_6H_8$.[12] This molecule has been crystallographically characterized[12] as a capped tricapped trigonal prism with the iron atoms within bonding distance of each other. The drastic departure from the usual 10-vertex geometry of a bicapped square antiprism may be attributed[12] to the presence of only $2n$ cage-valence electrons rather than the $(2n + 2)$ electrons normally found in a closed triangulated polyhedron (see Section II.B). A similar "electron-hyperdeficient"[141] molecule is discussed in Section IV.D.1.

The other known 10-vertex ferracarborane, bright yellow 2,1,4-$(CO)_3$-$Fe(CH_3)_2C_2B_7H_9$, has been characterized from nmr evidence[137] as a bicapped square antiprism, as expected for a $(2n + 2)$-electron 10-atom polyhedron. While the compound appears to have no particular structural distinction, its synthetic origin is remarkable; it was produced in $\sim 70\%$ yield from the reaction of solid $[(CH_3)_2C_2B_4H_4]_2FeH_2$ (Section IV.D.1) with carbon monoxide gas at 1 atm pressure and 150°, without solvent. At this writing, the reactivity of other metallocarboranes toward CO has not been examined.

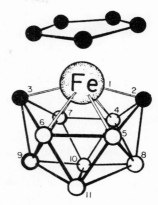

Figure 17. Proposed structure of 1,2,3-(η-C$_5$H$_5$)-FeC$_2$B$_8$H$_{10}$.[42] ●, CH; ○, BH. (From *J. Amer. Chem Soc.*, **95**, 4565 (1973). Copyright by the American Chemical Society.)

The polyhedral expansion of *closo*-1,6-C$_2$B$_8$H$_{10}$ via treatment with sodium naphthalide, FeCl$_2$, and NaC$_5$H$_5$ produced 1,2,3-(η-C$_5$H$_5$)FeC$_2$B$_8$H$_{10}$, an 11-atom cage system in which the metal is proposed to occupy the unique six-coordinate vertex with the carbons in four-coordinate vertices adjacent to the metal.[42] (Figure 17). The structurally analogous cobalt species and a number of its derivatives have been prepared (see Section IV.F.4 and Table I).

An 11-vertex cobaltaferracarborane is described in Section IV.F.3.

Experimental

Preparation of 1,4,6- *and* 4,1,8-(η-C$_5$H$_5$)FeC$_2$B$_6$H$_8$.[42] Under a nitrogen atmosphere, sodium (1.0 g, 44 mmol) and naphthalene (0.4 g, 3.2 mmol) were placed in a 100-ml flask, THF (20 ml) was added, and the solution was stirred 1 hr. Addition of 1,7-C$_2$B$_6$H$_8$ (0.9695 g, 10.0 mmol) by condensation into the flask cooled to $-78°$ caused the dark-green color of sodium naphthalide to disappear immediately. The flask was allowed to warm to $-20°$ in a cold room and stirring was continued for three days. The reduced carborane solution was added to a cooled solution of 30 mmol of freshly prepared NaC$_5$H$_5$. A gray slurry of FeCl$_2$ in THF, prepared by reaction of iron powder (1.2 g, 20 mmol) and FeCl$_3$ (3.2. g, 20 mmol) in 100 ml of THF at reflux for 3 hr, was added to the flask cooled to $-30°$. The flask was allowed to warm to room temperature and stirred overnight. After oxygen was bubbled through the solution for 30 min, it was filtered through Celite. Silica gel (150 ml) was added to the solution and the solvent removed. The remaining solids were placed in a Soxhlet thimble and extracted with 1 liter of hexane. The solution was evaporated to a small volume and placed on a column of silica gel (4 × 20 cm) in hexane. Among the several components that were separated on the column, three main bands were collected and characterized. The first fraction consisted of a yellow band characterized by its infrared spectrum as ferrocene. Next eluted was a red band that crystallized as the solvent was stripped by rotary evaporation. Recrystallization from hexane produced 0.228 g (1.05 mmol, 10.5%) of $4,1,8\text{-}(C_5H_5)FeC_2B_6H_8$, mp 132–133°. Elution with hexane–dichloromethane (9:1 v/v) resulted in a brown fraction. $1,4,6\text{-}(C_5H_5)FeC_2B_6H_8$ (0.170 g, 0.79 mmol, 7.9%) was recrystallized from hexane.

Preparation of $1,2,3\text{-}(\eta\text{-}C_5H_5)FeC_2B_8H_{10}$.[42] $1,6\text{-}C_2B_8H_{10}$ (0.8865 g, 7.05 mmol) was reduced with sodium as described above. Freshly prepared NaC_5H_5 (35 mmol) and a gray slurry of $FeCl_2$ (82 mmol) in THF, prepared as previously described, were added, and the reaction mixture was stirred overnight. Air was bubbled through the reaction mixture for 30 min, 20 ml silica gel was added, and the solvent was removed. The solids were placed on a silica gel column (5 × 30 cm) in hexane. A red-brown band was eluted with hexane–CH_2Cl_2 (8.5:1 v/v). $(C_5H_5)FeC_2B_8H_{10}$, mp 275–277° (1.17 g, 4.8 mmol, 68%), was recrystallized from hexane–CH_2Cl_2 as red-brown needles.

3. Twelve-Atom Cages

Icosahedral ferracarboranes have been prepared in several ways, the most common being the reaction of the 1,2- or $1,7\text{-}C_2B_9H_{11}{}^{2-}$ ion with anhydrous ferrous chloride in nonaqueous solvents,[100,101] and the interaction of 1,2- or $1,7\text{-}C_2B_9H_{12}^-$ salts with ferrous chloride in hot concentrated NaOH[91,210] (the general features of both types of reactions are discussed in Section III). The former method is analogous to the synthesis of ferrocene from NaC_5H_5 and $FeCl_2$. The formal iron(II) species formed in both routes

$$C_2B_9H_{11}^{2-} + FeCl_2 \longrightarrow (C_2B_9H_{11})_2Fe^{2-} \underset{Na}{\overset{O_2}{\rightleftarrows}} (C_2B_9H_{11})_2Fe^-$$

$$C_2B_9H_{12}^- + FeCl_2 \xrightarrow{\quad OH^- \quad}$$

is oxidized in air to the corresponding iron(III) anion.[100,101,210] The iron(II) species has been isolated under air-free conditions following reduction of the $(C_2B_9H_{11})_2Fe^-$ with sodium amalgam.

If the reaction with $C_2B_9H_{11}^{2-}$ in nonaqueous media is conducted in the presence of $C_5H_5^-$ ion, the paramagnetic cyclopentadienyl iron(III) compound is formed; this material can be readily reduced to the diamagnetic iron(II) species[95,100]:

$$C_5H_5^- + C_2B_9H_{11}^{2-} + FeCl_2 \xrightarrow[\text{reflux}]{THF} (\eta\text{-}C_5H_5)FeC_2B_9H_{11}$$

$$\downarrow Na/Hg$$

$$(\eta\text{-}C_5H_5)FeC_2B_9H_{11}^-$$

All of these reactions may be conducted with either the 1,2- or 1,7-isomer of the respective carborane anion. An improved method for the preparation of cyclopentadienyl metallocarboranes of iron and cobalt, which utilizes *in situ* formation of $C_2B_9H_{12}^-$ directly from $1,2-C_2B_{10}H_{12}$ (*o*-carborane), has recently been described.[116,164] In this procedure, *o*-carborane is degraded by alkaline methanolysis to the $C_2B_9H_{12}^-$ ion, which is not isolated but is allowed to react with $FeCl_2$ and C_5H_6 to produce $3,1,2-(\eta-C_5H_5)FeC_2B_9H_{11}$. The principal advantage of this modification is that the formation of the $(C_5H_5)_2Fe$ and $(C_2B_9H_{11})_2Fe^-$ species is considerably reduced and the yield of the desired cyclopentadienyl metallocarborane is increased to 50% or greater on a 100-mmol scale.[164] In comparison, the preparation from $C_2B_9H_{11}^{2-}$ described above gave a 25% yield of $3,1,2-(\eta-C_5H_5)FeC_2B_9H_{11}$ and 42% of ferrocene.

A novel synthesis of the $(1,2-C_2B_9H_{11})_2Fe^-$ ion by the reaction of the thallium species $TlC_2B_9H_{11}^-$ with $FeCl_2$ has been reported without details.[191]

A four-carbon 12-vertex ferracarborane, $(\eta-C_5H_5)Fe(CH_3)_4C_4B_7H_8$ has been prepared from $(CH_3)_4C_4B_8H_8$[136] and structurally characterized from an X-ray study.[11] The molecule has an open-cage geometry roughly similar to $B_{10}H_{14}$, with one bridge hydrogen atom on the open six-membered face; the iron atom occupies a six-coordinate vertex not on the open rim.

Experimental

Preparation of $3,1,2-(\eta-C_5H_5)FeC_2B_9H_{11}$.[164] Solid KOH (56 g, 1 mol) was added to a suspension of $1,2-C_2B_{10}H_{12}$ (14.4 g, 0.1 mol) in 100 ml of methanol. After cessation of a vigorous spontaneous reaction, the mixture was refluxed for 2 hr. The flask with the resulting mixture was cooled and connected by means of a bend (provided with a side stopcock) with a flask containing a two-layer mixture consisting of $FeCl_2$ (0.15 mol) in 60 ml methanol and freshly distilled cyclopentadiene (13.2 g, 0.2 mol). Both flasks were cooled to $-50°$ and evacuated to 10 torr; the temperature was allowed to rise to $-10°$ and the two-layer mixture was added to the reaction flask. The mixture was stirred and refluxed for 2.5 hr. Upon cooling to room temperature, the crude mixture was diluted with 200 ml of water, and methanol and excess cyclopentadiene were distilled off *in vacuo*. The reaction mixture was filtered and the solid residue washed three times with 100 ml of water. From the combined filtrates, methanol was evaporated off *in vacuo* and the clear red solution was oxidized by 20 ml of 30% hydrogen peroxide. The resulting crystals were filtered off and dissolved in three 100-ml portions of benzene; the combined benzene solution was filtered, evaporated to a volume of 50 ml *in vacuo*, and covered carefully with a layer of 100 ml of hexane. Overnight, the product separated as violet needles (13.2 g, 51%).

124 **Russell N. Grimes**

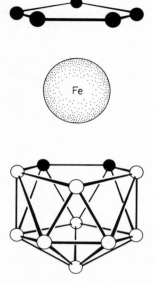

Figure 18. Structure of $3,1,2\text{-}(\eta\text{-}C_5H_5)FeC_2B_9H_{11}$.[100,245] ●, CH; ○, BH. (From *J. Amer. Chem. Soc.*, **90**, 879 (1968). Copyright by the American Chemical Society.)

    The ferrocene-like structure of $3,1,2\text{-}(\eta\text{-}C_5H_5)FeC_2B_9H_{11}$ has been established in an X-ray study[245] (Figure 18); paramagnetic resonance,[132] Mössbauer,[7,102] and magnetic susceptibility[64,65] studies on this molecule, as well as on $(1,2\text{-}C_2B_9H_{11})_2Fe^-$ and other icosahedral iron(III) metallo-carboranes, are basically in agreement with a ferrocene-like model of the bonding between the metal and the remainder of the cage system. It is important to note that the Fe–B and Fe–C distances in $3,1,2\text{-}(\eta\text{-}C_5H_5)FeC_2B_9H_{11}$ (Figure 18) are equal[245] and the molecule has pseudo-C_{5v} symmetry, in common with all known MC_2B_9 icosahedra in which the metal atom has fewer than seven d-shell electrons. Metals having d^8, d^9, or d^{10} configurations produce distorted icosahedra, as in the cases of certain copper and nickel species discussed later in this review (Sections IV.H.3 and IV.J.).

    The reaction of $1,2\text{-}C_2B_9H_{11}^{2-}$ salts with $Fe(CO)_5$ in refluxing THF generates a dimeric species, $(C_2B_9H_{11})_2Fe_2(CO)_4^{2-}$, linked by a pair of Fe-CO-Fe bridges.[97] The structure of this ion, as established from X-ray data,[72] is shown in Figure 19. This molecule is structurally analogous to $[(\eta\text{-}C_5H_5)Fe(CO)_2]_2$, except for the interesting difference that the latter species exists as the *trans* isomer whereas the carborane analog is of the *cis* form (Ref. 97, footnote 4) as shown in Figure 19. No explanation for this variance has been given but both species undoubtedly exist in solution as equilibrium mixtures of *cis,trans* and possibly a third form involving direct Fe–Fe bonding with all six carbonyls present as terminal groups.[97] The

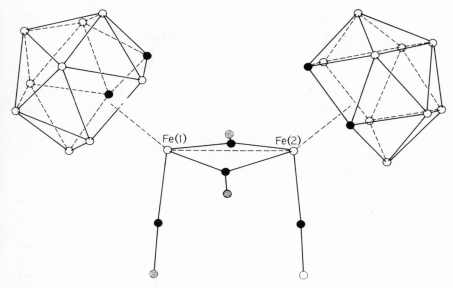

Figure 19. Structure of the $(1,2\text{-}C_2B_9H_{11})_2Fe_2(CO)_4^{2-}$ ion.[72] (From *Inorg. Chem.*, **9**, 1464 (1970). Copyright by the American Chemical Society.)

carborane dimer is diamagnetic, suggesting a direct metal–metal interaction which permits spin-pairing between the formal iron(I) atoms.

The pink biscarboranyl ion $3\text{-}(1,2\text{-}C_2B_9H_{11})_2Fe^{2-}$, a diamagnetic formal iron(II) species analogous to ferrocene, undergoes protonation by aqueous HCl or $HClO_4$ to form the orange $(C_2B_9H_{11})_2FeH^-$ ion which has been isolated as the trimethyl phosphonium salt and characterized from spectroscopic and electrochemical data.[98] The spectra are consistent with an iron-bonded hydrogen atom, but other locations for the extra hydrogen cannot be ruled out (for example, the infrared band at 1885 cm^{-1} is possibly indicative of an Fe–H group, but also falls in the region of B–H–B stretching frequencies). The protonated ion reacts easily with dialkyl sulfides, releasing a mole equivalent of hydrogen and forming B-substituted metallocarborane derivatives:

$$[(C_2B_9H_{11})_2Fe^{II}H]^- \xrightarrow{R_2S} H_2 + (C_2B_9H_{11})Fe^{II}(C_2B_9H_{10}SR_2)^-$$

$$+e^- \Big\uparrow\Big\downarrow -e^-$$

$$(C_2B_9H_{11})Fe^{III}(C_2B_9H_{10}SR_2)^0$$

$$R = CH_3,\ C_2H_5,\ n\text{-}C_3H_7$$

The sulfide derivatives are oxidized by air, I_2, or $FeCl_3$ to the corresponding air-stable iron(III) compounds, which are dark red, paramagnetic crystalline solids. The oxidized species are reducible to the original iron(II) compounds by treatment with sodium amalgam in solution.[98]

Treatment of the biscarboranyl iron(III) ion, $3\text{-}(1,2\text{-}C_2B_9H_{11})_2Fe^-$, with diethyl sulfide in the presence of anhydrous HCl gave both mono- and disubstituted products, the former being identical to the compound obtained by oxidation of the $(C_2B_9H_{11})Fe^{II}(C_2B_9H_{10}S[C_2H_5]_2)^-$ ion, described above:

$$[(C_2B_9H_{11})_2Fe^{III}]^- + (C_2H_5)_2S \xrightarrow[-H_2]{HCl} \begin{cases} [(C_2B_9H_{11})Fe^{III}(C_2B_9H_{10}S[C_2H_5]_2)]^0 \\ + \\ [(C_2B_9H_{10}S[C_2H_5]_2)_2Fe^{II}]^0 \end{cases}$$

$$[(C_2B_9H_{10}S[C_2H_5]_2)_2Fe^{III}]^+ \xleftarrow[CH_3CN]{-e^-}$$

The disubstituted neutral species, obtained as a minor product, contains formal iron(II); electrochemical oxidation of this compound produced a cation which was isolated as a shock-sensitive, red perchlorate salt.[98]

Although the position of sulfur attachment on the cages in these species is not certain, nmr and chemical evidence suggest B(8), the unique boron adjacent to the metal, as a probable location. A mechanism involving nucleophilic substitution has been tentatively postulated,[98] in which the H^+ on the protonated metallocene effectively removes H^- from the cage and thereby creates a location susceptible to nucleophilic attack by the sulfide. Several possibilities exist, however, for the structure of the intermediate, which at this time is unknown.

It is interesting that the analogous protonated cobalt(III) anion, $(C_2B_9H_{11})_2Co^-$, has not been isolated despite repeated attempts. However, acid-catalyzed substitution of diethyl sulfide on the cage proceeds readily, suggesting that a protonated cobalt intermediate is involved[98] (small cobalta-carboranes protonated at the metal are known: see Section IV.F.2). The reaction of the $(1,2\text{-}C_2B_9H_{11})_2Fe^-$ anion with CS_2 in protic media forms a derivative containing an HCS_2 unit bridging the two carborane cages.[52] A number of cobalt derivatives of this type have been prepared and are discussed in Section IV.F.5 below.

The oxidation of $[3,1,2\text{-}(\eta\text{-}C_5H_5)Fe^{II}C_2B_9H_{11}]^-$ and $[3\text{-}(1,2\text{-}C_2B_9H_{11})_2\text{-}Fe^{II}]^{2-}$ by ferric and ferricenium ions in aqueous media has been examined kinetically,[162] with rather inconclusive results. The reactions of the bis-carbollyl species occurred too rapidly ($k \geqslant 10^8 \text{ mol}^{-1}\text{sec}^{-1}$) to measure, but the oxidation of $[(\eta\text{-}C_5H_5)Fe^{II}C_2B_9H_{11}]^-$ by Fe^{3+} was found to have a rate constant of $(4.9 \pm 0.3) \times 10^6 \text{ mol}^{-1} \text{ sec}^{-1}$. These results are in agreement with theoretical predictions and kinetic findings on ferrocene oxidations, leading the authors to suggest a possible similarity in mechanism and implying a "side-by-side" configuration in the transition state.[162]

The electronic properties of the iron(II) biscarboranyl anion $(1,2\text{-}C_2B_9H_{11})_2Fe^{2-}$ have been explored[1] via the ^{19}F nmr spectra of its C-substituted m- and p-fluorophenyl derivatives, utilizing the method of Taft.[193,194] A large $+I$ effect was found, indicating an inductive electron-donating capability of similar magnitude to that of the $C_2B_9H_{11}^{2-}$ anions; in comparison, the cobalt(III) analog (a monoanion), exhibits only a small, if any, inductive effect, probably due to the high formal charge on the metal ion which tends to cancel the $+I$ effect that might be expected on the basis of the overall negative charge. Both the iron and cobalt species, however, were found to have negative Taft σ_R^0 values, indicative of extensive ground-state π-delocalization of the electron density of the cage system into the aryl substituent.[1] The limitations of the Taft approach, and the fact that only a handful of metallocarborane species were examined, preclude the drawing of any firm conclusions regarding the general nature of metallocarborane electronic structures (for example, only C-substituted fluorophenyl derivatives were examined).

Zakharkin and Bikkineev[237] have conducted a similar study on derivatives in which the substituent is attached to the cyclopentadienyl ring, rather than the polyhedral cage. From ^{19}F nmr studies of $(m\text{- and }p\text{-}FC_6H_4\text{-}\eta\text{-}C_5H_4)\text{-}CoC_2B_9H_{11}$, it was concluded that the $\text{-}(\eta\text{-}C_5H_4)CoC_2B_9H_{11}$ group is a more powerful inductive electron acceptor than is the $(\eta\text{-}C_5H_5)CoC_2B_9H_{10}\text{-}$ unit in which substitution is on the cage carbon atom.

4. Thirteen-Atom Cages

The reductive cage opening of icosahedral $1,2\text{-}C_2B_{10}H_{12}$ with sodium generates an open-cage $1,2\text{-}C_2B_{10}H_{12}^{2-}$ anion, which in turn complexes with transition metal ions to give 13-vertex metallocarboranes.[36] This polyhedral expansion yields biscarboranyl sandwich species or cyclopentadienyl–carboranyl compounds, the latter products requiring the presence of $C_5H_5^-$ during reaction with the metal ion.

$$1,2\text{-}C_2B_{10}H_{12} \xrightarrow[\text{C}_{10}\text{H}_8,\ \text{THF}]{\text{Na}} C_2B_{10}H_{12}^{2-} \xrightarrow{M^n} [(C_2B_{10}H_{12})_2M]^{n-4}$$

$$M = Fe^{II},\ Co^{III},\ \text{or } Ni^{II}$$

$$\downarrow MCl_2,\ C_5H_5^-$$

$$[(\eta\text{-}C_5H_5)M^{II}C_2B_{10}H_{12}]^-$$

$$\downarrow -e^-$$

$$(\eta\text{-}C_5H_5)M^{III}C_2B_{10}H_{12}$$

$$M = Fe,\ Co$$

The structure of $(\eta\text{-}C_5H_5)CoC_2B_{10}H_{12}$ (see Figure 32) has been established by the X-ray method, and the same cage geometry is assumed for the biscarboranyl system, $(C_2B_{10}H_{12})_2Co^-$, and for the iron analogs of these species.[36] The cobalt compounds have been studied in greater detail, including, for example, their fluxional behavior in solution, and are discussed in Section IV.F.6. Some work on the 13-vertex iron species has, however, been reported. The violet, paramagnetic $(\eta\text{-}C_5H_5)Fe^{III}C_2B_{10}H_{12}$ is readily reduced with aqueous $NaBH_4$ to the orange iron(II) anion. The tetramethylammonium salt of the biscarboranyl $(C_2B_{10}H_{12})_2Fe^{2-}$ ion undergoes reversible oxidation by cyclic voltammetry, and like its cobalt(III) and nickel(II) analogs, is slowly solvolyzed by acid or base. From nmr evidence, it appears that the $(C_2B_{10}H_{12})_2Fe^{2-}$ anion in solution is fluxional in the manner described below for the $(\eta\text{-}C_5H_5)CoC_2B_{10}H_{12}$ system, which involves an intermediate or transition state bisected by a mirror plane (the static structure is totally asymmetric).[36] Zakharkin[239] has published data suggesting that the $(C_2B_9H_{11})_2Fe^{2-}$ prepared under the conditions given in Ref. 36 is in fact a mixture of isomers (section IV.F.6).

A 13-atom cobaltaferracarborane has been prepared, and is described in Section IV.F.6.

5. Fourteen-Atom Cages

The only 14-vertex ferracarboranes known at this writing are several isomers of the $(\eta\text{-}C_5H_5)_2Fe_2(CH_3)_4C_4B_8H_8$[136,142] system which were obtained by reduction of the tetracarbon carborane $(CH_3)_4C_4B_8H_8$ with $NaC_{10}H_8$ followed by reaction with $FeCl_2$ and NaC_5H_5. Crystallographic studies[142] of two of the isomers revealed, surprisingly, *nido* structures in both cases; one compound has an open five-membered face and the other a four-sided open face. Both isomers undergo thermal isomerization, converting to a common isomer at 150°.[136]

Experimental

Preparation of $4,1,7\text{-}(\eta\text{-}C_5H_5)FeC_2B_{10}H_{12}$.[36] A solution of 20 mmol of $Na_2[C_2B_{10}H_{12}]$ and 40 mmol of NaC_5H_5 was placed in a 1-liter flask fitted with a mechanical stirrer and a nitrogen inlet. The flask was cooled to 0° and a gray-white slurry of $FeCl_2$ in THF, prepared by the reaction of 5.2 g (32 mmol) of $FeCl_3$ and 4.0 g (72 mg-atoms) of iron powder in 100 ml of THF at reflux for 3 hr, was added slowly. The reaction mixture was stirred for 15 hr while slowly warming to room temperature. A stream of oxygen was passed through the solution for 0.5 hr after which the solution was filtered through Celite. Silica gel (30 g) was added to the filtrate and the solvent removed by rotary evaporation. The solid was placed atop a silica gel chromatography column (4 × 30 cm) prepared with hexane. The column was eluted with hexane that was slowly enriched with dichloromethane. Four major bands were separated that contained, in order of elution, a mixture of ferrocene and naphthalene, green $2,1,7\text{-}(C_5H_5)FeC_2B_9H_{11}$, violet $(C_5H_5)FeC_2B_{10}H_{12}$ (0.8342 g, 3.1 mmol, 15%), and a purple band which moved extremely slowly on the column. This band was not isolated but was presumed to be the $(C_2B_{10}H_{12})_2Fe^{2-}$ complex.

E. Ruthenium Compounds

The only reported metallocarboranes containing ruthenium are the 12-vertex systems $3,1,2\text{-}[(C_6H_5)_3P]_2(CO)RuC_2B_9H_{11}$ and $3,1,2\text{-}(CO)_3RuC_2\text{-}B_9H_{11}$, obtained from reactions of the $1,2\text{-}C_2B_9H_{11}{}^{2-}$ ion with $[(C_6H_5)_3P]_2\text{-}Ru(CO)_2Cl_2$ and $[Ru(CO)_3Cl_2]_2$, respectively.[185] No chemistry has been described for these species.

F. Cobalt Compounds

1. General Comments

Of the dozen or more transition elements that have been incorporated into metallocarborane cage systems, by far the most extensive known chemistry is that of the cobalt species. Furthermore, nearly all of the characterized cobalt metallocarboranes, other than the biscarborane-type species, contain the formal $(\eta\text{-}C_5H_5)Co$ moiety which may be considered a two-electron donor that functions as a replacement for :BH in a boron framework. [An equivalent qualitative description is to regard $(\eta\text{-}C_5H_5)Co^{2+}$ and BH^{2+} as bonding to a dinegative borane or carborane fragment.] The cobalt atoms in such systems acquire an 18-electron filled-shell configuration, resulting in diamagnetic species which are readily studied by nmr spectroscopy. The ability of the $(\eta\text{-}C_5H_5)Co$ unit to replace BH groups and to stabilize borane or carborane cage fragments[114,208] is quite remarkable and as yet not fully explained. For example, the reaction of the $B_5H_8^-$ ion with $CoCl_2$, and $C_5H_5^-$ followed by air oxidation generates a group of novel cobaltaboranes[147,149] including two isomers of $(\eta\text{-}C_5H_5)CoB_4H_8$, a $(\eta\text{-}C_5H_5)_2Co_2B_4H_6$, and $(\eta\text{-}C_5H_5)CoB_9H_{13}$, which are isoelectronic analogs of B_5H_9, $B_6H_6^{2-}$, and $B_{10}H_{14}$, respectively. [In the $B_6H_6^{2-}\text{-}(\eta\text{-}C_5H_5)_2Co_2B_4H_6$ analogy, each $(C_5H_5)Co$ group replaces B^-; in the other cases, $(C_5H_5)Co$ replaces BH.] While there is as yet no direct evidence as to the mechanism of this particular reaction, it appears that cobalt functions as a scavenger for unstable polyborane species which are formed initially. The versatility of the $(\eta\text{-}C_5H_5)Co$ to function as a stabilizing agent has been widely utilized in metallocarborane synthesis, particularly in reactions of $CoCl_2$ and NaC_5H_5 with carborane anions generated by the reduction of $C_2B_{n-2}H_n$ polyhedra with sodium (Section III). Another method of incorporating the $(\eta\text{-}C_5H_5)Co$ unit into carborane frameworks utilizes the commercially available reagent cyclo-pentadienylcobalt(I) dicarbonyl, as described in the following paragraphs.

2. Six- and Seven-Atom Cages

The gas-phase reaction of $(\eta\text{-}C_5H_5)Co(CO)_2$ with the trigonal bi-pyramidal $1,5\text{-}C_2B_3H_5$ at $230°$ generates primarily $1,2,4\text{-}(\eta\text{-}C_5H_5)CoC_2B_3H_5$, plus small amounts of two isomers of a dicobalt system, $1,7,2,4\text{-}$ and $1,2,3,5\text{-}$ $(\eta\text{-}C_5H_5)_2Co_2C_2B_3H_5$.[151] The proposed[151] octahedral and pentagonal bi-pyramidal cage structures of these molecules are shown in Figure 11. The octahedral species and its analog $(CO)_3FeC_2B_3H_5$, described above, are the smallest known closed polyhedral metallocarboranes. The structures of these six- and seven-atom cobalt species have not yet been established by X-ray studies, but from their isoelectronic relationship to molecules of known geometry such as $1,6\text{-}C_2B_4H_6$, $CH_3GaC_2B_4H_6$, $1,2,3\text{-}(\eta\text{-}C_5H_5)Co(CH_3)_2C_2\text{-}$ B_4H_4,[215] and $1,7,2,3\text{-}(\eta\text{-}C_5H_5)_2Co_2C_2B_3H_5$, as well as from symmetry arguments based on [11]B and ^{1}H nmr spectra, the geometries shown can be regarded as reasonably certain. (The structural simplicity of the small polyhedra often permits assignment of a unique or highly probable geometry; this is less frequently the case in the larger and more complex metallo-carboranes where, commonly, a large number of possible isomers must be considered.)

The $1,7,2,4\text{-}(\eta\text{-}C_5H_5)Co_2C_2B_3H_5$ species, which can be viewed as a triple-decked metallocene,[3,79] is also generated in reactions of $1,6\text{-}C_2B_4H_6$, described below. The 1,2,3,5-isomer of this system is analogous to $1,2,3,5\text{-}$ $(CO)_6Fe_2C_2B_3H_5$, discussed earlier in this review. A third isomer, $1,7,2,3\text{-}$ $(\eta\text{-}C_5H_5)_2Co_2C_2B_3H_5$, is obtained in other reactions to be discussed in this section, and a fourth, identified as the 1,2,4,5 system, is formed in the pyrolysis (of the 1,7,2,3 species[77,148] as shown in Figure 20.

As expected, the direct gas-phase treatment of the octahedral $1,6\text{-}C_2B_4H_6$ with $(\eta\text{-}C_5H_5)Co(CO)_2$ produces a seven-vertex monocobalt species, $1,2,4\text{-}$ $\eta\text{-}C_5H_5)CoC_2B_4H_6$, in excellent yield.[151]

The reductive cage opening of $1,6\text{-}C_2B_4H_6$ with sodium naphthalide, followed by addition of $CoCl_2$ and air oxidation, generates the previously mentioned green $1,7,2,4\text{-}(\eta\text{-}C_5H_5)_2Co_2C_2B_3H_5$ (Figure 20) in 4% yield[3,79] with traces of an eight-vertex $(\eta\text{-}C_5H_5)_2Co_2C_2B_4H_6$ and a σ-bonded naphthyl derivative of the 1,7,2,4 system.[79] In view of the low yields this does not appear likely to be a synthetically important reaction, and direct gas-phase

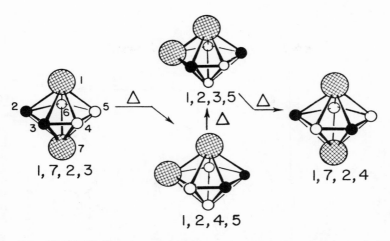

Figure 20. Schematic diagram of a pathway for thermal conversion of 1,7,2,3-$(\eta\text{-}C_5H_5)_2Co_2C_2B_3H_5$ to the 1,7,2,4 isomer via the intermediate 1,2,4,5 and 1,2,3,5 species.[77,148] The isomerization occurs above 200°; the structures of the 1,7,2,3 and 1,7,2,4 systems are established from X-ray studies[3,169]; the others have been deduced from [11]B and ^{1}H nmr and other spectroscopic data.[3,77,79,148] ●, CH; ○, BH; ◉, $Co(C_5H_5)$.

metal insertion seems the method of choice for $1,6\text{-}C_2B_4H_6$ and $1,5\text{-}C_2B_3H_5$, as discussed earlier in Section III.

The *nido*-carborane $2,3\text{-}C_2B_4H_8$, on the other hand, is a highly useful starting reagent for the preparation of small cobalt metallocarboranes. Bridge deprotonation of this material with sodium hydride to give the $C_2B_4H_7^-$ ion, followed by reaction of the latter species with $CoCl_2$ and O_2 and workup in aqueous media, gives three main products[79]: $1,2,3\text{-}(\eta\text{-}C_5H_5)\text{-}CoC_2B_4H_6$, $1,7,2,3\text{-}(\eta\text{-}C_5H_5)_2Co_2C_2B_3H_5$, and *nido*-$1,2,3\text{-}(\eta\text{-}C_5H_5)CoC_2B_3H_7$, as depicted in Figure 21. The same products are obtained by treatment of $2,3\text{-}C_2B_4H_8$ with sodium naphthalide, $CoCl_2$, NaC_5H_5, and O_2, but in this case $(\eta\text{-}C_5H_5)CoC_2B_4H_6$ is formed only in trace amount. In the synthesis via $C_2B_4H_7^-$, the relative yields of metallocarboranes are dependent upon workup conditions; thus, in acid media (1 M HCl), $1,2,3\text{-}(\eta\text{-}C_5H_5)CoC_2B_4H_6$ is obtained in 60% yield with much smaller amounts of the other two species, but workup of the products in neutral water results in base degradation of much of the $1,2,3\text{-}(\eta\text{-}C_5H_5)CoC_2B_4H_6$ to the three-boron species.[79] Clearly, the principal product formed initially is $1,2,3\text{-}(\eta\text{-}C_5H_5)CoC_2B_4H_6$, with the other compounds arising from base degradation of this material during workup in aqueous media. This hypothesis is supported by detailed studies of

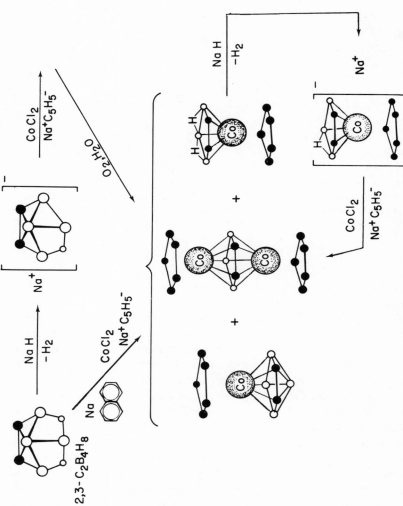

Figure 21. Synthesis of cobaltacarboranes from 2,3-$C_2B_4H_8$, showing established structures of 1,2,3-(η-C_5H_5)CoC$_2$B$_4$H$_6$[215] and 1,7,2,3-(η-C_5H_5)$_2$Co$_2$C$_2$B$_3$H$_5$,[3] and proposed structures[79] of 1,2,3-(η-C_5H_5)-CoC$_2$B$_3$H$_7$ and the (η-C_5H_5)CoC$_2$B$_3$H$_6^-$ ion. ●, CH; ○, BH.

reactions of $1,2,3\text{-}(\eta\text{-}C_5H_5)CoC_2B_4H_6$ and its *C*-alkyl derivatives in KOH solution, which give the above three-boron species or their derivatives in substantial yields.[150]

The structure of $1,2,3\text{-}(\eta\text{-}C_5H_5)CoC_2B_4H_6$ indicated in Figure 21 is based on ^{11}B and ^1H nmr data[79] and is distinguishable from the 1,2,4 (nonadjacent cage carbon) isomer by the independent synthesis of the latter from $1,6\text{-}C_2B_4H_6$ and from $2,4\text{-}C_2B_5H_7$ (see below). In addition, an X-ray diffraction study[215] of $2,3\text{-}(CH_3)_2\text{-}1,2,3\text{-}(\eta\text{-}C_5H_5)CoC_2B_4H_4$ has recently confirmed the postulated geometry of this cage system.

The structure of $1,2,3\text{-}(\eta\text{-}C_5H_5)CoC_2B_4H_6$ indicated in Figure 21 is based on ^{11}B and ^1H nmr data[79] and is distinguishable from the 1,2,4 derivatives,[3,169] and are analogous to triple-decked metallocene species (i.e., two metal ions sandwiched between three planar rings, as in the unique dinickel system mentioned below). If the formal charges of cobalt and the cyclopentadienyl ligand are taken to be $+3$ and -1, respectively, as usual in diamagnetic cobalt metallocarboranes, the formal charge associated with the central $C_2B_3H_5$ ring is -4. Since the formal $C_2B_3H_5{}^{4-}$ planar system is isoelectronic with $C_5H_5{}^-$, $1,7,2,3\text{-}(\eta\text{-}C_5H_5)_2Co_2C_2B_3H_5$ is an analog of the hypothetical triple-decked $Co_2(\eta\text{-}C_5H_5)_3{}^{3+}$ ion. No such cobalt triple-decked species is known, although the analogous dinickel ion, $Ni_2(\eta\text{-}C_5H_5)_3^+$, has been reported[182,221] and a triple-decked sandwich structure has been established in an X-ray study.[33] Since this latter species is the only reported example of an isolable triple-decked tris(cyclopentadienyl)metallocene, and is itself unstable in water and certain other solvents, the stability of the analogous but electrically neutral species $1,7,2,3\text{-}(\eta\text{-}C_5H_5)_2Co_2C_2B_3H_5$ is significant. We have suggested[3,169] that the formal $C_2B_3H_5^-$ planar ring system is highly electron delocalized, and a detailed examination[216] of substituent nmr effects has supported this model for the 1,7,2,4 isomer, which has nonadjacent central-ring carbon atoms. The same study, however, suggested that in the 1,7,2,3 species the central ring is less delocalized and contains an ethylene-like carbon–carbon bond which participates in a localized π-interaction with the metal atoms. In both isomers, strong inter- and intraannular electronic effects were observed.[216]

It seems reasonable to suppose that the large negative charge and the presence of three borons with relatively large bonding orbitals (compared to carbon) serve to stabilize metal sandwich complexes to an extent not matched by the isoelectronic $C_5H_5{}^-$ ion. Thus, an extensive family of metallocarboranes incorporating formal planar $C_2B_3H_5^{4-}$ might be anticipated, possibly including quadruple- or multiple-decked sandwiches containing several such cyclocarboranyl ring systems and end-capped by cyclopentadienyl rings. (Mass spectroscopic evidence for the existence of a quadruple-decked compound has been obtained but no such species has been isolated.[217])

The *nido*-metallocarborane $1,2,3-(\eta-C_5H_5)CoC_2B_3H_7$ (Figure 21) has been structurally characterized[79] from ^{11}B and ^1H nmr data, and is analogous to $1,2,3-(CO)_3FeC_2B_3H_7$, whose structure is known from an X-ray study[10] (see above and Figure 2a). The molecule is also an analog of $2,3-C_2B_4H_8$, in which $(\eta-C_5H_5)Co$ replaces the apex BH group, and hence might be expected to undergo bridge deprotonation on treatment with sodium hydride. This is indeed the case, as illustrated in Figure 21. Deprotonation generates the $(\eta-C_5H_5)CoC_2B_3H_6^-$ anion, into which a second cobalt can be inserted to give the previously described triple-decked sandwich, $1,7,2,3-(\eta-C_5H_5)_2$-$Co_2C_2B_3H_5$ in moderate yield.[79] (Although the last step shown involves loss of the remaining bridge hydrogen atom from the metallocarborane anion, its fate has not been determined.)

The reaction of the *C,C'*-dimethyl derivative of the $C_2B_4H_7^-$ ion with $CoCl_2$ in the *absence* of $C_5H_5^-$ has been studied recently and gives interesting and somewhat unexpected results. Instead of the anticipated biscarborane anion $[(CH_3)_2C_2B_4H_4]_2Co^-$, a red neutral species, $[(CH_3)_2C_2B_4H_4]_2CoH$, is obtained, in which the "extra" hydrogen is assumed from nmr evidence to be metal bound (Figure 22).[140] The proposed structure of this material is analogous to $1,2,3-(\eta-C_5H_5)CoC_2B_4H_6$ but contains two seven-vertex polyhedra fused at a common cobalt atom. In addition to the red compound, a pale yellow species formulated as $[(CH_3)_2C_2B_3H_5]CoH[(CH_3)_2C_2B_4H_4]$ (Figure 22) is obtained; the latter metallocarborane is presumably analogous to $1,2,3-(\eta-C_5H_5)CoC_2B_4H_6$ with a cyclo-$(CH_3)_2C_2B_3H_5$ ring replacing C_5H_5. Further work has shown that the red species, which is exceedingly reactive with moisture, is hydrolytically degraded with removal of a BH unit to yield the yellow species.[140] The yellow compound, which is an interesting example of a *nido–closo*-metallocarborane having a metal atom common to both an open and a closed polyhedral cage, is highly stable toward both hydrolysis and thermal degradation.

The chemistry of these new species has been explored in some detail[140] and the results are summarized schematically in Fig. 22. As shown, insertion of additional cobalt atoms in the monocobalt species (I) or (II) generates di- and tricobalt systems for which the structures drawn have been postulated from ^{11}B and ^1H nmr evidence. The red compound $[(CH_3)_2C_2B_4H_4]_2CoH$ (I) is air oxidized to yield a novel tetracarbon carborane, $(CH_3)_4C_4B_8H_8$, a colorless air-stable solid.[138,139] This compound, in turn, is a precursor to a variety of tetracarbon metallocarboranes incorporating Mo, W, Fe, Co, or Ni atoms (see Sections IV.B, IV.D, and IV.H.).

The previously mentioned seven-vertex species $1,2,4-(\eta-C_5H_5)CoC_2B_4H_6$, which has nonvicinal cage carbon atoms, is formed in substantial (20%) yield in the reductive cage opening of *closo*-$2,4-C_2B_5H_7$ with sodium, followed by reaction with $CoCl_2$ and NaC_5H_5.[146] In fact, this species is the main

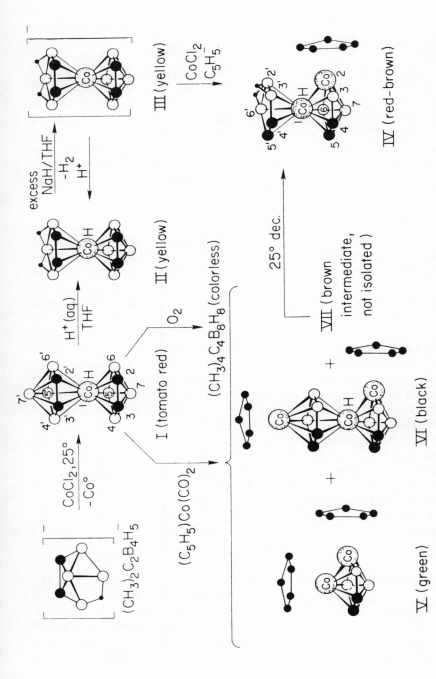

Figure 22. Preparation and chemistry of $[(CH_3)_2C_2B_4H_4]_2CoH$. Metal-bound hydrogen atoms, shown schematically, are proposed[140] to undergo rapid tautomerism through face-bonded locations on the polyhedral surface. ○, BH; ●, CCH₃; ●, CH; ● H.

metallocarborane product isolated following air oxidation and treatment with a water–acetone mixture. Other products, none obtained in more than ca. 2% yield, include several mono- and dicobalt species of 8–10 vertices plus σ-bonded naphthyl and carboranyl derivatives of 1,2,4-(η-C_5H_5)-$CoC_2B_4H_6$. Possible mechanisms for this reaction have been suggested,[146] but it is not clear whether the process occurs primarily via the initial formation of a nido-$C_2B_5H_7^{2-}$ ion (presently unknown), or whether 2,4-$C_2B_5H_7$ is degraded to a four-boron nido species such as the hypothetical 2,4-$C_2B_4H_6^{2-}$, which then undergoes metal insertion to yield the observed 1,2,4-(η-C_5H_5)-$CoC_2B_4H_6$. At the moment, available evidence[152,230] tends to support the former type of pathway, but this is not a settled question.

Experimental

Preparation of 1,2,4-(η-C_5H_5)$CoC_2B_3H_5$ *and* (η-C_5H_5)$_2Co_2C_2B_3H_5$ *Isomers by Direct insertion.*[151] The cylindrical hot–cold reactor described in Section IV.D.1. was charged with 3.59 mmol of $C_2B_3H_5$ and 3.90 mmol of (η-C_5H_5)Co(CO)$_2$. The central portion of the tube was heated to 230°, while the lower end was maintained at 60° in an oil bath. After 14 hr the tube was cooled in liquid nitrogen, the noncondensables were removed, and an additional 2.14 mmol of $C_2B_3H_5$ was added. The reactor was again heated for 24 hr, after which the above procedure was repeated and another 2.14 mmol of $C_2B_3H_5$ was added. The reaction was then allowed to continue for 48 hr with periodic interruptions to permit removal of noncondensables. At this point the reactor was opened to the vacuum line and the volatiles removed. Repeated fast distillation of the volatiles into a 0° trap allowed unreacted (η-C_5H_5)Co(CO)$_2$ to pass. No $C_2B_3H_5$ was recovered. The material retained in the 0° trap was combined with a solution obtained by extracting the residue in the reactor with CH_2Cl_2, and the resulting mixture was separated by preparative thin-layer chromatography. Development on silica gel plates with carbon tetrachloride as eluent gave yellow-orange 1,2,4-(η-C_5H_5)-$CoC_2B_3H_5$ (135 mg, 0.750 mmol), dark-green 1,2,3,5-(η-C_5H_5)$_2Co_2C_2B_3H_5$ (0.9 mg, 0.003 mmol), and dark-green 1,7,2,4-(η-C_5H_5)$_2Co_2C_2B_3H_5$ (10.2 mg, 0.033 mmol). The yields were 9.5, 0.04, and 0.42%, respectively, based on the total quantity of $C_2B_3H_5$ employed.

Preparation of 1,2,3-(η-C_5H_5)$CoC_2B_4H_6$, nido-1,2,3-(η-C_5H_5)$CoC_2B_3H_7$, *and* 1,7,2,3-(η-C_5H_5)$_2Co_2C_2B_3H_5$ *from* $NaC_2B_4H_7$.[79] A filtered solution of $NaC_2B_4H_7$ prepared from 5.06 mmol of $C_2B_4H_8$ and 6 mmol of NaH in THF was added to a filtered solution of NaC_5H_5 obtained from 19.6 mmol of C_5H_6 and 21.6 mmol of divided sodium in THF, and the combined solution was added dropwise over 20 min to a stirred solution of 25.1 mmol anhydrous

$CoCl_2$ in 50 ml of THF. The dark-blue $CoCl_2$ solution immediately turned dark brown. After stirring for 12 hr at 25°, the solvent was distilled off under reduced pressure and the residue was suspended in 60 ml of H_2O and stirred for 2 hr under a stream of air. After filtration of the solution, the residue was extracted with methylene chloride, followed by acetone. Thin-layer separation of the combined extracts on silica gel plates with CCl_4 gave orange 1,2,3-$(\eta-C_5H_5)CoC_2B_4H_6$ [R_f = 0.5, 104 mg (0.524 mmol, 10.3% yield)]; yellow nido-1,2,3-$(\eta-C_5H_5)CoC_2B_3H_7$ [R_f = 0.85, 156 mg (0.826 mmol, 16.3% yield)]; and red-brown 1,7,2,3-$(\eta-C_5H_5)_2Co_2C_2B_3H_5$ [R_f = 0.25, 40 mg (0.129 mmol, 2.5% yield)].

Modification of the above procedure to include workup in aqueous HCl gave substantially increased yields of 1,2,3-$(\eta-C_5H_5)CoC_2B_4H_6$. In a typical experiment, 5 ml of a THF solution of 0.66 mmol of $NaC_2B_4H_7$ and 2.4 mmol of NaC_5H_5 was added to 2.96 mmol of $CoCl_2$ in THF and stirred for 2.5 hr at 25°, after which the solvent was removed at reduced pressure, 20 ml of 1 M HCl was added and the solution stirred for another hour. Treatment with air, extraction with solvents, and tlc separation as described above gave 83 mg (0.42 mmol, 63% yield) of $(\eta-C_5H_5)CoC_2B_4H_6$, 5 mg (0.027 mmol, 4% yield) of $(\eta-C_5H_5)CoC_2B_3H_7$, and 4 mg (0.012 mmol, 2% yield) of $(\eta-C_5H_5)_2$-$Co_2C_2B_3H_5$.

3. Eight-, Nine-, and Ten-Atom Cages

The only known eight-vertex cobalt metallocarboranes are a mono-cobalt and a dicobalt species, 3,1,7-$(\eta-C_5H_5)CoC_2B_5H_7$ and 3,5,1,7-$(\eta-C_5H_5)_2$-$Co_2C_2B_4H_6$. Both compounds are minor products (ca. 1% yield) of the polyhedral expansion of 2,4-$C_2B_5H_7$ via sodium naphthalide reduction in THF,[146] but the monocobalt species is formed in larger yield (25%) in the direct thermal reaction of 2,4-$C_2B_5H_7$ with $(\eta-C_5H_5)Co(CO)_2$.[151] Several nine- and ten-vertex metallocarboranes having one to three cobalt atoms are also produced in these reactions of $C_2B_5H_7$; in several cases, identical isomers are produced despite the very dissimilar experimental conditions (see Table II; yields based on $C_2B_5H_7$ consumed are in parentheses). The fact that identical CoC_2B_5 and $Co_2C_2B_5$ systems are obtained in these two reactions is significant, and suggests that basically similar factors dictate the final polyhedral cage structure in the two reaction systems. Other similar examples of the formation of identical cage species under different conditions will be mentioned at appropriate points in the text.

The 1,8,5,6- and 1,7,5,6-$(\eta-C_5H_5)_2Co_2C_2B_5H_7$ isomers have been shown[148] to undergo reversible thermal interconversion in the vapor phase at elevated temperature, with K_{eq} = [1,7,5,6]/[1,8,5,6] = 1.00 ± 0.05 at 340° and ΔH = −2 ± 1 kcal for the [1,8,5,6] → [1,7,5,6] conversion. The

TABLE II
Reactions of $C_2B_5H_7$ and Their Products

$C_2B_5H_7 + Na + C_{10}H_8 + CoCl_2 + NaC_5H_5$, THF, 25° (Ref. 146)	$C_2B_5H_7 + (\eta\text{-}C_5H_5)Co(CO)_2$, vapor phase, 260° (Ref. 151)
Products (percent yield)	
$1,2,4\text{-}(\eta\text{-}C_5H_5)CoC_2B_4H_6$ (20)	$1,2,4\text{-}(\eta\text{-}C_5H_5)CoC_2B_4H_6$ (3.6)
$3,1,7\text{-}(\eta\text{-}C_5H_5)CoC_2B_5H_7$ (0.5)	$3,1,7\text{-}(\eta\text{-}C_5H_5)CoC_2B_5H_7$ (25)
$1,2,6\text{-}(\eta\text{-}C_5H_5)CoC_2B_6H_8$ (2)	
$3,5,1,7\text{-}(\eta\text{-}C_5H_5)Co_2C_2B_4H_6$ (0.2)	
$1,8,5,6\text{-}(\eta\text{-}C_5H_5)_2Co_2C_2B_5H_7$ (0.2)	$1,8,5,6\text{-}(\eta\text{-}C_5H_5)_2Co_2C_2B_5H_7$ (25)
	$1,7,5,6\text{-}(\eta\text{-}C_5H_5)_2Co_2C_2B_5H_7$ (2.4)
$2,4,3,10\text{-}(\eta\text{-}C_5H_5)_2Co_2C_2B_6H_8$ (0.2)	
	$2,3,8,1,6\text{-}(\eta\text{-}C_5H_5)_3Co_3C_2B_5H_7$ (2.4)^a
	$2,3,4,1,10\text{-}(\eta\text{-}C_5H_5)_3Co_3C_2B_5H_7$ (0.3)

^{a} Yield in sealed-bulb reactions was $\sim 20\%$.

structures originally proposed for these isomers[146,151] have been confirmed by X-ray crystallography[82] and consist of tricapped trigonal prisms with the cage-carbon atoms occupying two of the three capping vertices in each case; the cobalt atoms are located in vicinal (1,7) positions in the 1,7,5,6 isomer and nonvicinal (1,8) vertices in the 1,8,5,6 system (the numbering of the polyhedron is the same as that shown in Figures 10 and 16). The measured[82] Co–Co distance of 2.444(2) Å in the 1,7,5,6 compound and the calculated value of 2.07 Å for the nearest H–H intramolecular contact between the C_5H_5 rings in that molecule are consistent with an earlier suggestion[148] that the reversible rearrangement occurs because of opposing tendencies toward (1) metal–metal bond formation, which favors the 1,7,5,6 isomer, and (2) steric repulsion of the C_5H_5 rings, which favors the 1,8,5,6 species. The latter effect would presumably be stronger at high temperature, due to greater vibrational motion of the rings, in agreement with the slightly negative ΔH value. Aside from the less well-defined case of 1,7,2,3- and 1,2,4,5-$(\eta\text{-}C_5H_5)_2$-$Co_2C_2B_3H_5$ species that appear to be interconvertible,[148] these $(\eta\text{-}C_5H_5)_2$-$Co_2C_2B_5H_7$ isomers represent the only known example of reversible thermal rearrangement of metallocarborane isomers.

The tricobalt species listed earlier merit special mention since, with the exception of $[(CO)_3Fe]_2(\eta\text{-}C_5H_5)Co(CH_3)_2C_2B_4H_4$ (Section IV.D.2), they are the only known subicosahedral metallocarboranes* having more than two metal atoms in the cage [an icosahedral $(\eta\text{-}C_5H_5)_3Co_3C_2B_7H_9$ system has been isolated from the polyhedral expansion of 2,1,6-$(\eta\text{-}C_5H_5)CoC_2B_7H_9$, described in Section IV.F.5]. On the basis of ^{11}B and ^1H nmr spectra com-

---

* Cobaltaboranes (without cage carbon atoms) having three or four metal atoms have been prepared in the author's laboratory.[149]

bined with certain assumptions of the kind discussed in Section II (e.g., that the cage-CH groups will be nonadjacent as they are in the original $C_2B_5H_7$ system), the structures in Figure 23 have been suggested as likely.[151] Other structures cannot be excluded, but the patterns of fine splitting in the proton nmr signals arising from H–C–B–H spin–spin coupling[77,151] strongly support the geometries indicated in Figure 23.

A series of mono- and dicobalt species having 9 or 10 vertices has been

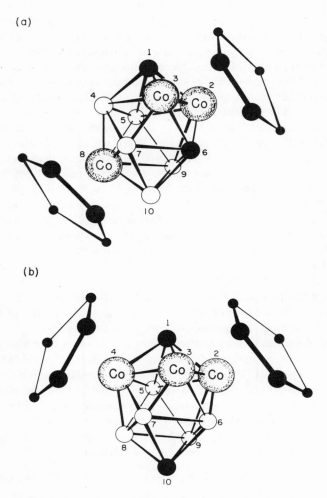

Figure 23. Proposed structures of $(\eta\text{-}C_5H_5)_3Co_3C_2B_5H_7$ isomers.[151] One C_5H_5 ring has been omitted in each drawing. ●, CH; ○, BH. (From *J. Amer. Chem. Soc.*, **96**, 3090 (1974). Copyright by the American Chemical Society.)

obtained by the reductive cage opening of $1,7\text{-}C_2B_6H_8$ and $4,5\text{-}C_2B_7H_9$
followed by cobalt insertion in the presence of cyclopentadienide ion.[34,42]
These reactions result in polyhedral expansion, the products from $C_2B_6H_8$
being $1,4,5\text{-}(\eta\text{-}C_5H_5)CoC_2B_6H_8$ and $2,6,1,10\text{-}(\eta\text{-}C_5H_5)_2Co_2C_2B_6H_8$ in 4 and
7% yield, respectively. [In the absence of $C_5H_5^-$, the $(C_2B_6H_8)_2Co^-$ ion was
obtained in 4% yield as the tetraphenylarsonium salt.] The metal atoms in
the $2,6,1,10\text{-}(\eta\text{-}C_5H_5)_2Co_2C_2B_6H_8$ species occupy adjacent vertices,[104] but
thermal rearrangement in refluxing hexadecane produces the 2,7,1,10 isomer
with nonadjacent cobalt.[49,50] Although the Co–Co link is broken in this
rearrangement and in those of $1,6,2,4\text{-}(\eta\text{-}C_5H_5)_2Co_2C_2B_7H_9$ and $3,7,1,2\text{-}$
$(\eta\text{-}C_5H_5)_2Co_2C_2B_8H_{10}$, the movement of cobalt to nonadjacent vertices is not
a completely general phenomenon, nor is it irreversible. Studies[148] of the
thermal rearrangement of dimetallic cobaltacarboranes of seven, eight, or
nine vertices have shown that in some instances migration of cobalt atoms
toward each other occurs, as in the conversion of $1,7,2,3\text{-}(C_5H_5)_2Co_2C_2B_3H_5$
to the 1,7,2,4 isomer (Figure 20). Although in most cases the metal atoms will
probably occupy nonadjacent vertices in the thermally most stable isomer
(again refer to Figure 20), even this rule is not without exception, as shown
by the 1,7,5,6- and $1,8,5,6\text{-}(\eta\text{-}C_5H_5)_2Co_2C_2B_5H_7$ systems discussed above.

The sodium naphthalide–cobalt insertion treatment of $C_2B_7H_9$[42]
generates the "expanded" product $2,3,10\text{-}(\eta\text{-}C_5H_5)CoC_2B_7H_9$ in 9% yield,
but the major product, obtained in 12% yield, is the nonexpanded cage
$1,4,6\text{-}(\eta\text{-}C_5H_5)CoC_2B_6H_8$, which has the same number of vertices as the
original carborane. It is interesting that an 11-vertex species, $1,2,3\text{-}(\eta\text{-}C_5H_5)\text{-}$
$CoC_2B_8H_{10}$, was found as a minor product, even when chromatographically
pure $C_2B_7H_9$ was employed.[42] Similar evidence for "boron expansion"
has been observed in reactions of $1,10\text{-}C_2B_8H_{10}$ and of $2,4\text{-}C_2B_5H_7$, described
elsewhere in this review. The mechanism responsible for this effect is obscure,
particularly in the absence of any information as to the nature of the intermediate ions in most of these cage-opening reactions.

Small quantities of nine and ten-vertex cobalt species are also produced
in the reductive cage opening of 1,6- and $1,10\text{-}C_2B_8H_{10}$, in which the major
products are 11-vertex "expanded" cages as discussed in Section IV.F.5. The
ten-vertex systems $(\eta\text{-}C_5H_5)CoC_2B_7H_9$ and $(C_2B_7H_9)_2Co^-$ are also formed
in two additional kinds of reactions, and the various isomers of these species
have received considerable attention. Treatment of the $1,3\text{-}C_2B_7H_{11}^{2-}$
anion (Section III) with $CoCl_2$, or with $CoCl_2$ and NaC_5H_5, yields respectively
the $2\text{-}(1,6\text{-}C_2B_7H_9)_2Co^-$ anion and $2,1,6\text{-}(\eta\text{-}C_5H_5)CoC_2B_7H_9$, (Figure 24).[63,93]

$$1,3\text{-}C_2B_7H_{13} \xrightarrow[\text{Et}_2\text{O}]{\text{NaH}} C_2B_7H_{11}^{2-} \xrightarrow{\text{Co}^{2+}} Co^0 + H_2 + (C_2B_7H_9)_2Co^-$$

$$\xrightarrow[\text{NaC}_5\text{H}_5]{\text{CoCl}_2} 2,1,6\text{-}(\eta\text{-}C_5H_5)CoC_2B_7H_9$$

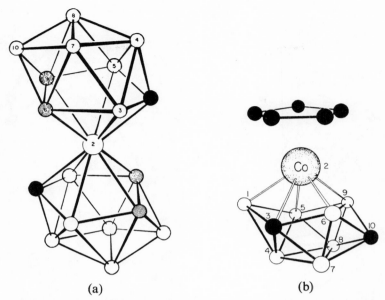

(a)   (b)

Figure 24. Structure of the $2\text{-}(1,6\text{-}C_2B_7H_9)_2Co^-$ ion[176] (a) and proposed structure of $2,1,6\text{-}(\eta\text{-}C_5H_5)CoC_2B_7H_9$ (b). In (a): $\bigcirc$, Co; $\bigcirc$, B; $\bullet$, C; ⊘, disordered B and C. In (b): $\bullet$, CH; $\bigcirc$, BH. (The first drawing is from *Inorg. Chem.*, **11**, 377 (1972). Copyright by the American Chemical Society.)

The structure depicted for the biscarboranyl ion has been established in an X-ray investigation.[176]

The polyhedral contraction (Section III) of $1,2,4\text{-}(\eta\text{-}C_5H_5)CoC_2B_8H_{10}$, an 11-vertex species with adjacent framework carbon atoms, affords a novel *nido*-cobaltaborane intermediate $(\eta\text{-}C_5H_5)CoC_2B_7H_{11}$ (an analog of $C_2B_8H_{12}$) and traces of a *closo* isomer, believed to be $2,6,9\text{-}(\eta\text{-}C_5H_5)CoC_2B_7H_9$[113] (Figure 25). Recently, the structure of $(\eta\text{-}C_5H_5)CoC_2B_7H_{11}$ has been established from X-ray diffraction.[14] The molecule has the *nido* shape shown in Figure 25, but both carbon atoms are on the open face and the two bridging hydrogens are present as Co–H–B and B–H–B groups.

On heating the $(\eta\text{-}C_5H_5)CoC_2B_7H_{11}$ intermediate at 70°, H_2 is evolved and $2,1,6\text{-}(\eta\text{-}C_5H_5)CoC_2B_7H_9$ is generated, very likely via initial formation of the 2,6,9 isomer (formerly[113] numbered 2,6,7) as indicated in Figure 25. [An apparent inconsistency arises, however, from an earlier report[63] which indicated that the $2\text{-}(1,6\text{-}C_2B_7H_9)_2Co^-$ and $2\text{-}(6,9\text{-}C_2B_7H_9)_2Co^-$ ions are *not* thermally interconvertible. This does not preclude rearrangement of $2,6,9\text{-}(\eta\text{-}C_5H_5)CoC_2B_7H_9$ to the 2,1,6 isomer, but casts some doubt on the identity of the proposed 2,6,9 species]. Further rearrangement of the 2,1,6 isomer

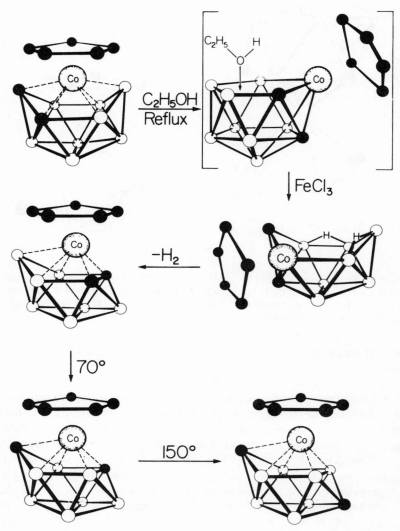

Figure 25. Polyhedral contraction of 1,2,4-$(\eta$-C$_5$H$_5)$CoC$_2$B$_8$H$_{10}$ showing proposed structures of *nido* and *closo* species formed.[113] A recent X-ray study[14] has shown that the *nido*-$(\eta$-C$_5$H$_5)$CoC$_2$B$_7$H$_{11}$ species has a different structure from that depicted, and contains a Co–H–B bridge. ●, CH; ○, BH. (From *J. Amer. Chem. Soc.*, **94**, 8391 (1972). Copyright by the American Chemical Society.)

occurs at 150°, yielding 2,1,10-$(\eta$-$C_5H_5)CoC_2B_7H_9$ as shown.[62,63,113] The sequence illustrates the general observation in polyhedral carborane chemistry that the framework carbon atoms tend to separate on thermal rearrangement, the driving force presumably being electrostatic repulsion of the relatively positive carbon nuclei.[38,75] It also supports another empirical rule, that the most stable polyhedra tend to place the carbons in low-coordinate vertices.[38] The biscarboranyl anions 2-$(1,6$-$C_2B_7H_9)_2Co^-$ and 2-$(6,9$-$C_2B_7H_9)_2Co^-$ undergo similar cage rearrangements at 250–315°, in both cases forming the 2-$(1,10$-$C_2B_7H_9)_2Co^-$ ion.[62,63]

A fourth isomer of $(\eta$-$C_5H_5)CoC_2B_7H_9$ is formed in ca. 2% yield in the polyhedral contraction of the icosahedral compound 3,1,2-$(\eta$-$C_5H_5)CoC_2B_9$-H_{11},[113,114] a reaction which is discussed in greater detail below. Although the nmr data eliminate a number of possibilities, this isomer has not been structurally assigned.[113]

The chemistry of 2,1,6-$(\eta$-$C_5H_5)CoC_2B_7H_9$ has been examined to a limited extent.[66] Bromination in CCl_4 produces a dibromo and traces of a tribromo derivative, and acylation with acetyl chloride occurs at a boron position exclusively. Unlike ferrocene and other metallocenes, no acetyl substitution occurs on the C_5H_5 ring. The acetyl derivative decomposes at 250° and no cage rearrangement occurs, but it is soluble in aqueous HCl, in contrast to the parent species. (Significantly, the 2,1,10 isomer could not be acylated at all under the conditions used to produce the 2,1,6 acetyl derivative.) In addition, the 2,1,6 isomer reacts with concentrated nitric acid to produce a nitro derivative which exhibits extreme shock sensitivity.[66]

The reduction of 2,1,6-$(\eta$-$C_5H_5)CoC_2B_7H_9$ with sodium followed by metal ion insertion in the presence of NaC_5H_5 (polyhedral expansion) generates 11-vertex dimetallic species, such as 1,8,2,3-$(\eta$-$C_5H_5)_2Co_2C_2B_7H_9$[43] or the mixed-metal system $(\eta$-$C_5H_5)_2Co^{III}Fe^{III}C_2B_7H_9$,[37] depending on the metal reagent employed. The cobalt insertion also produces other materials,[43] notably the tricobalt cage system $(\eta$-$C_5H_5)_3Co_3C_2B_7H_9$, described in Section IV.F.5.c.

A novel nine-vertex *closo*-metallocarborane having only one framework carbon atom has been prepared from the orange 13-atom $(\eta$-$C_5H_5)CoC_2$-$B_{10}H_{12}$ isomer,[39,41] which has been tentatively assigned a 4,1,8 structure.[38] The latter compound, described more fully in Section IV.F.6, is degraded in refluxing ethanolic KOH to give the $(\eta$-$C_5H_5)CoCB_7H_8^-$ ion, whose X-ray determined structure[15] is depicted in Figure 26. This particular polyhedral contraction reaction is unique in the loss of four skeletal atoms from the parent polyhedron, and is the first report of extraction of carbon from a metallocarborane. Traces of other 11- to 13-vertex metallocarboranes were also obtained. The analogous base-degradation of the *C*-monomethyl derivative of 4,1,8-$(\eta$-$C_5H_5)CoC_2B_{10}H_{12}$, present as an inseparable mixture

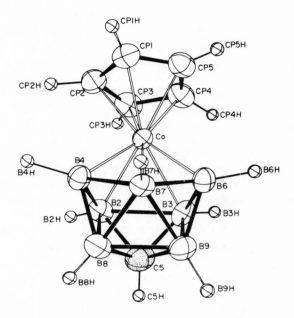

Figure 26. Structure of the $1,5\text{-}(\eta\text{-}C_5H_5)CoCB_7H_8^-$ ion.[15]

of two isomers, as expected gave two products, identified as the $(\eta\text{-}C_5H_5)\text{-}$ $CoCB_7H_8^-$ ion and its *C*-methyl derivative, respectively.[39,41] While the mechanism of this degradation is unknown, it is significant that a different 13-atom cobaltacarborane isomer, red-orange $4,1,13\text{-}(\eta\text{-}C_5H_5)CoC_2B_{10}H_{12}$, is *not* degraded by ethanolic KOH; interestingly, the same is true of the *C,C'*-dimethyl derivative of the 4,1,8 isomer. In the latter case steric factors may be responsible,[39] but the difference in the behavior of the two parent $(\eta\text{-}C_5H_5)CoC_2B_{10}H_{12}$ isomers is attributable to electronic effects within the cage structure.

The $(\eta\text{-}C_5H_5)CoCB_7H_8^-$ ion is oxidized by $FeCl_3$ to a neutral $(\eta\text{-}C_5H_5)\text{-}$ $CoCB_7H_8$ species, which is notable in that cobalt achieves the unusual $+4$ formal oxidation state. (The carborane ligand in this compound is regarded as $nido\text{-}CB_7H_8^{3-}$, although no such species is known as a free ion.) The *C*-methyl derivative of the ion undergoes an analogous reaction. Both the parent and methylated cobalt(IV) compounds are green paramagnetic materials which decompose at room temperature in the dark, even when stored *in vacuo* or under nitrogen.[39]

A different cobalt(IV) metallocarborane, the $(CB_{10}H_{11})_2Co^{2-}$ ion, has been prepared by Knoth,[119] who noted that monocarbon *nido*-carborane ligands tend to stabilize high formal metal oxidation states. This point is discussed further in Section V.B.

Reductive cage opening of the $(\eta\text{-}C_5H_5)CoCB_7H_8^-$ ion with sodium naphthalide in THF, followed by treatment with $NiBr_2 \cdot 2C_2H_4(OCH_3)_2$ and NaC_5H_5 gave several isomers of a mixed-metal ten-vertex polyhedron, $(\eta\text{-}C_5H_5)CoCB_7H_8Ni(\eta\text{-}C_5H_5)$.[178,180] One species, formed in 6% yield, has been tentatively assigned the structure of a bicapped square antiprism with the metal atoms in the 2 and 4 positions and the lone cage carbon occupying the 10-(apex) vertex. The predominant isomer (25% yield), is proposed to be $1,6,10\text{-}(\eta\text{-}C_5H_5)_2NiCoCB_7H_8$, i.e., with nickel and carbon in the apex positions and cobalt adjacent to carbon.[178,180] The 1,6,10 isomer rearranges slowly to the 2,4,10 species at 300° in a sealed tube or in refluxing hexane; if the original synthesis is conducted at reflux temperature, the 2,4,10 isomer predominates among the products.[178,180] The 2,4,10 compound in turn isomerizes at 450° in a quartz tube, forming a mixture of the 1,2,10 and 2,3,10 species; the 1,2,10 isomer converts at 450° to the 2,3,10.[180] These findings support the view that metal atoms in the thermally most stable isomer tend to occupy high-coordinate vertices although, as with most empirical observations concerning metalloboron cage systems, exceptions to this rule are to be expected. Indeed, many species containing metals in low-coordinate polyhedral positions have been isolated and some of these may well prove to be thermodynamically favored.

Experimental

Preparation of $1,2,4\text{-}(\eta\text{-}C_5H_5)CoC_2B_4H_6$, $3,1,7\text{-}(\eta\text{-}C_5H_5)CoC_2B_5H_7$, $(\eta\text{-}C_5H_5)_2Co_2C_2B_5H_7$ *Isomers, and* $(\eta\text{-}C_5H_5)_3Co_3C_2B_5H_7$ *Isomers by Direct Metal Insertion into* $2,4\text{-}C_2B_5H_7$.[151] A mixture of 2.90 mmol of $C_2B_5H_7$ and 2.70 mmol of $(\eta\text{-}C_5H_5)Co(CO)_2$ was allowed to react in the hot–cold apparatus (See Section IV.D.1) at 260/70° for 50 hr. The more volatile materials were distilled onto a vacuum line while cooling the reaction vessel at 0° and were separated using trap-to-trap fractionation to give 2.1 mmol of $C_2B_5H_7$ and 1.8 mmol of $(\eta\text{-}C_5H_5)Co(CO)_2$. The reaction vessel was opened to the air and the CH_2Cl_2 extract was evaporated onto a small amount of silica gel, placed on a silica gel column (100 ml of dry solid) prepared in hexanes, and eluted with a benzene–hexanes mixture. The first band (44 mg, 0.21 mmol, 26%) was crystallized from pentane at −78° and identified as the known[146] $3,1,7\text{-}(\eta\text{-}C_5H_5)CoC_2B_5H_7$ (previously[146] numbered 4,1,8-). The second band (5.8 mg, 0.029 mmol, 3.6%) was identified as $1,2,4\text{-}(\eta\text{-}C_5H_5)CoC_2B_4H_6$ from its ir and mass spectra.[146] The third band (69 mg, 0.20 mmol, 25%) was crystallized from warm hexanes and identified as the known $1,8,5,6\text{-}(\eta\text{-}C_5H_5)_2 Co_2C_2B_5H_7$ (formerly[146] numbered 3,8,1,9-). The fourth band (7.1 mg, 0.020

mmol, 2.5%) was characterized as green 1,7,5,6-(η-C_5H_5)$_2$$Co_2$$C_2$$B_5$$H_7$ and was purified by crystallization from pentane at $-78°$. The fifth band (8.7 mg, 0.019 mmol, 2.4%) was characterized as brown 2,3,8,1,6-(η-C_5H_5)$_3$$Co_3$$C_2$$B_5$$H_7$, and was crystallized by slowly evaporating a hexanes solution. The sixth band (1.3 mg, 0.0028 mmol, 0.3%) was characterized as brown 2,3,4,1,10-(η-C_5H_5)$_3$$Co_3$$C_2$$B_5$$H_7$, and was purified by preparative tlc on silica gel using 50% benzene in hexanes. All yields given are based on $C_2B_5H_7$ consumed.

Preparation of Cobalt Metallocarboranes from $1,7$-$C_2B_6H_8$ *via Sodium Naphthalide Reduction.*[42] The carborane (20.1 mmol) was reduced with 44 mmol sodium and 3.2 mmol naphthalene in 20 ml THF under a nitrogen atmosphere at $-78°$. The reaction mixture was warmed to $-20°$ and stirred three days. The amber solution which formed was syringed into a 500-ml, three-neck flask containing 20 mmol of previously prepared NaC_5H_5 cooled to $-50°$. A slurry of finely ground anhydrous $CoCl_2$ (6.0 g, 46 mmol) in 50 ml of THF was added dropwise to the stirred solution. After addition was complete the flask was allowed to warm to room temperature slowly and stirred for 15 hr. After a stream of oxygen was passed through the reaction mixture for 30 min, the brown solution was filtered into a 1-liter flask, 150 ml (dry volume) of silica gel was added, and the solvent was stripped using a rotary evaporator and a water aspirator. The last traces of solvent were removed under high vacuum. The solids were placed in a large Soxhlet thimble and extracted with 1 liter of hexane. The solvent was stripped to approximately 30 ml and the solution was passed through a column of silica gel (4 × 20 cm) in hexane. Two bands were separated. The deep-red solution comprising the first fraction was stripped to approximately 50 ml and placed in a flask attached to a sublimer. The solvent was evaporated, and dark red crystals were collected on the $0°$ cold finger, affording 0.1811 g of 1,4,5-(η-C_5H_5)$CoC_2B_6H_8$ (0.82 mmol, 4.1%), mp 61–64°. Continued elution with hexane produced a second fraction, which was recrystallized from hexane giving 0.508 g (1.47 mmol, 7.3%) of 2,6,1,10-(η-C_5H_5)$_2$$Co_2$$C_2$$B_6$$H_8$ as green crystals, mp 236–238°.

Preparation of nido-(η-C_5H_5)$CoC_2B_7H_{11}$ *by Polyhedral Contraction of* 1,2,4-(η-C_5H_5)$CoC_2B_8H_{10}$.[113] A sample of (η-C_5H_5)$CoC_2B_8H_{10}$ (Section IV.F.4) (1.0 g, 4.0 mmol) was dissolved in absolute ethanol (200 ml), ferric chloride (5.0 g, 31 mmol) was then added, and the mixture was heated to reflux for 45 min with stirring. After this time the reaction mixture was quenched with water (1000 ml), and the products were extracted into dichloromethane (3 × 300 ml). The combined extracts were dried over $MgSO_4$, filtered, and stripped* onto silica gel (ca. 40 ml). This material was mounted on a silica-gel column (400 ml), and the products were eluted with hexane–dichloromethane. The first band to elute was yellow and was identified as $2,6,9$-$(\eta$-$C_5H_5)CoC_2B_7H_9$ from ^1H nmr, ^{11}B nmr, and mass spectral measurements (20 mg, 2%). The second band to elute was red and, on evaporation of the solvent, afforded the red crystalline $(\eta$-$C_5H_5)CoC_2B_7H_{11}$ (380 mg, 40%) which was recrystallized from dichloromethane–hexane, mp 123°.

Pyrolysis of $(\eta$-$C_5H_5)CoC_2B_7H_{11}$. *Formation of* 2,1,6- *and* 2,1,10-$(\eta$-$C_5H_5)CoC_2B_7H_9$.[113] A sample of *nido*-$(\eta$-$C_5H_5)CoC_2B_7H_{11}$ (100 mg, 0.43 mmol) was added to cyclooctane (20 ml) and the mixture heated to reflux for 24 hr. After this time the bulk of the solvent was evaporated at reduced pressure, and the residual reaction products were separated using preparative thick-layer chromatography with hexane–dichloromethane eluent. The first band to elute was yellow and contained 2,1,10-$(\eta$-$C_5H_5)$-$CoC_2B_7H_9$ (84 mg, 85%) which was identified by ^1H nmr, ^{11}B nmr, and mass-spectral measurements. Traces of a red material with an R_f value equal to that of 2,1,6-$(C_5H_5)CoC_2B_7H_9$ were also observed.

The above reaction was repeated using *nido*-$(\eta$-$C_5H_5)CoC_2B_7H_{11}$ (101 mg, 0.43 mmol) and cyclooctane (5 ml) which were placed in a tube with a break-seal. The tube was evacuated and sealed off on the vacuum line, then heated to 150° for 24 hr. After this time the tube was vented into the vacuum line, and the volume of noncondensable ($-190°$) gas formed was measured using a Sprengel pump (0.42 mmol, 98%). The residual material in the tube was treated in the manner described in the previous paragraph and afforded 2,1,10-$(\eta$-$C_5H_5)CoC_2B_7H_9$ (80 mg, 80%) and 2,1,6-$(\eta$-$C_5H_5)CoC_2B_7H_9$ (6%).

Preparation of $(\eta$-$C_5H_5)Co^{III}CB_7H_8^-$ *and* $(\eta$-$C_5H_5)Co^{IV}CB_7H_8$.[39] To a 100-ml, three-necked, round-bottom flask fitted with a mechanical stirrer, reflux condenser topped with a nitrogen inlet and a thermometer, were added 0.30 g (1.1 mmol) of orange 1,6,7-$(\eta$-$C_5H_5)CoC_2B_{10}H_{12}$ and 75 ml of absolute ethanol. The solution was heated to the reflux temperature and 5.0 g of KOH was added directly as a solid. The mixture was allowed to reflux for 48 hr during which time the color changed from orange to green-brown. The solvent was removed by rotary evaporation and the residual oil was re-dissolved in 250 ml of distilled water. The aqueous solution was extracted with four 125-ml portions of benzene and then added to a solution of 5 g of tetramethylammonium chloride in 25 ml of water. The cloudy solution was

---

* Silica gel was added to the solution, which was then evaporated to dryness. The solid gel with adsorbed products was placed on the top of a silica gel column and then eluted as indicated.

clarified by the addition of 50 ml of acetone and then rotary-evaporated to a volume of about 25 ml by which time black crystals had deposited. The crystals were filtered and washed with ethanol to afford 0.21 g (0.71 mmol, 65%) of product.

A 50-mg (0.17 mmol) sample of $[(CH_3)_4N][(\eta-C_5H_5)CoCB_7H_8]$ was dissolved in 50 ml of dichloromethane. Anhydrous, sublimed ferric chloride (28 mg, 0.17 mmol) was added as a slurry in 25 ml of dichloromethane via an addition funnel. The solution was stirred under nitrogen for 0.5 hr during which time the color became intensely green and a white precipitate formed. The reaction mixture was transferred to a vacuum line where the solvent was removed. A dark-green, nearly black, solid was sublimed at room temperature to a $-80°$ cold finger. The sublimer was opened in a nitrogen-filled drybox yielding approximately 25 mg (0.11 mmol, 65%) of $(\eta-C_5H_5)CoCB_7H_8$.

4. Eleven-Atom Cages

As with other cobaltacarboranes of intermediate size, 11-vertex species have been prepared both by the polyhedral expansion of smaller systems and via the contraction of larger (usually icosahedral) frameworks. The expansion of the ten-vertex cages 1,6- and $1,10-C_2B_8H_{10}$ by reductive cage opening and subsequent reaction with $CoCl_2$ and NaC_5H_5 generates primarily $(\eta-C_5H_5)$-$CoC_2B_8H_{10}$ compounds,[42] whose assumed geometry is that shown in Figure 17. The main product obtained from $1,6-C_2B_8H_{10}$ at room temperature is the purple $1,2,3-(\eta-C_5H_5)CoC_2B_8H_{10}$ (44% yield), with much smaller amounts of $2,1,6-(\eta-C_5H_5)CoC_2B_7H_9$, $2,3,1,7-(\eta-C_5H_5)Co_2C_2B_8H_{10}$, a linked-cage species $[(\eta-C_5H_5)CoC_2B_8H_9(\sigma-C_2B_8H_9)]$, and the biscarboranyl ion $[1-(2,3-C_2B_8H_{10})_2Co^-]$. The latter species has also been prepared as the cesium salt, isolated in 48% yield from a reaction conducted under similar conditions but excluding NaC_5H_5.[42,44]

When the expansion of $1,6-C_2B_8H_{10}$ is conducted at $-80°$ the yield of $1,2,3-(\eta-C_5H_5)CoC_2B_8H_{10}$ is much lower than that conducted in the room-temperature reaction, the main products being two isomers of $(\eta-C_5H_5)_2$-$Co_2C_2B_8H_{10}$ and a $(\eta-C_5H_5)Co_2(C_2B_8H_{10})_2$ anion. An X-ray investigation of the latter species established the presence of a central 12-vertex $Co_2C_2B_8$ polyhedron, one of the cobalts also being part of an 11-vertex CoC_2B_8 cage.[51]

The corresponding reactions of $1,10-C_2B_8H_{10}$ in the presence of NaC_5H_5 yield a similar set of compounds.[42] The main product, interestingly, is $1,2,3-$ $(\eta-C_5H_5)CoC_2B_8H_{10}$ (44%), identical to the principal species derived from $1,6-C_2B_8H_{10}$. The minor products include $3,1,10-(\eta-C_5H_5)CoC_2B_7H_9$, $2,7,1,12-$ $(\eta-C_5H_5)_2Co_2C_2B_8H_{10}$ (an icosahedral species proposed to contain a Co–Co bond), $2,1,12-(\eta-C_5H_5)CoC_2B_9H_{11}$, $(\eta-C_5H_5)CoC_2B_8H_9(\sigma-C_2B_8H_9)$, and

2,3,1,7-$(\eta$-C$_5$H$_5)_2$Co$_2$C$_2$B$_8$H$_{10}$, the last two being identical to the products obtained in the 1,6-C$_2$B$_8$H$_{10}$ reaction.

Since the product distribution in both reactions is dominated by species having more framework atoms than the original C$_2$B$_8$H$_{10}$ isomers, the description of these as polyhedral expansion reactions[42] is reasonable. Indeed, in terms of yields of "expanded" metallocarborane products, these C$_2$B$_8$H$_{10}$ syntheses are more successful than any of the corresponding reactions of their lower homologs with sodium naphthalide, CoCl$_2$, and NaC$_5$H$_5$, described above. (Curiously, the attempted polyhedral expansion of the 11-vertex C$_2$B$_9$H$_{11}$ gave only a 2.5% yield of the well-known $(\eta$-C$_5$H$_5)$CoC$_2$B$_9$H$_{11}$ species, with minute amounts of other products.[42]) Thus, for reasons not yet clear, the two C$_2$B$_8$H$_{10}$ isomers are particularly amenable to expansion via sodium naphthalide reduction and subsequent metal insertion.

Plesek, Štíbr, and Hermanek[164] have prepared 1,2,3-$(\eta$-C$_5$H$_5)$CoC$_2$B$_8$H$_{10}$ in 63% yield by the reaction of the disodium salt of C$_2$B$_8$H$_{10}$$^{2-}$ with methanolic KOH followed by CoCl$_2$ and cyclopentadiene (not NaC$_5$H$_5$). Presumably, the intermediate is an *arachno*-type anion with two open faces, although this is not stated.

Another excellent preparative route to 11-atom cobaltacarboranes is the polyhedral contraction of 12-atom cage systems, which are readily derived from 1,2-C$_2$B$_{10}$H$_{12}$ (o-carborane). The contraction (Section III) of 3,1,2-$(\eta$-C$_5$H$_5)$CoC$_2$B$_9$H$_{11}$ via treatment with strong base and oxidation of the resulting *nido* anion produced 1,2,4-$(\eta$-C$_5$H$_5)$CoC$_2$B$_8$H$_{10}$ in 63% yield,[113,115] plus two isomers of $(\eta$-C$_5$H$_5)$CoC$_2$B$_7$H$_9$, described in Section IV.F.3. A thermal process with similar results is the formation of 1,2,3-$(\eta$-C$_5$H$_5)$-CoC$_2$B$_8$H$_{10}$ in the pyrolysis of 2,3,1,7-$(\eta$-C$_5$H$_5)_2$Co$_2$C$_2$B$_8$H$_{10}$,[43] a 12-vertex dimetallic system.

Two isomeric four-carbon 11-vertex cobaltacarboranes, $(\eta$-C$_5$H$_5)$Co-(CH$_3)$C$_4$B$_6$H$_6$, have been prepared from the tetracarbon carborane (CH$_3)_4$-C$_4$B$_8$H$_8$.[136] The structures have not been determined, but electron-counting considerations (Section II.B) suggest open-cage geometries.

Experimental

Preparation of 1,2,3-$(\eta$-C$_5$H$_5)$CoC$_2$B$_8$H$_{10}$.[164] A solution of 10 g KOH in 20 ml methanol was added to 5 mmol of crude Na$_2$[C$_2$B$_8$H$_{10}$]. To this mixture were added, at $-10°$, a solution of 5.0 g CoCl$_2 \cdot$6H$_2$O in 20 ml methanol and 5.0 g cyclopentadiene previously mixed at $-50°$. The reaction mixture was then refluxed under nitrogen for 2 hr, the cyclopentadiene and a part of the methanol were distilled off *in vacuo*, and the residue was isolated as in the case of 3,1,2-(C$_5$H$_5$)CoC$_2$B$_9$H$_{11}$ (Section IV.F.5.a). The final extraction from metallic cobalt was performed by three 50-ml portions of benzene. The

solution was filtered, evaporated to a volume of 50 ml *in vacuo*, covered carefully with a layer of 100 ml of hexane, and allowed to stand overnight. Dark-
blue prisms of $(\eta\text{-}C_5H_5)CoC_2B_8H_{10}$ (0.77 g, 63%) separated and were filtered
off and dried in air.

Preparation of $1,2,4\text{-}(\eta\text{-}C_5H_5)CoC_2B_8H_{10}$ *by Polyhedral Contraction of*
$3,1,2\text{-}(\eta\text{-}C_5H_5)CoC_2B_9H_{11}$.[113] A sample of $(\eta\text{-}C_5H_5)CoC_2B_9H_{11}$ (Section
IV.F.5) (2.0 g, 7.8 mmol) was added to a mixture of ethylene glycol–ethanol–
water (150 ml of a mixture in the proportions 100:40:10) containing potassium
hydroxide (12.0 g, 210 mmol). The mixture was then heated to reflux for 25 hr,
after which time the reaction mixture was quenched with water (500 ml) to
give a red solution. Careful addition of hydrogen peroxide (10 ml of a 30%
aqueous solution) resulted in the formation of a blue precipitate; this was
extracted into dichloromethane (4 × 200 ml), and the combined extracts were
dried over $MgSO_4$. After filtration the purple extract was stripped onto silica
gel (ca. 60 ml) and mounted on a silica gel column (400 ml).

Elution with hexane–dichloromethane developed a yellow band con-
taining a mixture of $(\eta\text{-}C_5H_5)CoC_2B_7H_9$ isomers, followed by a purple band
containing $1,2,4\text{-}(\eta\text{-}C_5H_5)CoC_2B_8H_{10}$. This band was collected; evaporation
of most of the solvent resulted in the precipitation of the product (1.2 g,
63%) which was recrystallized from dichloromethane–hexane to give deep-
purple needles, mp 168°.

The polyhedral contraction of the biscarboranyl anion $(1,2\text{-}C_2B_9H_{11})_2Co^-$
with OH^- and H_2O_2 produces a $(1,2\text{-}C_2B_9H_{11})Co(2,4\text{-}C_2B_8H_{10})^-$ anion in
which the cobalt atom occupies a vertex common to both a 12- and an
11-atom closed polyhedron.[113] Treatment of this species with pyridine generates a red adduct $(C_2B_9H_{11})Co(C_5H_5N\cdot C_2B_8H_{10})^-$, whose C_2B_8 cage has
been shown in an X-ray study[22,24] to have an open face with the pyridine
appended to one edge (Figure 27).[113,115] The formation of this *nido* species
can be viewed as a reductive cage-opening process in which the two electrons
contributed by pyridine to the polyhedral framework force the latter to
open, as predicted by the $(2n + 2)$ rule (Section II). The terminal hydrogen
originally bonded to B(9′), and displaced by pyridine, was not located in the
X-ray study but is assumed to move to a B–H–B bridging position on the
open face.[22,24,113,115]

The neutral cyclopentadienyl species $1,2,4\text{-}(\eta\text{-}C_5H_5)CoC_2B_8H_{10}$ also
forms a pyridine adduct, in this case electrically neutral, to which has been
assigned a *nido* structure analogous to that in Figure 27, with $C_5H_5^-$ replacing
the $C_2B_9H_{11}^{2-}$ ligand.[114,115] A piperidine adduct of presumably analogous
structure has also been prepared.[114] The pyridine adducts are oxidized by
$FeCl_3$, forming products which are postulated to have closed polyhedral
structures with the base ligand attached to B(7) on the 11-vertex cage[114,115]:

$$(\eta\text{-}C_5H_5)CoC_2B_8H_{10}\cdot C_5H_5N \xrightarrow{\ FeCl_3\ } [(\eta\text{-}C_5H_5)CoC_2B_8H_9\cdot C_5H_5N]^+$$

$$nido \qquad\qquad\qquad\qquad\qquad\qquad closo$$

$$[(C_2B_9H_{11})CoC_2B_8H_{10}\cdot C_5H_5N]^- \xrightarrow{\ FeCl_3\ } (C_2B_9H_{11})CoC_2B_8H_9\cdot C_5H_5N$$

$$nido \qquad\qquad\qquad\qquad\qquad\qquad closo$$

The cyclopentadienyl cation is unstable in polar solvents or moist air, probably as a consequence of its net positive charge.[114] A neutral and stable piperidine-substituted analog has therefore been prepared by treatment of the piperidine adduct of $1,2,4\text{-}(\eta\text{-}C_5H_5)CoC_2B_8H_{10}$ with $FeCl_3$ followed by deprotonation with K_2CO_3:

$$(\eta\text{-}C_5H_5)CoC_2B_8H_{10}\cdot C_5H_{10}NH \xrightarrow{\ FeCl_3\ } [(\eta\text{-}C_5H_5)Co(C_2B_8H_9\cdot C_5H_{10}NH)]^+$$

$$nido \qquad\qquad\qquad\qquad\qquad\qquad closo$$

$$(\eta\text{-}C_5H_5)Co(C_2B_8H_9\cdot C_5H_{10}N) \xleftarrow{\ K_2CO_3\ }$$

$$closo$$

The neutral product has been characterized[114] as structurally analogous to the pyridine-substituted cation above; both species are formally derived from $closo\text{-}(\eta\text{-}C_5H_5)CoC_2B_8H_{10}$ by replacement of H^- with the respective bases C_5H_5N and $C_5H_{10}N^-$.

The fact that the oxidation of the base adducts of $(\eta\text{-}C_5H_5)CoC_2B_8H_{10}$ and $(C_2B_9H_{11})Co(C_2B_8H_{10})^-$ proceeds via *nido* intermediates as described above, suggested to Hawthorne and co-workers that the polyhedral contraction of icosahedral $(C_2B_9H_{11})_2Co^-$ and $(C_5H_5)CoC_2B_9H_{11}$ species by base degradation and subsequent oxidation with H_2O_2 involves intermediate species containing the $nido\text{-}7,8\text{-}C_2B_8H_{11}^-$ moiety. In support of this view, the product of the reaction of $(1,2\text{-}C_2B_9H_{11})_2Co^-$ with aqueous hydroxide ion was isolated as the brown tetramethylammonium salt of the dianion$[(1,2\text{-}C_2B_9H_{11})\text{-}3,9'\text{-}Co(7',8'\text{-}C_2B_8H_{11})]^{2-}$. The similarity of the [11]B nmr spectrum of this species to that of the *nido* adduct $[(C_2B_9H_{11})Co(C_2B_8H_{10}\cdot C_5H_5N)]^-$, described above, is taken as evidence that the two structures are analogous except for replacement of H^- by C_5H_5N.[114] One interesting distinction between the two species is that the B–H–B bridging proton in the unsubstituted dianion evidently moves to a B–H–Co bridging position in the pyridine adduct, as deduced from the presence of a bridge-proton peak at high field in the ^{1}H nmr spectrum of the latter compound.[114] The appearance of such resonances in the region above $+5$ ppm relative to tetramethylsilane has been taken by Grimes and co-workers as evidence for B–H–M bonding in metalloboranes[147,149] and metallocarboranes,[189,190] and is consistent with the observed high-field ^{1}H nmr peaks associated with transition metal hydrides.

Both *nido* compounds are apparent analogs of the $nido\text{-}C_2B_9H_{12}^-$

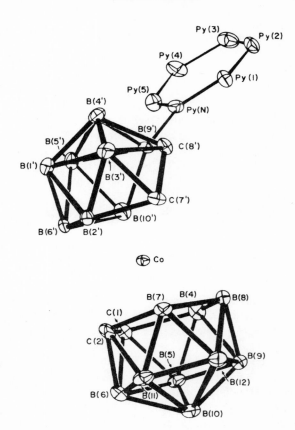

Figure 27. Structure of the $(1,2\text{-}C_2B_9H_{11})Co(9'\text{-}C_5H_5N\text{-}7',8'\text{-}C_2B_8H_{10})^-$ ion.[22] (From *Inorg. Chem.*, **12**, 1157 (1973). Copyright by the American Chemical Society.)

anion, the species obtained from base degradation of icosahedral $C_2B_{10}H_{12}$. Accordingly, since protonation of $C_2B_9H_{12}^-$ yields the neutral *nido*-$C_2B_9H_{13}$, treatment of $[(1,2\text{-}C_2B_9H_{11})\text{-}3,9'\text{-}Co(7',8'\text{-}C_2B_8H_{11})]^{2-}$ with HCl reversibly forms the $[(1,2\text{-}C_2B_9H_{11})\text{-}3,9'\text{-}Co(7',8'\text{-}C_2B_8H_{12})]^-$ monoanion, an analog of $C_2B_9H_{13}$ in which a $(C_2B_9H_{11})Co^-$ moiety replaces a BH unit. The monoanion is believed to be structurally analogous to the dianion, but with a B–H–Co bridging proton (Figure 28).[114]

Oxidation of the dianion with H_2O_2 produces the *closo* ion $[(1,2\text{-}C_2B_9H_{11})\text{-}3,1'\text{-}Co(2',4'\text{-}C_2B_8H_{10})]^-$, which is identical to the product of the polyhedral contraction of $3\text{-}(1,2\text{-}C_2B_9H_{11})_2Co^-$, thereby demonstrating that the dianion is indeed an intermediate in the contraction process.[113,114] The protonation of the dianion is reversible and does not result in separation

$$(1,2\text{-}C_2B_9H_{11})_2Co^- \xrightarrow{\ OH^-\ } [(1,2\text{-}C_2B_9H_{11})\text{-}3,9'\text{-}Co(7',8'\text{-}C_2B_8H_{11})]^{2-}$$

$$[(1,2\text{-}C_2B_9H_{11})\text{-}3,9'\text{-}Co(7',8'\text{-}C_2B_8H_{12})]^-$$

$$\xrightarrow[\ H_2O_2\ ]{\ OH^-\ } [(1,2\text{-}C_2B_9H_{11})\text{-}3,1'\text{-}Co(2',4'\text{-}C_2B_8H_{10})]^-$$

of the framework carbon atoms which are proposed to occupy adjacent positions in the open face (Figure 28), as shown by the fact that the protonated species can be oxidized by alkaline H_2O_2 to give the same *closo* species as was produced by oxidation of the dianion directly.[114]

Pyrolysis of the *nido* monoanion $[(1,2\text{-}C_2B_9H_{11})\text{-}3,9'\text{-}Co(7',8'\text{-}C_2B_8H_{12})]^-$ at 150° in cyclooctane results in loss of H_2, generating in 80% yield the green *closo* species $[(1,2\text{-}C_2B_9H_{11})\text{-}3,1'\text{-}Co(2',3'\text{-}C_2B_8H_{10})]^-$ which is an isomer of the monoanion discussed above; in the 2',3'-isomer, however, the carbon atoms in the 11-vertex polyhedron are proposed to occupy nonadjacent (2' and 3') vertices while in the 2',4' isomer they are in adjacent locations.[114] The process is analogous to the thermal conversion of *nido*-$C_2B_9H_{13}$ to *closo*-$C_2B_9H_{11}$ and H_2. The direct thermal conversion of the 2',4' to the 2',3' isomer occurs at 150° (for numbering of the 11-vertex polyhedron, see Figure 17).

$$[(1,2\text{-}C_2B_9H_{11})\text{-}3,1'\text{-}Co(2',4'\text{-}C_2B_8H_{10})]^- \xrightarrow{\ \Delta\ } [(1,2\text{-}C_2B_9H_{11})\text{-}3,1'\text{-}Co(2',3'\text{-}C_2B_8H_{10})]^-$$

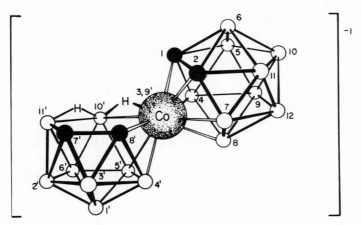

Figure 28. Proposed structure of the $(1,2\text{-}C_2B_9H_{11})\text{-}3,9'\text{-}Co(7',8'\text{-}C_2B_8H_{12})^-$ ion.[114] ●, CH; ○, BH. (From *J. Amer. Chem. Soc.*, **95**, 7633 (1973). Copyright by the American Chemical Society.)

Scholer and co-workers have studied the rearrangement of 1,2,4-(η-C$_5$H$_5$)Co(CH$_3$)$_2$C$_2$B$_8$H$_8$ to the 1,2,3 isomer, and isolated an intermediate species formulated as 10,2,3-(η-C$_5$H$_5$)Co(CH$_3$)$_2$C$_2$B$_8$H$_8$ with the cobalt atom in a five-coordinate vertex bound exclusively (in the cage) to boron atoms.[143]

The polyhedral contraction of the icosahedral 3,1,2-(η-C$_5$H$_5$)CoC$_2$B$_9$H$_{11}$ to 1,2,4-(η-C$_5$H$_5$)CoC$_2$B$_8$H$_{10}$ has been examined in detail,[114] and it appears to proceed via intermediates which are structurally analogous to those found in the contraction of 3-(1,2-C$_2$B$_9$H$_{11}$)$_2$Co$^-$. Thus, the reaction of 3,1,2-(η-C$_5$H$_5$)CoC$_2$B$_9$H$_{11}$ with hydroxide ion produces a red solution containing a mixture of anionic species. The addition of (CH$_3$)$_4$NCl to this solution precipitates a red solid which could not be purified, but whose major component has been shown to be the *nido*-[9,7,8-(η-C$_5$H$_5$)CoC$_2$B$_8$H$_{11}$]$^-$ ion, analogous to the previously mentioned *nido*-9,7,8-(η-C$_5$H$_5$)CoC$_2$B$_8$H$_{10}$·NC$_5$H$_5$ species with H$^-$ replacing NC$_5$H$_5$. As in the case of the corresponding [(C$_2$B$_9$H$_{11}$)Co(C$_2$B$_8$H$_{11}$)]$^{2-}$ ion and its pyridyl derivative (Figure 27), it appears that the parent compound has a B–H–B bridging proton which on pyridyl substitution moves to a B–H–Co location.

If HCl instead of (CH$_3$)$_4$NCl is added to the red solution, a yellow solid containing two components is obtained, but these have not been separated. However, by indirect means the major product has been found to be the neutral species 9,7,8-(η-C$_5$H$_5$)CoC$_2$B$_8$H$_{12}$, analogous to the [(1,2-C$_2$B$_9$H$_{11}$)-3,9'-Co(7',8'-C$_2$B$_8$H$_{12}$)]$^-$ ion mentioned above but with C$_5$H$_5$$^-$ replacing the C$_2$B$_9$H$_{11}^{2-}$ ligand. The oxidation of 9,7,8-(η-C$_5$H$_5$)CoC$_2$B$_8$H$_{12}$ by H$_2$O$_2$ in basic media generates 1,2,4-(η-C$_5$H$_5$)CoC$_2$B$_8$H$_{10}$ in 81% yield, and the pyrolysis of the same *nido* species at 120° in cyclooctane produces a mixture of 1,2,3- and 1,2,4-(η-C$_5$H$_5$)CoC$_2$B$_8$H$_{10}$ together with decomposition products. (At 150°, not surprisingly, the more stable 1,2,3 isomer with nonvicinal cage carbon atoms is formed exclusively.) From these and other data summarized in the following scheme, the intermediate in the polyhedral contraction of 3,1,2-(η-C$_5$H$_5$)CoC$_2$B$_9$H$_{11}$ is evidently the *nido*-[9,7,8-(η-C$_5$H$_5$)-CoC$_2$B$_8$H$_{11}$]$^-$ ion[114]:

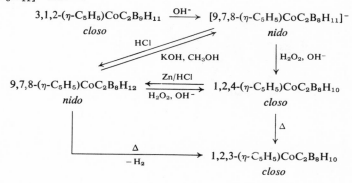

When the base degradation of $3,1,2-(\eta-C_5H_5)CoC_2B_9H_{11}$ or $[3-(1,2-C_2B_9H_{11})_2Co]^-$ is conducted in the presence of cobalt, icosahedral dicobalt species containing the $Co_2C_2B_8H_{10}$ polyhedron are obtained, presumably as a consequence of removal of B(6)–H from the original metallocarborane and its subsequent replacement by cobalt. This pathway, however, requires a rearrangement of the intermediate *nido* species $[(\eta-C_5H_5)CoC_2B_8H_{11}]^-$ or $[(C_2B_9H_{11})Co(C_2B_8H_{11})]^{2-}$ *prior* to the incorporation of the second cobalt, since the intermediates are believed to have cobalt in the open face (see above). Experiments involving delay time prior to the addition of Co^{2+} ions indicate that unless the initial degradation product is quickly complexed with cobalt, it converts to a species which does not react with cobalt. Hawthorne *et al.*[114] suggest that this inertness is caused by protonation of the open face to create a hydrogen bridge which prevents reaction with cobalt. That such protonation can in fact occur even under strongly basic conditions was demonstrated by allowing the above intermediate species to stand in THF solution containing excess sodium hydride; the hydrogen evolved was less than 10% of theory even after a five-day period at room temperature, indicating the relative inertness of the hydrogen bridges in basic media.

One general observation to be drawn from these extensive studies of 11-atom cobalt metallocarboranes is the striking parallel with carborane chemistry, e.g., the degradation of *closo*-$1,2-C_2B_{10}H_{12}$ to *nido*-$C_2B_9H_{12}^-$ and the subsequent conversion of this intermediate to *closo*-$C_2B_9H_{11}$ or the insertion of metal atoms to generate MC_2B_9-type species. This work thus further supports the observation made above that the $(\eta-C_5H_5)Co$ or $[(\eta-C_5H_5)Co^{2+}]$ group possesses a remarkably versatile ability to replace the BH (or BH^{2+}) unit in borane cage structures. Despite the extensive structural analogies between the carboranes and cobaltacarboranes, the similarity in chemical behavior is more limited; thus, Hawthorne[114] has noted that while the bridge hydrogen in $C_2B_9H_{12}^-$ is protic in character, in the analogous $(\eta-C_5H_5)CoC_2B_8H_{11}^-$ and $[(C_2B_9H_{11})Co(C_2B_8H_{11})]^{2-}$ species it is essentially hydridic (it might be noted, however, that the bridge hydrogen in $C_2B_4H_7^-$ is also hydridic).

The reactions of $1,2,4-(\eta-C_5H_5)CoC_2B_8H_{10}$ with several carbanions have been studied. Treatment of this material with η-butyllithium resulted in butyl substitution on the C_5H_5 ligand, and no reaction occurred with the carborane anions $C_2B_{10}H_{11}^-$ and $CH_3C_2B_{10}H_{10}^-$. The same metallocarborane reacts rapidly with piperidine and diethylamine to form red adducts in benzene solution; pyridine and 2-methylpyridine react more slowly, and triethylamine fails to react at all under the same conditions. In view of the Lewis base strength of the latter reagent, the authors postulate that the C_5H_5 ligand sterically prevents attack on the reactive boron by bulky amines or carbanions.[114]

The 11-vertex cobaltacarboranes are useful precursors to icosahedral bimetallic systems, by at least three general routes. Thermolysis of 1,2,4-$(\eta$-$C_5H_5)CoC_2B_8H_{10}$ generates six isomers of $(\eta$-$C_5H_5)_2Co_2C_2B_8H_{10}$ (see Table I). Five of these are also obtained on heating the cobalticinium salt $[(\eta$-$C_5H_5)_2Co][1$-$Co(2,3$-$C_2B_{10}H_{12})_2].^{45}$

The polyhedral expansion of the 11-vertex species 1,2,3-$(\eta$-$C_5H_5)$-$CoC_2B_8H_{10}$ by successive treatments with sodium naphthalide and a NaC_5H_5–$CoCl_2$ mixture gives an icosahedral $(\eta$-$C_5H_5)_2Co_2C_2B_8H_{10}$ metallocarborane in 26% yield.[43,46] Originally a 5,12,1,7 icosahedral structure was postulated,[43,46] but later the 2,3,1,7-$(\eta$-$C_5H_5)_2Co_2C_2B_8H_{10}$ geometry containing a Co–Co bond was favored [42] because the presence of cobalt atoms in neighboring positions had been established in an X-ray study of the ten-vertex 2,6,1,10-$(\eta$-$C_5H_5)_2$-$Co_2C_2B_6H_8$ cage,[42] described in Section IV.F.3 (the 2,3,1,7 geometry has recently been confirmed crystallographically[16]). The 2,3,1,7-$(\eta$-$C_5H_5)_2Co_2$-$C_2B_8H_{10}$ species has been obtained in other reactions (for example, in the polyhedral expansion of 1,6- and 1,10-$C_2B_8H_{10}$ described earlier), and at least two other isomers of this icosahedral system, characterized as the 3,6,1,2[116,164] and 2,7,1,12[42] cages, have been reported. The polyhedral expansion of 1,2,4-$(\eta$-$C_5H_5)CoC_2B_8H_{10}$ generates a mixture of products, including the asymmetric icosahedral isomer 3,4,1,2-$(\eta$-$C_5H_5)_2Co_2C_2B_8H_{10}.^{43}$ Further details on the icosahedral species are presented in the following section.

Stone and co-workers have prepared a mixed cobalt–platinum metallo-carborane by the direct-insertion technique. The reaction of 1,2,4-$(\eta$-$C_5H_5)$-$CoC_2B_8H_{10}$ with $[(C_2H_5)_3P]_2Pt(trans$-stilbene) produces a 12-vertex icosa-hedral species, 4,7,1,2-$(\eta$-$C_5H_5)Co[(C_2H_5)_3P]_2PtC_2B_8H_{10}$, whose gross structure has been confirmed in an X-ray investigation.[71] A corresponding unstable nickel–platinum complex has also been reported without definitive structural characterization.[71]

An 11-vertex mixed-metal Co–Fe compound was described in Section IV.F.3.

5. Twelve-Atom Cages

a. Synthesis. In common with the analogous iron species described above, icosahedral cobalt metallocarboranes have been prepared in most cases either from the $C_2B_9H_{11}^{2-}$ (dicarbollide) ion in nonaqueous media[90,100] or from the $C_2B_9H_{12}^-$ ion in hot aqueous NaOH solution[100,210]; additional synthetic approaches have been reported recently and are described later in this section. Further paralleling the iron reactions, the main products are biscarboranyl cobalt species unless cyclopentadienide ion is present; in the latter case, cyclopentadienyl cobaltacarboranes are obtained. A detailed description of the synthesis of the $(1,2$-$C_2B_9H_{11})_2Co^-$ ion is available.[91]

$$2 \ C_2B_9H_{11}^{2-} + \tfrac{3}{2} \ CoCl_2 \longrightarrow (C_2B_9H_{11})_2Co^- + \tfrac{1}{2} \ Co^0 + 3 \ Cl^-$$

$$\xrightarrow[\text{NaC}_5\text{H}_5]{\text{CoCl}_2} (\eta\text{-}C_5H_5)CoC_2B_9H_{11}$$

$$2 \ C_2B_9H_{\overline{12}} + CoCl_2 \xrightarrow[\text{aq.}]{\text{NaOH}} (C_2B_9H_{11})_2Co^{2-}$$

$$\downarrow \text{air}$$

$$(C_2B_9H_{11})_2Co^- + Co^0$$

A number of C-substituted derivatives[134,231,240] of the $(C_2B_9H_{11})_2Co^-$ and $(\eta\text{-}C_5H_5)CoC_2B_9H_{11}$ species (Table I) have been prepared in analogous fashion, via base degradation of the corresponding derivative of 1,2- or 1,7-$C_2B_{10}H_{12}$. The only known cobaltacarborane carbonyl species is the yellow, air-sensitive $[3,1,2\text{-}(CO)_2CoC_2B_9H_{11}]^-$ ion, prepared from 1,2-$C_2B_9H_{11}^{2-}$ and $Co_2(CO)_8$ in refluxing THF.[97] The final products of these reactions are diamagnetic and contain formal cobalt(III), the oxidation from the cobalt(II) stage apparently occurring both by disproportionation to Co(III) and cobalt metal,[210] and also by air oxidation. The corresponding Co(II) species presumably form initially on insertion of Co^{2+} into the icosa-hedral framework,[210] but are extremely unstable toward oxidation. [The facile air oxidation of $(\eta\text{-}C_5H_5)_2Co$ to the cobalticinium ion, $(\eta\text{-}C_5H_5)_2Co^+$, is a related observation.]

The reduced Co(II) species, $(1,2\text{-}C_2B_9H_{11})_2Co^{2-}$, has been obtained as a purple air-sensitive ion by reduction of $(1,2\text{-}C_2B_9H_{11})_2Co^-$ with n-butyl-lithium,[90] and is also apparently formed in the electrochemical reduction of the latter metallocarborane.[100] The icosahedral geometry of the $(1,2\text{-}C_2B_9H_{11})_2Co^-$ ion[244] and several derivatives[21,25,30] has been confirmed in X-ray structural investigations.

An improved method for the preparation of icosahedral cyclopenta-dienyl cobaltacarboranes has been developed independently by two groups. The procedure utilizes the reaction of the $C_2B_9H_{\overline{12}}$ ion with cyclopentadiene and $CoCl_2$ in aqueous[164] or alcoholic[116,164] media. The $C_2B_9H_{\overline{12}}$ salt need not be isolated, and can be prepared *in situ* directly from 1,2-$C_2B_{10}H_{12}$ (*o*-carborane)[116]; one report[164] indicates that the optimal solvent is methanol, in which, for unknown reasons, the side formation of $(\eta\text{-}C_5H_5)_2Co^+$ and $(C_2B_9H_{11})_2Co^-$ species is minimized:

$$1,2\text{-}C_2B_{10}H_{12} \xrightarrow[\text{CH}_3\text{OH}]{\text{KOH}} (C_2B_9H_{12})^- \xrightarrow[\text{C}_5\text{H}_6]{\text{CoCl}_2} 3,1,2\text{-}(\eta\text{-}C_5H_5)CoC_2B_9H_{11}$$
$$79\%$$

A minor product of this reaction, characterized as $3,6,1,2\text{-}(\eta\text{-}C_5H_5)_2Co_2C_2\text{-}B_8H_{10}$, has been obtained in high yield by treatment of the monocobalt

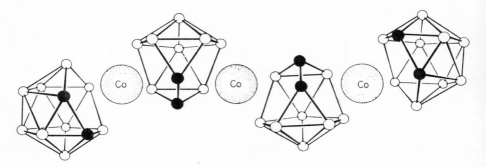

Figure 29. Structure of the $[(1,2\text{-}C_2B_9H_{11})Co(1,2\text{-}C_2B_8H_{10})]_2Co^{3-}$ ion.[53] ●, CH; ○, BH. (From *Inorg. Chem.*, **10**, 863 (1971). Copyright by the American Chemical Society.)

compound with $CoCl_2$ and C_5H_6 in methanolic KOH at reflux[164]:

$$3,1,2\text{-}(\eta\text{-}C_5H_5)CoC_2B_9H_{11} + CoCl_2 + C_5H_6 \xrightarrow[\text{KOH}]{\text{CH}_3\text{OH}} 3,6,1,2\text{-}(\eta\text{-}C_5H_5)_2Co_2C_2B_8H_{10}$$
$$57\%$$

If the latter reaction is conducted in the absence of cyclopentadiene, the major product (75%) is a tricobalt anionic species, $Co[C_2B_8H_{10}Co(\eta\text{-}C_5H_5)]_2^-$, which is proposed to have a central cobalt common to two icosahedra, each of which also incorporates a $(\eta\text{-}C_5H_5)Co$ moiety.[116] The analogous species in which the "end" ligands are $C_2B_9H_{11}^{2-}$ rather than $C_5H_5^-$ has been characterized by X-ray diffraction[28] and is shown in Figure 29.

The direct-insertion technique described earlier (Sections III,IV.D, and IV.F) has been applied to the preparation of $(\eta\text{-}C_5H_5)CoC_2B_9H_{11}$ species, and involves the reaction of $1,2\text{-}C_2B_{10}H_{12}$ with $(\eta\text{-}C_5H_5)Co(CO)_2$ in the gas phase at 300°.[151] The initial step is probably the formation of an unstable 13-vertex $(\eta\text{-}C_5H_5)CoC_2B_{10}H_{12}$ species which loses boron to form $(\eta\text{-}C_5H_5)CoC_2B_9H_{11}$. However, due to thermal rearrangement,[118] many, and perhaps all nine, of the possible $(\eta\text{-}C_5H_5)CoC_2B_9H_{11}$ isomers are formed.[151]

Another novel synthetic approach utilizes the thallium species $Tl_2C_2B_9$-H_{11} or its *C*-alkyl derivatives, which are assumed to be ionic $Tl^+[TlC_2B_9H_{11}]^-$ salts containing an icosahedral anion incorporating a thallium atom in the cage. Displacement of the thallium with $CoCl_2$ in THF is reported to form $(1,2\text{-}C_2B_9H_{11})_2Co^-$ species, although details and yields are not given.[191]

Several 12-vertex tetracarbon cobaltacarboranes of composition $(\eta\text{-}C_5H_5)Co(CH_3)_4C_4B_7H_7$ have been prepared from $(CH_3)_4C_4B_8H_8$ via reduction to the dianion.[136] The structures have not been determined, but are probably similar to that of the isoelectronic species $(\eta\text{-}C_5H_5)Fe(CH_3)_4C_4\text{-}B_7H_8^{11}$ (Section IV.D.3), which consists of an open 12-vertex cage.

Experimental

Preparation of 3,1,2-$(\eta$-$C_5H_5)CoC_2B_9H_{11}$.[164] A mixture of KOH (56 g, 1 mol) and 1,2-$C_2B_{10}H_{12}$ (14.4 g, 0.1 mol) in 100 ml of methanol was allowed to react and then refluxed for 2 hr; a two-layer mixture of $CoCl_2 \cdot 6H_2O$ (36 g, 0.15 mol) in 60 ml of methanol and freshly distilled cyclopentadiene (13.2 g, 0.2 mol) was added as described in the preparation of 3,1,2-$(C_5H_5)CoC_2B_9$-H_{11}[164] given in Section IV.D.3. The combined mixture was diluted with 200 ml of water, and the insoluble residue was filtered off and washed with two 100-ml portions of water and two 100-ml portions of 10% HCl. A mixture of the product and metallic cobalt remained on the filter. The product was dissolved in three 50-ml portions of acetone, the combined solutions were filtered, diluted with 50 ml of water, and acetone was distilled off *in vacuo*. Separated crystal flakes were filtered off, washed three times with 30 ml of chloroform to remove brown impurities, and dried in air. The product (18.7 g, 72%) was chromatographically pure. The chromatography of the chloroform extract on silica gel yielded another 1.8 g (7%) of product, which was crystallized as yellow needles from benzene solution and was covered carefully with a layer of an equal volume of hexane.

Preparation of 3,6,1,2-$(\eta$-$C_5H_5)_2Co_2C_2B_8H_{10}$.[164] The above compound, 3,1,2-$(\eta$-$C_5H_5)CoC_2B_9H_{11}$ (8.0 g, 0.031 mol) was added to a suspension of KOH (60 g, 1.1 mol) in 80 ml of methanol. To this mixture a solution of $CoCl_2 \cdot 6H_2O$ (14.0 g, 0.06 mol) in 15 ml of methanol together with freshly distilled cyclopentadiene (6.6 g, 0.1 mol) were added as in the preparation in the preceding paragraph. The reaction mixture was refluxed for 10 hr under nitrogen and the product isolated in the same manner as above. The final mixture of metallocarborane product, unreacted starting material, and metallic cobalt was washed with two 30-ml portions of acetone to remove starting carborane. The undissolved residue was extracted with boiling acetone as long as the run off was red. In the extraction flask, the product separated in the form of bright red flakes which, after drying in air, weighed 6.5 g (57%).

b. Electronic Studies. The icosahedral CoC_2B_9 cage has received more detailed examination than other metallocarborane systems, including studies designed to probe the electronic structure and properties of the cage framework. A [59]Co nuclear quadropole resonance investigation[84] of (1,2-C_2B_9-$H_{11})_2Co^-$ indicated virtually identical boron–cobalt and carbon–cobalt bonding, consistent with the uniformity of bond distances as determined by the X-ray method.[244] The electronic nature of the 1,2 and 1,7 isomers of the $(C_2B_9H_{11})_2Co^-$ anion and of $(\eta$-$C_5H_5)CoC_2B_9H_{11}$ has been examined by applying the method of Taft[193,194] to the ^{19}F nmr spectra of derivatives of these species in which a *m*- or *p*-fluorophenyl ligand is bonded to a cage carbon atom[1] or to the C_5H_5 ring.[237] The conclusions drawn from this work have been discussed in Section IV.D.3. A detailed study of the [11]B nmr spectra of the $(1,2\text{-}C_2B_9H_{11})_2Co^-$ ion and a number of substituted derivatives has appeared,[187] but interpretation of the data in terms of electronic effects within the cage was not attempted. A recent investigation of electron delocalization in paramagnetic species was mentioned in Section II.A.

   c. *Cage Degradation Reactions.* Dicobalt and Tricobalt Icosahedra. The synthesis of icosahedral cobaltacarboranes from 1,2- or $1,7\text{-}C_2B_9H_{11}$ by base degradation to generate an open-faced anion followed by insertion of Co^{2+} ions has been described above. Repetition of this procedure with the $(1,2\text{-}C_2B_9H_{11})_2Co^-$ ion results in the introduction of a second cobalt into the icosahedron and leads to polyicosahedral species, as shown schematically[53]:

$$(C_2B_9H_{11})_2Co^- \xrightarrow{\ OH^-\ } [(C_2B_9H_{11})Co(C_2B_8H_{10})]^{3-}$$

I (*unobserved*)

$$[(C_2B_9H_{11})Co(C_2B_8H_{10})Co(C_2B_9H_{11})]^{2-} \qquad \{[(C_2B_9H_{11})Co(C_2B_8H_{10})]_2Co\}^{3-}$$

II (15%)   III (6%)

The intermediate I formed initially has not been isolated, but is the species expected from the degradation of the $(C_2B_9H_{11})_2Co^-$ ion. Reaction of two equivalents of this ion with Co^{2+} leads to III with four fused icosahedra; alternatively, reaction of one equivalent of the intermediate with Co^{2+} and $C_2B_9H_{11}^{2-}$ [formed by the *in situ* degradation of $(C_2B_9H_{11})_2Co^-$] produces II.[28,53,54] The structures of II and III have been established by X-ray investigations,[28,174] and that of III is illustrated in Figure 29. The 12-atom cages in these species are significantly distorted from regular icosahedral symmetry, particularly where two vertices in the same cage are occupied by cobalt.

   A small quantity of III has been isolated in the preparation of $(C_2B_9\text{-}H_{11})_2Co^-$ itself,[54] presumably formed by the sequence indicated above.

   The incorporation of two cobalt atoms in an icosahedral cage via polyhedral expansion of the 11-atom system $(\eta\text{-}C_5H_5)CoC_2B_8H_{10}$ has been mentioned in Section IV.F.4. Recently, an icosahedral cage containing three cobalts has been isolated from the expansion of $2,1,6\text{-}(\eta\text{-}C_5H_5)CoC_2B_7H_9$.[43,47]

$$(\eta\text{-}C_5H_5)CoC_2B_7H_9 \xrightarrow[C_{10}H_8]{Na} \xrightarrow[Na^+C_5H_5]{CoCl_2} (\eta\text{-}C_5H_5)_2Co_2C_2B_7H_9 + (\eta\text{-}C_5H_5)_3Co_3C_2B_7H_9$$

22%   (3%)

The proposed structure[43,47] of the cobalt species, $2,3,5,1,7\text{-}(\eta\text{-}C_5H_5)_3\text{-}Co_3C_2B_7H_9$, is given in Figure 30. This compound decomposes in solution or in silica gel, yielding $(C_5H_5)_2Co_2C_2B_7H_9$, believed to be the 1,8,2,3 isomer.

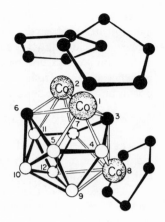

Figure 30. Proposed structure of $(\eta\text{-}C_5H_5)_3Co_3C_2\text{-}$ B_7H_9.[43] ●, CH; ○, BH. (From *Inorg. Chem.*, **13**, 869 (1974). Copyright by the American Chemical Society.)

Although this is the only reported icosahedral trimetallocarborane, two 10-vertex species having three cobalts have been prepared (see Section IV.F.3).

d. Substitution on the Icosahedral Cage. The direct introduction of ligand groups at carboranyl carbon positions has been reported by Zakharkin and Bikkineev,[234] who demonstrated that $3,1,2\text{-}(\eta\text{-}C_5H_5)CoC_2B_9H_{11}$ can be C-metallated by treatment with *n*-butyllithium in ether at $-35°$; the resulting C-lithio or C,C'-dilithio species can be carboxylated to yield the mono- or di-C-carboxylic acid. When the metallation is conducted with either *n*-butyllithium or phenylethynyllithium *in THF*, substitution occurs exclusively on the cyclopentadienyl ring to give $3,1,2\text{-}(n\text{-}RC_5H_4)CoC_2B_9H_{11}$.[234] The C-metallation technique has been widely used in the preparation of C-substituted derivatives of the $C_2B_{10}H_{12}$ isomers and other carboranes,[75] but nearly all of the known C-substituted cobaltacarboranes were obtained from derivatives of 1,2- or $1,7\text{-}C_2B_{10}H_{12}$ containing ligands on carbon.

Alternative routes[236] to C-substituted derivatives of $3,1,2\text{-}(C_5H_5)\text{-}$ $CoC_2B_9H_{11}$ utilize the treatment of derivatives of the thallium carborane $Tl_2C_2B_9H_{11}$ with $CoCl_2$ and C_5H_5Tl, and the reaction of $CoCl_2$ and C_5H_6 with $K^+[RC_2B_9H_{10}]^-$ salts in ethanol. Electrophilic bromination of $(1,2\text{-}C_2B_9H_{11})_2Co^-$ produces the $(8,9,12\text{-}Br_3\text{-}1,2\text{-}C_2B_9H_8)_2Co^-$ ion, as established in an X-ray study[30] of the product.[100] This result parallels the bromination of $1,2\text{-}C_2B_{10}H_{12}$ itself, in that the halogenation occurs at the borons furthest from carbon, presumably the most electronegative positions on the cage.[100] Similar treatment of $3,1,2\text{-}(C_5H_5)CoC_2B_9H_{11}$ generates mono-, di-, and tribromo derivatives; the C-methyl derivative behaves comparably.[235]

The attack of halogens other than bromine has not been reported.

In contrast to the $(1,2\text{-}C_2B_9H_{11})_2Fe^{2-}$ species that readily undergoes protonation in strong acids,[98] no direct evidence for a corresponding protonated derivative of the $(1,2\text{-}C_2B_9H_{11})_2Co^-$ ion has been found[98] although metal-protonated seven-vertex cobalt systems are known (Section IV.F.2). However, acid-catalyzed substitution reactions of ferracarboranes described in Section IV.D.3 *are* paralleled by the cobalt(III) species.[98] Thus, in the presence of anhydrous HCl, diethyl sulfide replaces a hydrogen bonded to boron and yields a neutral derivative:

$$(C_2B_9H_{11})_2Co^- + HCl + (C_2H_5)_2S \longrightarrow (C_2B_9H_{11})Co[C_2B_9H_{10}S(C_2H_5)_2] + H_2 + Cl^-$$

As was discussed above for the iron compounds, the position of sulfur attachment is uncertain; it may be at B(8), the unique boron adjacent to cobalt but not bonded to carbon.[98]

The reaction of the $(1,2\text{-}C_2B_9H_{11})_2Co^-$ anion with carbon disulfide in the presence of $AlCl_3$ and HCl generates a CS_2-bridged species in which the bridging carbon atom is protonated.[25,52] An X-ray diffraction study[21] has confirmed this remarkable structure, shown in Figure 31:

$$(1,2\text{-}C_2B_9H_{11})_2Co^- + CS_2 + H^+ \longrightarrow (1,2\text{-}C_2B_9H_{10})_2CoS_2CH + H_2$$

The $(1,7\text{-}C_2B_9H_{11})_2Co^-$ anion undergoes an analogous reaction, as does $(1,2\text{-}C_2B_9H_{11})_2Fe^-$. However, the bis ($C,C'$-dimethyl) derivative of $(1,2\text{-}C_2B_9H_{11})_2Co^-$ failed to give an isolable product, probably due to steric interference of the methyl groups in achieving the required cisoid arrangement of the two carborane cages (Figure 31).[52]

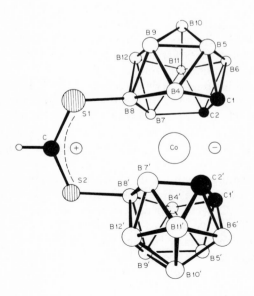

Figure 31. Structure of $(1,2\text{-}C_2B_9\text{-}H_{10})_2CoS_2CH$.[21] (From *J. Amer. Chem. Soc.*, **91**, 1222 (1969). Copyright by the American Chemical Society.)

A presumably analogous bridged species is formed in the reaction of $(1,2\text{-}C_2B_9H_{11})_2Co^-$ with 4:1 acetic acid–acetic anhydride in the presence of $HClO_4$. The orange product has the formula $(C_2B_9H_{10})_2CoO_2CCH_3$ and is proposed to have a bridged structure like that in Figure 31, with a CH_3CO_2 group replacing the HCS_2 moiety.[52] The initial process in this reaction is suggested to be addition of acetic acid to form an acetate, followed by attack of the bound acetate group on the unsubstituted cage. Since no hydrogen is evolved (in contrast to the CS_2 reaction above), and acetic anhydride is required, it has been concluded that some of the acetic acid is reduced, forming water which terminates the reaction unless it is continuously removed. The bridging carbon atoms are electron-deficient carbonium ion centers and this is reflected in their chemical behavior. Reduction of $(1,2\text{-}C_2B_9H_{10})_2CoS_2CH$ to $(1,2\text{-}C_2B_9H_{10})_2CoS_2CH_2$, assumed to contain an $-S\text{-}CH_2\text{-}S-$ linking group, occurs easily with BH_4^- ion, and the $(1,2\text{-}C_2B_9H_{10})_2CoS_2CH$ and $(1,2\text{-}C_2B_9H_{10})_2CoO_2CCH_3$ species are hydrolyzed in refluxing aqueous ethanol to a dithiol and a diol, respectively.[52]

The reaction of $[(CH_3)_2SOH]^+[(1,2\text{-}C_2B_9H_{11})_2Co]^- \cdot (CH_3)_2SO$ with S_2Cl_2 in methylene chloride, followed by alkaline methanolysis affords a unique sulfur-bridged neutral product, $8,8'\text{-}CH_3S\text{-}(1,2\text{-}C_2B_9H_{10})_2Co$, which features a three-membered $B\text{-}S\text{-}B$ bridge between the two cages.[163]

A closely related type of reaction of the $(1,2\text{-}C_2B_9H_{11})Co^-$ ion, that has been investigated, involves the formation of an apparent diazonium salt $[C_6H_5N_2][(1,2\text{-}C_2B_9H_{11})_2Co]$ by reaction of $[C_6H_5N_2][BF_4]$ with $K^+[(C_2B_9H_{11})_2Co]^-$ at $0°$ in water.[55] The salt is an unstable yellow material which, however, dissolves in warm benzene; on heating the solution above $60°$, both H_2 and N_2 are evolved and a $[(C_6H_4)(C_2B_9H_{10})_2Co]^-$ species is formed. This compound is believed to contain a benzene ring bridging the two carborane cages,[55] analogous to the HCS_2 bridged species depicted in Figure 31. The bridging C_6H_4 unit is derived from the *solvent*, not from the diazonium function, as shown by two additional experiments. When this yellow benzene-diazonium salt is heated in toluene solution, the product is exclusively a toluyl bridged $[(CH_3C_6H_3)(1,2\text{-}C_2B_9H_{10})_2Co]^-$ species; similarly, reaction of the corresponding p-toluenediazonium salt $[p\text{-}CH_3C_6H_4N_2][(1,2\text{-}C_2B_9H_{10})_2Co]$, with hot benzene yields only the phenyl-bridged species. In both of the mixed reactions, methylbiphenyl is obtained as a side product but dimethylbiphenyl and biphenyl have not been detected, indicating that the reactions probably do not involve the formation of aryl radicals from the solvent; such radicals would be expected to produce symmetrical biphenyls. However, a radical mechanism involving decomposition of the original diazonium salt to a $[(1,2\text{-}C_2B_9H_{11})Co(1,2\text{-}C_2B_9H_{10})]^{\mp}$ intermediate has been proposed. A cationic mechanism has also been considered but is deemed less likely.[55]

Experimental

Preparation of $(1,2\text{-}C_2B_9H_{10})_2CoS_2CH$.[52] To a dry flask containing a stirring bar, 100 ml of CS_2 and 1.5 g of $AlCl_3$, was added 2 g (5.5 mmol) of $[(1,2\text{-}C_2B_9H_{11})_2Co]^-K^+$. A slow stream of HCl (equivalent to one or two bubbles/sec) was passed over the stirred solution for 5 hr. By this time, gas evolution had nearly ceased, and the solution was a deep brownish-orange. The mixture was filtered, the solid was washed with 100 ml of CS_2, and the combined filtrates were reduced to dryness. The orange and red product was dissolved in 100 ml of dichloromethane, and the solution was run through a column (2-cm i.d.) of about 5 g of silica gel. About 50 ml of heptane was added to the orange solution which was reduced to a volume of 50 ml. This produced a thick slurry of the voluminous yellow product; it amounted to 1.6 g (4 mmol, 70%). It was contaminated with some sulfur-containing compounds, and in this form was easily electrified, making transfers difficult. Recrystallization from boiling 1:1 heptane–benzene afforded large brownish-yellow crystals containing 30% by weight of benzene. These were more easily handled, and the benzene could be removed easily by washing with hexane or drying *in vacuo*. The pure material melted at 302–304°.

e. Substitution on the Cyclopentadienyl Ring. Derivatives 3,1,2-$(\eta\text{-}C_5H_5)CoC_2B_9H_{11}$ containing substituents on the C_5H_5 ligand have been prepared by the reaction of the parent compound with Grignard reagents, $RMgBr$, where $R = C_6H_5$, $m\text{-}FC_6H_4$, or $p\text{-}FC_6H_4$.[237] Electronic studies based on ^{19}F nmr spectra are discussed in Section IV.D.3. A series of cyclopentadienyl ring-substituted derivatives of the seven-vertex 1,7,2,3- and 1,7,2,4-$(\eta\text{-}C_5H_5)_2Co_2C_2B_3H_5$ triple-decked complex species (Section IV.F.2) has been prepared[216] from $Na^+RC_5H_4^-$ salts and the $[1,2,3\text{-}(\eta\text{-}C_5H_5)CoC_2B_3H_6]^-$ ion; the compounds reported are listed in Table I.

The incorporation of a C_6H_5B group *into* the C_5H_5 ring in 3,1,2-$(\eta\text{-}C_5H_5)CoC_2B_9H_{11}$ was mentioned in Section IV.F.6.

f. Thermal Rearrangement of $(\eta\text{-}C_5H_5(CoC_2B_9H_{11}$ Species. The cage isomerization of the 3,1,2-$(\eta\text{-}C_5H_5)CoC_2B_9H_{11}$ icosahedral system in the vapor phase has been examined in detail.[94,118] Sublimation of the parent compound through glass wool at 380–495° generated four additional isomers, and one of these, 4,1,7-$(\eta\text{-}C_5H_5)CoC_2B_9H_{11}$, produced two more isomers at 600–700°. Thus, seven of the nine possible CoC_2B_9 icosahedral systems (disregarding enantiomers) were generated, the only missing ones being the 8,1,2 and 9,1,2 species in which the cage carbons are nonadjacent to the metal *and* adjacent to each other. The latter two systems were, however, produced by rearrangement of 3,1,2-$(\eta\text{-}C_5H_5)Co[\mu\text{-}1,2\text{-}(CH_2)_3C_2B_9H_9]$ in which the cage carbons are tied together by a trimethylene bridge.[161] Hence, all nine possible CoC_2B_9 isomers are represented among the products of the thermal rearrangement of parent 3,1,2-$(\eta\text{-}C_5H_5)CoC_2B_9H_{11}$ or its *C*-methyl, *C,C'*-

dimethyl, or C,C'-μ-trimethylene derivatives.[118] The overall trends observed in the distribution of isomers as a function of reaction temperature are primarily (1) a tendency of the cage carbons to achieve a nonvicinal configuration (although one product, the 4,1,2 isomer, does have adjacent carbons), and (2) apparently increasing thermal stability as the number of carbons directly bonded to the metal decreases, i.e., the isomers having a B_5 face fused to cobalt are most stable. The latter trend is paralleled by electrochemical measurements which indicate that the isomers having no Co–C bonds are the least readily reduced to the cobalt(II) species.[118] A similar conclusion regarding the greater ability of a B_4C face than a B_3C_2 face to stabilize a high formal oxidation state has been independently reached from studies of monometallocarboranes of the $MCB_{10}H_{11}$ type.[119]

The mechanism by which these cage rearrangements occur is not understood, and the authors of the study just described[118] reach no firm conclusions except for the observation that the *simplest* form of the diamond-square-diamond, or dsd mechanism of Lipscomb is in itself inadequate to explain the results. This pathway, originally proposed for the rearrangement of icosahedral $C_2B_{10}H_{12}$ carboranes, and described elsewhere in detail,[75,126] would account for five of the observed cobaltacarborane products; it is incapable, however, of converting the original 3,1,2-CoC_2B_9 cage into the observed 5,1,7, 9,1,2, or 2,1,12 isomers. More complicated schemes,[117,154] involving rotation of triangular clusters of atoms on the icosahedral surface, or mutual rotation of two halves of the icosahedron (Ref. 75, p. 158) do allow interconversion of all possible isomers. As in the case of the thermal rearrangement of 1,2-$C_2B_{10}H_{12}$ and its derivatives, the observed isomer distribution suggests no clear preference for any single proposed mechanism and may indeed reflect a mixture or hybrid of several different processes.[118]

6. Thirteen- and Fourteen-Atom Cages

The $C_2B_{10}H_{12}{}^{2-}$ ion, prepared by reductive cage opening of 1,2-$C_2B_{10}H_{12}$ with sodium naphthalide in THF, reacts with cobaltous chloride and cyclopentadienide ion to produce 4,1,7-(η-C_5H_5)$CoC_2B_{10}H_{12}$.[35,36] The reaction is analogous to the syntheses of 13-vertex metallocarboranes of iron and nickel described elsewhere in this review. The geometry[19,20] of the cobalt compound, given in Figure 32, was determined by X-ray. It differs somewhat from the structure originally proposed[35] and has the metal in a unique six-coordinate vertex on the 13-atom polyhedron:

$$1,2\text{-}C_2B_{10}H_{12} \xrightarrow[\text{THF}]{\text{Na, }C_{10}H_8} [C_2B_{10}H_{12}]^{2-} \xrightarrow[\text{CoCl}_2]{\text{NaC}_5H_5}$$

$$(\eta\text{-}C_5H_5)CoC_2B_9H_{11} + (\eta\text{-}C_5H_5)CoC_2B_{10}H_{12} + (C_2B_{10}H_{12})_2Co^-$$
$$1\% \qquad\qquad \text{red, } 50\% \qquad\qquad 12\%$$

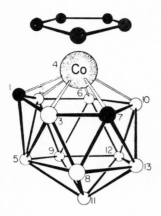

Figure 32. Structure of $4,1,7\text{-}(\eta\text{-}C_5H_5)CoC_2B_{10}\text{-}H_{12}$.[20,36] ●, CH; ○, BH. (From *J. Amer. Chem. Soc.*, **95**, 1109 (1973). Copyright by the American Chemical Society.)

The red $4,1,7\text{-}(\eta\text{-}C_5H_5)CoC_2B_{10}H_{12}$ undergoes thermal rearrangement in refluxing hexane, yielding an orange isomer; this compound rearranges further in refluxing benzene to give a third isomer which is red-orange. The structures of the latter two species have not been assigned but are believed to retain the 13-vertex polyhedral cage.[36]

The red 4,1,7 isomer exhibits a [11]B nmr spectrum which is inconsistent with the known solid-state geometry[20] and is indicative of fluxional behavior in solution, such that the enantiomeric 4,1,7- and 4,3,10- configurations undergo rapid interconversion via an intermediate containing a symmetry plane.[36] This equilibrium exists at temperatures as low as $-30°$, but at $-90°$ the nmr spectrum is more complex and indicates that the molecules are locked into the asymmetric 4,1,7-structure (Figure 32) and its enantiomer. The [11]B nmr spectra of the other isomers of $(\eta\text{-}C_5H_5)CoC_2B_{10}H_{12}$, and of the biscarboranyl $(C_2B_{10}H_{12})_2Co^-$ species, all indicate similar fluxional behavior in solution which generates a time-averaged plane of symmetry.[36]

The green bis complex, $(1,2\text{-}C_2B_{10}H_{12})_2Co^-$, is formed by reaction of $CoCl_2$ with the $C_2B_{10}H_{12}^{2-}$ ion in the absence of NaC_5H_5, and is slowly solvolyzed in acidic or basic media. In the solid state decomposition occurs at $230°$, but in refluxing CH_2Cl_2 an apparent thermal rearrangement to an unidentified red isomer is reported to occur.[36] However, two groups of Russian workers have independently determined that, in fact, isomerization of the $(1,2\text{-}C_2B_{10}H_{12})_2Co^-$ ion can happen at a much lower temperature. Thus, Brattsev and Stanko[8] have reported that this complex isomerizes at temperatures as low as $20°$, and that oxidation of the isomerized species with O_2, Cu^{2+}, $KMnO_4$, or other agents in anhydrous THF or in H_2O produces a mixture of the neutral carboranes 1,2-, 1,7-, and $1,12\text{-}C_2B_{10}H_{12}$. The ratio of carborane isomers obtained depends upon the reaction time and the temperature. Oxidation of the $(1,7\text{-}C_2B_{10}H_{12})_2Co^-$ and $(1,12\text{-}C_2B_{10}H_{12})_2Co^-$ species following isomerization in refluxing THF also gave mixtures of all three $C_2B_{10}H_{12}$ isomers, and the suggestion given is that these 13-vertex metallocarborane complexes have lower activation energy for isomerization than do the uncomplexed $C_2B_{10}H_{12}^{2-}$ anions, since the latter do not re-arrange even at $150°$.[8] The mechanism proposed by these workers involves an unstable cobalt(IV) intermediate:

$$[(C_2B_{10}H_{12})_2Co^{III}]^- \xrightarrow{-e} [(C_2B_{10}H_{12})_2Co^{IV}]^0 \longrightarrow 2\ C_2B_{10}H_{12} + Co^0$$

Zakharkin and Kalinin[238] report generally similar results, although they found only $1,7\text{-}C_2B_{10}H_{12}$ as a product of the isomerization of $(1,2\text{-}C_2B_{10}H_{12})_2Co^-$ at $60\text{-}70°$ in THF followed by oxidation with $CuCl_2$. (In contrast, Brattsev and Stanko[8] found 63% $1,2\text{-}C_2B_{10}H_{12}$, 36% of the 1,7 isomer, and 0.4% of the 1,12 isomer formed under presumably similar conditions.) Clearly, these experiments involve at least three different types of reactions— namely, metallocarborane formation, isomerization, and oxidative degrada-tion—and the complexities involved in conducting these processes in sequence could easily lead to different carborane product distributions depending on the exact procedures employed. Even the length of time in which the metallo-carborane ion is stored has a substantial effect on the carborane isomer ratio produced.[8]

In recent work,[239] Zakharkin et al. have investigated the oxidation of the $[1,2\text{-}(C_2B_{10}H_{12})_2Co^{III}]^-$, $[(1,2\text{-}C_2B_{10}H_{12})_2Fe^{II}]^{2-}$ and $[(1,2\text{-}C_2B_{10}H_{12})_2\text{-}Ni^{II}]^{2-}$ complexes and their C-substituted derivatives which were prepared at low temperature $(5\text{-}15°)$ and found that under these conditions $1,2\text{-}C_2B_{10}\text{-}H_{12}$ is produced almost quantitatively. However, when the same metallo-carborane complexes were prepared under the conditions originally employed by Hawthorne and co-workers[36] $(20°$ over 12 hr$)$, subsequent oxidation generated mixtures of 1,2- and $1,7\text{-}C_2B_{10}H_{12}$; on this basis it is claimed[239] that the complexes originally reported[36] must have been mixtures of isomers rather than pure compounds.

These reactions appear to have potential utility in facilitating the con-version of $1,2\text{-}C_2B_{10}H_{12}$ to the 1,7 and 1,12 isomers, which have been in-corporated into heat-stable polymers.[75] For comparison, the direct thermal conversion of the 1,2 isomer in the vapor phase requires temperatures of $425\text{-}500°$ for $1,7\text{-}C_2B_{10}H_{12}$ formation and $600\text{-}700°$ for the preparation of the 1,12 isomer.

The reactions of the 13-vertex cobaltacarboranes other than isomeriza-tion have been little studied, except for a novel base degradation which results in extraction of three borons and a carbon from the cage (Section IV.F.3). Recently, several 13-vertex bimetallic species, including isomers of $(\eta\text{-}C_5H_5)_2\text{-}Co_2C_2B_9H_{11}$ and a mixed metal system, $(\eta\text{-}C_5H_5)_2Co^{III}\ Fe^{III}\ C_2B_9H_{11}$, have been obtained from $(\eta\text{-}C_5H_5)CoC_2B_{10}H_{12}$ via base degradation.[37,40] The polyhedral expansion of $4,1,13\text{-}(\eta\text{-}C_5H_5)CoC_2B_{10}H_{12}$ by the sodium naphthalide method yields two 14-vertex dicobalt systems, $(\eta\text{-}C_5H_5)_2Co_2C_2B_{10}H_{12}$, which are proposed to have the geometry of a bicapped hexagonal antiprism with the metal atoms occupying the six-coordinate capping vertices, the isomers differing in the location of cage carbon atoms.[48]

Dicobalt 13-vertex cages have been obtained in the polyhedral expansion of $3,1,2\text{-}(\eta\text{-}C_5H_5)CoC_2B_9H_{11}$ utilizing sodium in THF followed by metal insertion;[43] the two $(\eta\text{-}C_5H_5)_2Co_2C_2B_9H_{11}$ isomers thus formed have not been structurally characterized as of this writing. An attempt[124] to prepare a 13-vertex monocobaltacarborane by sodium reduction of $3,1,2\text{-}(\eta\text{-}C_5H_5)\text{-}CoC_2B_9H_{11}$, followed by reaction with $C_6H_5BCl_2$ was expected to effect insertion of a boron atom into the cage, forming $(\eta\text{-}C_5H_5)CoC_2B_{10}H_{11}(C_6H_5)$. Instead, however, the boron insertion occurred exclusively in the cyclopentadienyl ring, giving the phenylborinato complex $3,1,2\text{-}[\eta^6\text{-}(C_6H_5)\text{-}BC_5H_5]CoC_2B_9H_{11}$.

A 13-vertex *tetracarbon* cobaltacarborane has been prepared from $(CH_3)_4C_4B_8H_8$[139] by reduction to the dianion and reaction with $CoCl_2$ and NaC_5H_5.[136] Although the structure could not be determined from [11]B and ^{1}H nmr spectra, an X-ray diffraction study[11] established the geometry as that of an open cage.

Experimental

Preparation of $4,1,7\text{-}(\eta\text{-}C_5H_5)CoC_2B_{10}H_{12}$.[36] To a solution of 25 mmol of $Na_2C_2B_{10}H_{12}$[36] was added a solution of 75 mmol of NaC_5H_5. The flask was cooled in an ice bath, and a slurry of 12 g (92 mmol) of finely ground anhydrous $CoCl_2$ in 50 ml of THF was added from an addition funnel. The contents of the flask were stirred for 15 hr while slowly warming to room temperature. The addition funnel and nitrogen inlet were removed and, in order to convert any cobaltocene formed to a cobalticinium salt, a stream of oxygen was passed through the solution with occasional stirring for 0.5 hr. The reaction mixture was then filtered through Celite to remove cobalt metal. Approximately 30 g of silica gel was added to the filtrate, and the solvent was removed at room temperature using a rotary evaporator and a water aspirator. The solid was then chromatographed on silica gel in the manner described for $(\eta\text{-}C_5H_5)FeC_2B_{10}H_{12}$ (Section IV.D.4). Seven major bands were separated consisting of, in order of elution: naphthalene, 1,2-$C_2B_{10}H_{12}$ (2.55 g, 17.7 mmol); orange $(\eta\text{-}C_5H_5)CoC_2B_{10}H_{12}$ (0.1054 g, 0.4 mmol); yellow $3,1,2\text{-}(\eta\text{-}C_5H_5)CoC_2B_9H_{11}$; red $(\eta\text{-}C_5H_5)CoC_2B_{10}H_{12}$ (2.16 g, 8 mmol), the yellow cobalticinium salt of $C_2B_9H_{12}^-$, a red-brown band which contained several isomers of $[(\eta\text{-}C_5H_5)_2Co]^+[Co(C_2B_{10}H_{12})_2]^-$ (1.73 g,

4.0 mmol). The orange and red isomers of $(\eta\text{-}C_5H_5)CoC_2B_{10}H_{12}$ were recrystallized from hexane–dichloromethane by slow removal of the solvent on a rotary evaporator.

Thermal Rearrangement of $4,1,7\text{-}(\eta\text{-}C_5H_5)CoC_2B_{10}H_{12}$.[36] To a 25-ml, round-bottom flask were added 0.268 g (1 mmol) of red $4,1,7\text{-}(\eta\text{-}C_5H_5)\text{-}CoC_2B_{10}H_{12}$ and 20 ml of hexane. A reflux condenser topped with a nitrogen inlet was connected, and the solution was heated at reflux for 15 hr. After cooling to room temperature, 10 g of silica gel was added and the solvent removed using a rotary evaporator and a water aspirator. The solid was placed on a silica-gel column (2.5 × 30 cm) and eluted with hexane gradually enriched with dichloromethane. The bands that were separated, in order of elution, were red-orange $(\eta\text{-}C_5H_5)CoC_2B_{10}H_{12}$, an orange isomer (0.1575 g, 0.6 mmol, 60%), and the original red isomer (0.1011 g, 0.37 mmol, 37%).

In a 200-ml, round-bottom flask were placed 1.0444 g (3.8 mmol) of the red isomer and 100 ml of benzene. A reflux condenser topped with a nitrogen inlet was connected and the flask was heated at reflux for 24 hr. The mixture was chromatographed as above yielding only one band. The solvent was evaporated and 0.7344 g (2.7 mmol, 71%) of the red-orange isomer was isolated.

G. Rhodium Compounds. Catalytic Activity of 2,1,2- and $2,1,7\text{-}[(C_6H_5)_3P]_2Rh(H)C_2B_9H_{11}$

The reaction of the tris(triphenylphosphine)rhodium(I) cation with the 1,2- or $1,7\text{-}C_2B_9H_{12}^-$ *nido*-carborane anions produces the corresponding icosahedral species 2,1,2- and $2,1,7\text{-}[(C_6H_5)_3P]_2Rh(H)C_2B_9H_{11}$, which are proposed to have a hydride (formally H^-) ligand σ-bonded to the rhodium atom.[160] These metallocarboranes have been reported (in two short communications[103,160]) to catalyze the isomerization of 1-hexene, the hydrogenation of 1-hexene to *n*-hexane, and the exchange of D_2 gas with terminal B–H bonds in boron cage compounds. This catalytic activity is potentially a major development in metallocarborane chemistry, and, if the early promise is borne out, may herald a significant practical role for metalloboron cage compounds.

A series of rhodacarboranes $[(C_6H_5)_3P]_2ClRhC_2B_9H_{11}$, $[(C_6H_5)_3P)\text{-}(C_6H_6)RhC_2B_9H_{11}]_2$, and $[(C_6H_5)_3P](H)RhC_2B_9H_{11}$ has been prepared from the $1,2\text{-}C_2B_9H_{11}^{2-}$ ion by treatment with organometallic rhodium reagents[185]; the formulation of the last-mentioned species suggests a 16-electron configuration on the metal atom, although this has not been established. A species analyzed as $[(C_6H_5)_4B][(C_6H_5)_3P]RhC_2B_9H_{11}$, in which the tetra-phenylboron group may occupy two coordination positions on the rhodium atom, was also isolated.[185]

H. Nickel Compounds

1. Seven- and Nine-Atom Cages

Methods for the preparation of small metallocarboranes from closo-1,6-$C_2B_4H_6$ and nido-2,3-$C_2B_4H_8$, described in preceding sections, afford nickel species containing the NiC_2B_4 system which is assumed to have pentagonal bipyramidal gross geometry. Thus, the reaction of ethylenebis-(triphenylphosphine)nickel with 1,6-$C_2B_4H_6$ in refluxing THF effects the replacement of ethylene with a carborane ligand[151]:

$$(\pi\text{-}C_2H_4)Ni[(C_6H_5)_3P]_2 + 1,6\text{-}C_2B_4H_6 \longrightarrow 1,2,4\text{-}[(C_6H_5)_3P]_2NiC_2B_4H_6$$

The same compound is ultimately obtained in a similar reaction between 2,4-$C_2B_5H_7$ and $(\pi\text{-}C_2H_4)Ni[(C_6H_5)_3P]_2$; an unstable violet compound which may be $[(C_6H_5)_3P]_2NiC_2B_5H_7$ is formed initially, but decomposes during thin-layer chromatography on silica gel to generate the above species.[151]

The reaction of the sodium salt of the nido-2,3-$C_2B_4H_7^-$ anion with bis-(diphenylphosphine)ethanenickel(II) chloride in THF generates 1,2,3-$[(C_6H_5)_2$-$PCH_2]_2NiC_2B_4H_6$.[79] Treatment of the same carborane anion with $NiBr_2$, NaC_5H_5, HCl, and air yields a $(\eta\text{-}C_5H_5)_2Ni_2C_2B_5H_7$ species (Figure 33). The phosphorus-substituted compound exhibits a [11]B nmr spectrum contain-ing a single broad peak, similar to that described above, but the $(\eta\text{-}C_5H_5)_2$-$Ni_2C_2B_5H_7$ has a normal spectrum and is evidently diamagnetic. This compound is interesting in that it has 22 cage valence electrons, two more than would be allowed by the $(2n + 2)$ rule for a closed nine-atom polyhedron (Section II), and hence some form of cage opening or distortion is expected. This author and his co-workers have suggested[79] that the distortion takes the form of a stretched Ni–Ni bond in the polyhedron, which could result from placement of the two "extra" electrons in a nickel–nickel antibonding orbital. Normally, boron cages having more than $(2n + 2)$ framework valence electrons adopt an open, or nido, geometry, but the placement of surplus electrons in metal–metal antibonding orbitals is common in metal cluster systems.[74] The structure indicated in Figure 33 is consistent with the nmr data, but other possibilities cannot be excluded even if a Ni–Ni bond is assumed.[79]

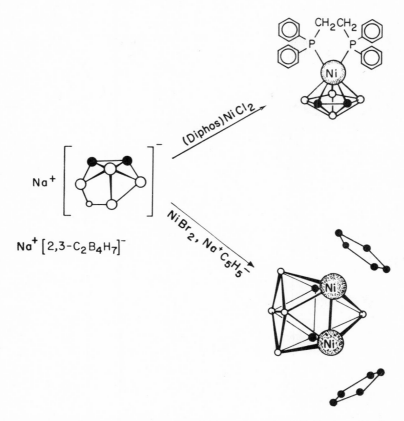

Figure 33. Preparation of nickel metallocarboranes from the $C_2B_4H_7^-$ ion, showing proposed structures of $1,2,3\text{-}[(C_6H_5)_2PCH_2]_2NiC_2B_4H_6$ and $(\eta\text{-}C_5H_5)_2Ni_2C_2B_5H_7$.[79] The latter molecule is probably distorted from the regular polyhedron shown (see text); Diphos = $[(C_6H_5)_2PCH_2]_2$.

Experimental

Reaction of $1,6\text{-}C_2B_4H_6$ *with* $(\pi\text{-}C_2H_4)Ni[P(C_6H_5)_3]_2$.[151] Under a nitrogen atmosphere, 1.83 mmol of $(\pi\text{-}C_2H_4)Ni[P(C_6H_5)_3]_2$ was placed in a 100-ml Schlenk tube which was connected to a vacuum line, and $1,6\text{-}C_2B_4H_6$ (1.83 mmol) and THF (30 ml) were added at $-196°$. The reaction mixture was allowed to warm to room temperature with stirring. During this period the solution changed from a rusty brown to a bright red color with substantially no H_2 evolution.

After a total reaction period of 3 hr, the solvent was removed and the product extracted with benzene. The benzene extract was rapidly passed through a silica-gel column with benzene as the eluent. The recovered red-orange fraction was chromatographed on a silica-gel column with a 50:50 hexane–benzene mixture. Recrystallization from methylene chloride–hexane yielded 120 mg (0.183 mmol, 10%) of red-orange plates of $1,2,4\text{-}[(C_6H_5)_3P]_2\text{-}NiC_2B_4H_6$.

Preparation of $1,2,3\text{-}[(C_6H_5)_2PCH_2]_2NiC_2B_4H_6$.[79] A solution of 3.0 mmol of $Na^+C_2B_4H_7^-$ in 20 ml of THF was taken into a dry box, filtered, and transferred to a pressure-equalized addition funnel. The carborane solution was added dropwise under nitrogen to a solution of 1.5 mmol of $[(C_6H_5)_2\text{-}PCH_2]_2NiCl_2$ in 20 ml of THF. The mixture was stirred at 26° for 12 hr. The resulting brown-violet solution was filtered, solvent was removed *in vacuo*, and the brown residue was transferred to a nitrogen glove bag and extracted with benzene. After removal of benzene *in vacuo*, the brown material was dissolved in a minimum of methylene chloride and cooled to 5° for 14 hr to form 100 mg of brown needle-like crystals. A second crop of crystals was obtained in the same manner from the mother liquor to give a total of 135 mg (0.255 mmol, 17% yield).

2. Ten-Atom Cages

The only other known subicosahedral metallocarboranes containing nickel are four isomers of a mixed cobalt–nickel species, $(\eta\text{-}C_5H_5)CoCB_7\text{-}H_8Ni(\eta\text{-}C_5H_5)$, discussed in Section IV.F.3, and a *nido* ten-vertex $[(CH_3)_2P]_2\text{-}Ni(CH_3)_2C_2B_7H_9$ species having $B_{10}H_{14}$-like geometry.[67,68] The latter compound was prepared from *nido*-$1,3\text{-}(CH_3)_2C_2B_7H_{11}$ by treatment with tetrakis(triethylphosphine)nickel.

3. Twelve-Atom Cages

a. Synthesis. The preparation of icosahedral metallocarboranes of nickel parallels those of the analogous iron and cobalt species; thus, reactions of nickel salts with the 1,2- or $1,7\text{-}C_2B_9H_{11}^{2-}$ ion (or with 1,2- or $1,7\text{-}C_2B_9\text{-}H_{12}^-$ in hot aqueous base) followed by air oxidation yield the corresponding $(C_2B_9H_{11})_2Ni^-$ ion in which the metal is in a formal oxidation state of $+3$.[100,210] Both the $(1,2\text{-}C_2B_9H_{11})_2Ni^-$ and $(1,7\text{-}C_2B_9H_{11})_2Ni^-$ ions are paramagnetic with one unpaired electron and are easily oxidized to the neutral, diamagnetic nickel(IV) species, which are unusual compounds and are discussed in detail below. The $(1,2\text{-}C_2B_9H_{11})_2Ni^-$ ion can be reduced to the formal nickel(II) species $(1,2\text{-}C_2B_9H_{11})_2Ni^{2-}$, which is extremely air-sensitive and paramagnetic with two unpaired electrons[100,210,212]:

$$[(C_2B_9H_{11})_2Ni^{II}]^{2-} \underset{\text{air}}{\overset{\text{Na/Hg}}{\rightleftarrows}} [(C_2B_9H_{11})_2Ni^{III}]^- \underset{\text{Cd}}{\overset{\text{FeCl}_3}{\rightleftarrows}} [(C_2B_9H_{11})_2Ni^{IV}]^0$$

The interconversion of species having different metal oxidation states is actually somewhat more complicated than the above sequence implies, and involves profound structural rearrangement of the molecule. These reactions are described more fully later in this section.

The cyclopentadienyl nickel(III) analog, $3,1,2-(\eta-C_5H_5)NiC_2B_9H_{11}$, has been obtained from $1,2-C_2B_9H_{11}^{2-}$, NaC_5H_5, and $NiBr_2 \cdot 2C_2H_4(OCH_3)_2$ in THF with subsequent air oxidation, and, as expected, is paramagnetic with one unpaired electron.[226] This neutral species can be electrochemically oxidized to the $[(\eta-C_5H_5)NiC_2B_9H_{11}]^+$ cation containing formal nickel(IV), but the latter ion has not been isolated. Although it is more stable than the $[(\eta-C_5H_5)_2Ni^{IV}]^{2+}$ nickelocene dication, whose existence at $-40°$ has been demonstrated electrochemically in the same study, the $[(\eta-C_5H_5)NiC_2B_9H_{11}]^+$ species decomposes slowly in acetonitrile at room temperature. Since the $(1,2-C_2B_9H_{11})_2Ni^{IV}$ compound is much more stable, a clear trend has been established which demonstrates the much greater ability of the $C_2B_9H_{11}^{2-}$ ligand to stabilize high formal metal oxidation states, as compared to the $\eta-C_5H_5^-$ ligand.[226]

A related icosahedral nickel metallocarborane, $1,7-(CH_3)_2-2,1,7-(1,5-C_8H_{12})NiC_2B_9H_9$, has obtained from the reaction of $Ni(1,5-C_8H_{12})_2$ with $closo-2,3-(CH_3)_2C_2B_9H_9$ in toluene; similarly, $1,7-(CH_3)_2-2,1,7-[(C_2H_5)_3P]_2NiC_2B_9H_9$ was prepared from $2,3-(CH_3)_2C_2B_9H_9$ and $[(C_2H_5)_3P]_2-Ni(1,5-C_8H_{12})$ in toluene.[71,191] The isolation of a mixed-metal species $(C_8H_{12})NiC_2B_8H_{10}Co(\eta-C_5H_5)$, from the reaction of $1,2,4-(\eta-C_5H_5)CoC_2-B_8H_{10}$ with $(1,5-C_8H_{12})Ni$ has been reported, but characterization of the product was incomplete.[71]

A *tetracarbon* 12-vertex nickelacarborane, $[(C_6H_5)_2PCH_2]_2Ni(CH_3)_4C_4-B_7H_7$, has been obtained from $(CH_3)_4C_4B_8H_8$[139] via reduction to the dianion.[136] Two isomers have been characterized from [11]B and ^{1}H nmr and mass spectra, but the structures have yet to be established; however, electron-counting arguments of the type discussed in Section II.B. predict open-cage geometries for these molecules.

Experimental

Nonaqueous Preparation of the $[(1,2-C_2B_9H_{11})_2Ni^{III}]^-$ *and* $[(1,7-C_2B_9H_{11})_2Ni^{III}]^-$ *Anions.*[100] A solution of nickel acetylacetonate (0.667 g, 2.59 mmol) in THF (35 ml) was added dropwise under nitrogen to a stirred solution of $Na_2[1,7-C_2B_9H_{11}]$ (5.17 mmol), prepared as in Ref. 100. The red mixture was heated at 35° under nitrogen for 3 hr with stirring. Oxygen was then bubbled through the solution until it turned green (ca. 10 min). The mixture was filtered, the solvent removed *in vacuo*, the residue extracted with ether, and the ether evaporated to dryness. The residue was taken up in 50 ml of water and the resulting green aqueous solution treated with tetramethyl-ammonium chloride. The product was recrystallized from aqueous acetone to give 0.804 g (2.02 mmol, 78%) of olive-green $[(CH_3)_4N][(1,7-C_2B_9H_{11})_2Ni]$, mp > 300°.

Using the procedure described above, 5.45 mmoles of $Na_2[1,2-C_2B_9H_{11}]$ gave 0.80 g (2.01 mmol, 74%) of yellow-brown $[(CH_3)_4N][(1,2-C_2B_9H_{11})_2Ni]$, mp > 300°.

Aqueous Preparation of the $[(1,2-C_2B_9H_{11})_2Ni^{III}]^-$ *Anion.*[100] A mixture of 10.0 g (51.8 mmol) of $[(CH_3)_3NH][1,2-C_2B_9H_{12}]$ (20.0 g, 84 mmol) of $NiCl_2 \cdot 6H_2O$, and 250 ml of freshly prepared 40% aqueous sodium hydroxide solution was treated as described in the aqueous iron dicarbollide preparation. The brown residue from the ether extraction was taken up in 300 ml of water filtered, warmed to ca. 70°, and treated with 5.0 g (41.3 mmol) of rubidium chloride. Upon cooling 7.99 g (19.6 mmol, 76%) of Rb $[(1,2-C_2B_9H_{11})_2Ni]$ was obtained as black crystals with a greenish luster.

Preparation of $(1,2-C_2B_9H_{11})_2Ni^{IV}$ *and* $(1,7-C_2B_9H_{11})_2Ni^{IV}$.[100] A solution of 1.32 g of $FeCl_3 \cdot 6H_2O$ (4.89 mmol) in 250 ml of water was added slowly to a freshly prepared solution of 2.00 g (4.89 mmol) of Rb $(1,2-C_2B_9H_{11})_2Ni$ in 1 liter of warm water. The yellow precipitate formed was filtered, washed with water, and dried under vacuum. Further purification of this crude material (1.58 g) was achieved by column chromatography on silica gel with a 1:1 benzene–hexane mixture as the eluent. The solvents were reduced to a low volume *in vacuo* and heptane was added. Slow removal of the remaining benzene from the solution at room temperature under reduced pressure left orange crystals of pure $(1,2-C_2B_9H_{11})_2Ni$, 1.43 g (4.43 mmol, 90.5%). The compound decomposed at ca. 265°. The pure compound was sublimed unchanged at 150° under high vacuum.

The 1,7-isomer was obtained in a similar manner. The $Rb[(1,7-C_2B_9-H_{11})_2Ni]$ (0.852 g, 2.09 mmol) was obtained from a nonaqueous preparation, substituting rubidium chloride for tetramethylammonium chloride as the precipitating agent. The Ni complex was then treated with $FeCl_3 \cdot 6H_2O$ (0.563 g, 2.08 mmol) as described above. The crude, dried product was purified by column chromatography on silica gel with *n*-pentane as the eluent to give 0.575 g (1.78 mmol, 85%) of $(1,7-C_2B_9H_{11})_2Ni$ as orange needles. Further purification was effected by rapid recrystallization from hot aceto-nitrile and by sublimation at 150° under high vacuum. The bis-1,7 compound was much less stable in solution than its bis-1,2 isomer.

   b. Structures. The bis(carboranyl)nickel(II) species, $(1,2-C_2B_9H_{11})_2Ni^{2-}$, has an electronic spectrum whose assigned *d-d* transitions yield ligand field parameters qualitatively in agreement with those of nickelocene.[159] Unlike the latter species, however, X-ray data[228] suggest that $(1,2\text{-}C_2B_9H_{11})_2Ni^{2-}$ has a "slipped" configuration in which the two carborane ligands are shifted so that the metal atom is closer to the borons than to the carbons in the bonding faces (Figure 34). Similar slipped $(1,2\text{-}C_2B_9H_{11})_2M$ structures have been found in X-ray studies of several other compounds,[227,228] in all of which the metal atom has eight or more d electrons as defined by its assignment of formal oxidation state, e.g., for M = Ni(II), Cu(II), Cu(III), Au(II), Au(III), and Pd(II).[92] Significantly, all $(1,2\text{-}C_2B_9H_{11})_2M$ species in which M has six or

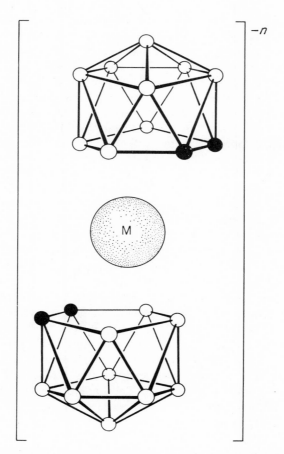

Figure 34. View of the slipped $(1,2\text{-}C_2B_9H_{11})_2M^{n-}$ structure, where M = Ni(II), Cu(II), Cu(III), Au(II), Au(III), or Pd(II).[92] ●, CH; ○, BH. (From *J. Amer. Chem. Soc.*, **90**, 4823 (1968). Copyright by the American Chemical Society.)

fewer d electrons, e.g., Cr(III), Fe(III), Fe(II), Co(III), and Co(II), and for which precise bond distances are available from X-ray crystallographic studies, have a symmetrical (nonslipped) geometry corresponding to that shown in Figure 8.[92] The cases of Ni(IV) and Pd(IV), which have d^6 electronic configurations, as well as that of Ni(III), a d^7 metal, are discussed below.

The problem of the slipped $(C_2B_9H_{11})_2M$ structures has been discussed at some length by Hawthorne and co-workers,[92,211] who proposed an explanation based on MO symmetry considerations, and also by Wing,[227,228] who viewed the slipped metallocarboranes as analogs of π-allylic metal complexes.

Wade[208] has suggested that the slippage is a form of distortion induced by an excess of delocalized valence electrons in the polyhedral cage systems beyond the $(2n + 2)$—or $(n + 1)$ pairs—which constitute a filled-shell molecular orbital configuration (Section II). Thus, if the metal atom utilizes six of its nine available bonding orbitals in binding to the two carborane ligands, there can be only three nonbonding orbitals[208] with a capacity of six electrons. A metal atom having more than six d electrons might then be expected to produce a distorted structure when incorporated into a $(C_2B_9-H_{11})_2M^q$-type species. On this basis, it would be predicted that $(C_2B_9H_{11})_2-M^q$ complexes of formal d^7 metals will exhibit structural anomalies in comparison with d^6 species. The only such metallocarborane for which X-ray data have been published is $(1,2\text{-}C_2B_9H_{11})_2Ni^-$, a formal Ni(III) anion, and the structure is close to that of the nonslipped species (Figure 8) with the carbons in the two cages in a *trans* orientation.[83] There are, however, some indications of slip distortion although certainly small compared to the d^8 complexes discussed above. The analogous palladium(III) complex, $(1,2\text{-}C_2B_9H_{11})_2Pd^-$, is claimed to be isostructural with the nickel species on the basis of chemical and spectroscopic similarities,[212] but no X-ray data have been reported.

The structure of the $(1,7\text{-}C_2B_9H_{11})_2Ni^{2-}$ ion, a d^8 metal system, has been examined by X-ray, and the results are particularly interesting.[229] Although the analogous $(1,2\text{-}C_2B_9H_{11})_2Ni^{2-}$ ion is slip-distorted,[228] as discussed above, the bis-1,7 isomer is much less slipped with the nickel shifted only 0.15 Å compared to 0.6 Å in the $(1,2\text{-}C_2B_9H_{11})_2Ni^{2-}$ and $(1,2\text{-}C_2B_9H_{11})_2Cu^{2-}$ species. Thus, the greater asymmetry of the $1,2\text{-}C_2B_9H_{11}^{2-}$ ligand with adjacent carbons in the bonding face does indeed appear to play some role in causing the slip-distortions, as proposed by Wing.[229] It is noteworthy, however, that significant distortion nevertheless occurs in $(1,7\text{-}C_2B_9H_{11})_2-Ni^{2-}$, that is, the five-atom bonding face is nonplanar, is irregular in shape, and one of the carbons is substantially further from the nickel atom than the other four atoms in the face (2.39 vs 2.14 Å).

The bis(carboranyl) complexes of nickel(IV), a d^6 metal, are neutral, diamagnetic, presumably filled MO systems containing $(2n + 2)$ electrons per cage and hence are expected to have nonslipped, symmetrical sandwich configurations analogous to the isoelectronic cobalt(III) species. An X-ray investigation[177] of $(1,2\text{-}C_2B_9H_{11})_2Ni^0$ confirmed the nonslipped arrangement but showed, surprisingly, that the cage carbon atoms in the two ligands are in a *cis*-orientation and are as close to each other as possible in the staggered configuration (Figure 35). This cisoid geometry, although nonslipped, contrasts with the "normal" centrosymmetric configuration (Figure 8) of $(1,2\text{-}C_2B_9H_{11})_2Co^-$ and other metallocarboranes which incorporate a formal d^6 metal in the polyhedral system. The implications of this novel structural arrangement, which is said to be paralleled in the isomorphous[92] $(1,2\text{-}C_2B_9H_{11})_2Pd^0$ species, will be discussed further in the context of the redox chemistry of the nickel(II), (III), and (IV) metallocarboranes.

c. Reactions and Interconversion. Nickel and palladium $(C_2B_9H_{11})_2M^q$ complexes, containing the metal in formal oxidation states of $+2$, $+3$, and $+4$, have a remarkable stereochemistry which is, in fact, unique in the metallocarborane family and has been thoroughly examined.[161,212] The $(C_2B_9H_{11})_2Ni^q$ species in which $q = -2, -1$, and 0 corresponding to formal Ni(II), Ni(III), and Ni(IV), each can exist in three different isomeric forms which are distinguished by the relative placement of their cage carbon atoms.

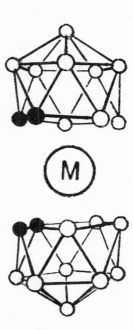

Figure 35. View of the nonslipped cisoid geometry of $(1,2\text{-}C_2B_9H_{11})_2M$; $M = Ni(IV)$ or Pd(IV).

Hawthorne and associates have designated these isomeric forms [studied as bis(C,C'-dimethyl) or trimethylene derivatives] as series A, B, and C, each series having three species corresponding to the three available metal oxidation states. The series A compounds have the slipped 1,2-1',2' structure depicted in Figure 34; the series B species have a 1,2-1',7' configuration in which one cage carbon in one of the carborane ligands has moved to the 7 position; and the series C isomers have a 1,7-1',7' arrangement in which one cage carbon in each ligand has moved to the seven-vertex in the icosahedron (this geometry gives rise to d,l and meso forms). The interrelationship between these species is complex and involves both redox reactions and thermal rearrangements of the polyhedron. Detailed discussions of these findings and their many ramifications have been given elsewhere[161,212] and will not be reiterated here. Scheme 2 summarizes the sequence of reactions involving the C,C'-dimethyl nickel species which is prepared from nickel acetylacetonate and $1,2\text{-}(CH_3)_2\text{-}1,2\text{-}C_2B_9H_9^{2-}$ ion.[92] In this scheme $[Ni^{II} - A]^{2-}$ stands for $[1,2\text{-}(CH_3)_2\text{-}1,2\text{-}C_2B_9H_9]_2Ni^{2-}$, $[Ni^{II} - B]^{2-}$ represents $[1,2\text{-}(CH_3)_2\text{-}1,2\text{-}C_2B_9H_9]Ni[1,7\text{-}(CH_3)_2\text{-}1,7\text{-}C_2B_9H_9]^{2-}$, and $[Ni^{II} - C]^{2-}$ indicates $[1,7\text{-}(CH_3)_2\text{-}1,7\text{-}C_2B_9H_9]_2Ni^{2-}$; the other bracketed symbols represent analogous nickel(III) and (IV) species. The redox voltages given in this scheme, taken from a recent review,[92] differ slightly from those presented earlier,[212] which were obtained in ether solution. The structures of the various isomers were assigned from nmr and other spectroscopic data, from an X-ray study[23] of $[1,2\text{-}(CH_3)_2\text{-}1,2\text{-}C_2B_9H_9]Ni[1,7\text{-}(CH_3)_2\text{-}1,7\text{-}C_2B_9H_9]^0$ ($[Ni^{IV} - B]$), and from an investigation of the rearrangement and electrochemistry of the analogous μ-trimethylene derivatives (Figure 36).[161] In these compounds, the framework carbon atoms in each ligand are held adjacent to each other by an external $-(CH_2)_3-$ chain; since the behavior of these nickel metallocarboranes is completely analogous to that of the C,C'-dimethyl derivatives, it is assumed that the cage carbons remain adjacent in the thermal rearrange-

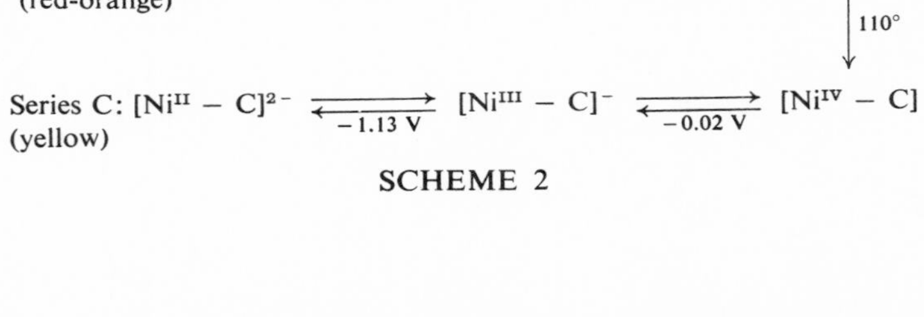

$$Ni(CH_3COCHCOCH_3)_2 + (CH_3)_2C_2B_9H_9^{2-} \xrightarrow{\text{benzene-ether}}$$

Series A: $[Ni^{II} - A]^{2-}$ $\underset{-0.51 \text{ V}}{\rightleftharpoons}$ $[Ni^{III} - A]^-$ $\underset{+0.50 \text{ V}}{\rightleftharpoons}$ $[Ni^{IV} - A]$
(yellow)
                                        $200°$ ↓ $[O]$ $0°$ ↓

Series B: $[Ni^{II} - B]^{2-}$ $\underset{-0.92 \text{ V}}{\rightleftharpoons}$ $[Ni^{III} - B]^-$ $\underset{+0.26 \text{ V}}{\rightleftharpoons}$ $[Ni^{IV} - B]$
(red-orange)
                                                                              $110°$ ↓

Series C: $[Ni^{II} - C]^{2-}$ $\underset{-1.13 \text{ V}}{\rightleftharpoons}$ $[Ni^{III} - C]^-$ $\underset{-0.02 \text{ V}}{\rightleftharpoons}$ $[Ni^{IV} - C]$
(yellow)

SCHEME 2

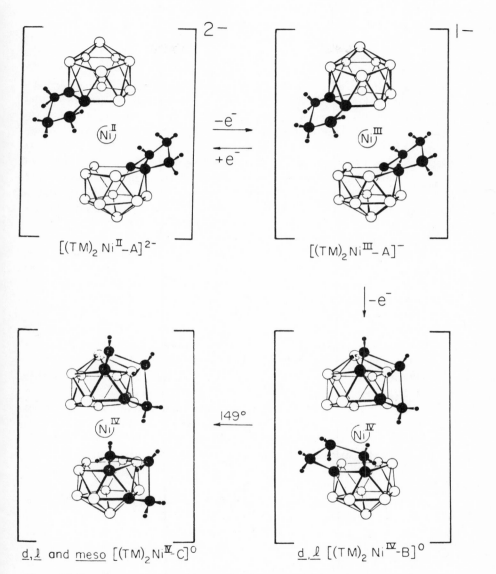

Figure 36. Reaction sequence and interconversions of bis(1,2-μ-trimethylene) derivatives (TM) of the $(1,2\text{-}C_2B_9H_{11})_2Ni^q$ system.[161] $\bigcirc$, BH; $\bullet$, C; $\bullet$, H. (From *J. Amer. Chem. Soc.*, **94**, 4882 (1972). Copyright by the American Chemical Society.)

ment of the latter species also.[212] This conclusion is supported by the X-ray investigation of [Ni^{IV} — B] mentioned earlier.

The rearrangement sequence

$$[Ni^{IV} - A] \xrightarrow{\Delta} (Ni^{IV} - B] \xrightarrow{\Delta} [Ni^{IV} - C]$$

has been rationalized on the basis of severe steric interaction between the four methyl groups in the A species, which is relieved somewhat in the B and C structures. These rearrangements occur with much greater facility than do those of the isoelectronic (d^6) iron(II) and cobalt(III) species, which require temperatures of 400–600°.[117] The unsubstituted orange $(1,2\text{-}C_2B_9H_{11})_2Ni^{IV}$ (Series A) species rearranges at 360–400° in the vapor phase to a mixture of yellow isomers which were tentatively identified from electrochemical data as the corresponding B and C series compounds. The higher temperatures required for the isomerization of the unsubstituted species have been rationalized on the basis of lesser steric requirements compared to the bis(C,C'-dimethylcarboranyl) compounds.[212]

The neutral $(1,2\text{-}C_2B_9H_{11})_2Ni$ species and its C-substituted derivatives are Lewis acids which form adducts with a number of "soft" Lewis bases such as halide ions, naphthalene, phenanthrene, pyrene, thiocyanate, and others. The bonding between the metallocarborane cage and the Lewis-base ligands has been proposed to be primarily of the dipole-induced dipole type, which is consistent with the large dipole moment of 6.16 D measured in cyclohexane.[212] Not unexpectedly, the larger, more polarizable base donors tend to produce the most stable adducts.

Treatment of the nickel(IV) metallocarboranes with "hard" Lewis bases such as hydroxide ion or amines effects a reduction to the corresponding nickel(III) anions; however, detailed study of these reactions has disclosed that the process actually involves a partial degradation of the metallocarborane, yielding fragments which reduce the remainder of the nickel(IV) compound to the nickel(III) species.[212] Similar reactions occur slowly in alcohols or in polar solvents containing traces of water, but in acidic ethanol–water solution, $(1,2\text{-}C_2B_9H_{11})_2Ni$ is stable indefinitely.

The species $(1,2\text{-}C_2B_9H_{11})_2Ni$ is strongly acidic with respect to lability of the C-H protons, reflecting high electron deficiency at the carbon atoms. This observation can be correlated with the unusually high +4 oxidation state of the metal atom, which tends to draw electron density away from the carborane ligands. (A further indication[161] of the same general effect may be the short metal–ligand bonding distance of 1.47 Å in $(1,2\text{-}C_2B_9H_{11})_2Ni^{177}$ (Figure 35). Consistent with these findings is the demonstration, via an ^{19}F nmr investigation[1] of m- and p-fluorophenyl derivatives of $(1,2\text{-}C_2B_9H_{11})_2Ni$, that this system exhibits a very strong inductive electron-withdrawing (-I) effect upon substituents bonded at carbon, comparable to -CN or -F, and much greater than that of the $-C_2B_{10}H_{11}$ icosahedral moiety. In comparison, the isoelectronic iron(II) and cobalt(III) analogs showed a large $+I$ and a nearly zero inductive effect, respectively, which are accountable in terms of the lower formal charges on the metal in these species (see Section IV.D.3). It should be emphasized that these inductive effects refer to substitution at cage carbon locations only, and quantitative measurements at other locations on the metallocarborane polyhedra are not available.

4. Thirteen-Atom Cages

A bis(carboranyl) nickel species, $(C_2B_{10}H_{12})_2Ni^{2-}$, containing formal nickel(II), has been prepared from nickel acetylacetonate and the sodium salt of the $C_2B_{10}H_{12}^{2-}$ ion in THF at room temperature.[36] The nickel atom is assumed to reside in a 13-vertex polyhedron analogous to that in $(\eta-C_5H_5)-CoC_2B_{10}H_{12}$ (Figure 32), except that in the present case two polyhedra are fused at the metal atom. Cyclic voltammetry measurements on the nickel ion indicated a two-electron oxidation to the corresponding nickel(IV) compound, $(C_2B_{10}H_{12})_2Ni^\circ$; this material is also obtainable by reaction of $(C_2B_{10}H_{12})_2-Ni^{2-}$ with iodine. The nickel(II) ion, like some other nickel(II) metallocarboranes discussed elsewhere in this review, is diamagnetic but displays a ^{11}B nmr spectrum containing rounded peaks suggestive of the presence of an equilibrium in solution involving a paramagnetic spin state.[36] However, Zakharkin[239] and co-workers have reported data on the oxidation of this complex to $C_2B_{10}H_{12}$ species which indicate that the $(C_2B_{10}H_{12})_2Ni^{2-}$ originally reported[36] was a mixture of isomers (Section IV.F.6).

The treatment of the tetracarbon carborane$(CH_3)_4C_4B_8H_8$[139] with organonickel reagents, following reduction by sodium naphthalide, has given a novel 13-vertex four-carbon nickelacarborane system, $[(C_6H_5)_2PCH_2]_2Ni(CH_3)_4-C_4B_7H_7$, three isomers of which have been isolated and characterized.[136] From electron-counting arguments (Section II.B), it can be predicted that these species will have open-cage structures, but X-ray investigations have not been conducted as yet.

Experimental

Preparation of $[(C_2H_5)_4N]_2[(C_2B_{10}H_{12})_2Ni^{II}]$.[36] To a 25 mmol solution of $Na_2[C_2B_{10}H_{12}]$ in 100 ml of THF was added 3.2 g (12.5 mmol) of nickel(II) acetylacetonate. The mixture was stirred at room temperature for 12 hr and then filtered through Celite. The solvent was removed from the filtrate by rotary evaporation using a water aspirator. The resulting dark-green oil was redissolved in a minimum amount of ethanol and added to an ethanol solution of excess tetraethylammonium bromide. The precipitate was redissolved in

20% acetonitrile in dichloromethane. The addition of hexane, followed by slow cooling to 0°, resulted in the formation of 2.4 g (4.0 mmol, 32%) of orange-brown crystals of $[(C_2H_5)_4N]_2[(C_2B_{10}H_{12})_2Ni^{II}]$.

I. Palladium and Platinum Compounds

The first metallocarboranes of palladium and platinum to be prepared were of the icosahedral type derived from the 1,2- or 1,7-$C_2B_9H_{11}^{2-}$ ligand.[211] Recently, however, Stone and co-workers have synthesized a number of smaller systems and determined the structures of several species via crystallographic studies.[2,69,70] The reaction of organometallic reagents with *closo*-carboranes in hydrocarbon solvents effects the insertion of one or two metal atoms into the cage, in a manner similar to the vapor-phase insertions of iron, cobalt, and nickel conducted in our laboratory[135,151,189] and described previously in detail.[151] The treatment of *closo*-1,6-$C_2B_4H_6$ with $[(CH_3)_3P]_2$-Pt(*trans*-stilbene) produces *closo*-4,5,6-$[(CH_3)_3P]_2Pt(CH_3)_2C_2B_6H_6$ (a tricapped trigonal prism in which the three capping vertices are occupied by the metal and two carbon atoms), and an isomer of this species which has an open structure approximating a distorted tricapped trigonal prism.[70] In both molecules the metal atom is coordinated to only four cage atoms; the *closo* species provides a further example of metal-cyclobutadiene-type bonding as was found earlier in 1-$(C_5H_5)CoB_4H_8$[147] and several metallocarboranes,[42,151,181] and more recently in 1-$(CO)_3FeB_4H_8$[73] as well.

A similar metal-insertion reaction[2] conducted at room temperature on *closo*-2,4-$C_2B_5H_7$ has given 1,7,4,6-$[(C_2H_5)_3P]_4Pt_2C_2B_5H_7$, a tricapped trigonal prism in which the platinum atoms occupy adjacent vertices in separate triangular faces of the prism, and the carbon atoms reside in two of the capping locations. It is intriguing that this structure differs from 1,7,5,6- and 1,8,5,6-$(\eta$-$C_5H_5)_2Co_2C_2B_5H_7$ isomers (Section IV.F.3) which were structurally characterized from ^{11}B and ^{1}H nmr data[146,151] and confirmed by crystal structure analyses,[82] in view of the fact that no other isomers of the cobalt system have been found under any conditions studied.[146,148,151] Furthermore, the placement of metal atoms in the diplatinum species is considerably different from the monoplatinum nine-vertex system described above. It is becoming increasingly apparent that metallocarborane geometry in many instances reflects subtle factors which may include intramolecular steric interactions,[148] metal-orbital symmetry, metal-atom size, and mechanism of insertion into the cage. Thermodynamic arguments alone, though useful, are clearly inadequate for dealing with cage structures prepared under low-energy conditions.

closo-1,6-$C_2B_8H_{10}$ reacts with $[(CH_3)_2P]_2Pt(1,5-C_8H_{12})]$ to generate an 11-vertex *nido* system, $\mu(4,8)$-$\{[(CH_3)_3P]_2Pt\}$-8,7,10-$[(CH_3)_3P]_2PtC_2B_8H_{10}$

which contains one metal atom in the open face of the cage and one metal in an *exo*-polyhedral Pt–Pt–B bridge.[2] On treatment with charcoal, this compound loses the bridging $[(CH_3)_3P]_2Pt$ group, forming the parent species, *nido*-8,7,10$[(CH_3)_3P]_2PtC_2B_8H_{10}$.

In an extension of this work, the reaction of *closo*-1,6-$(CH_3)_2C_2B_7H_7$ and $Pt[(CH_3)_2C_6H_5P]_4$ has been reported to yield a 10-vertex *nido* system[69] $[(C_2H_5)_3P]_2Pt(CH_3)_2C_2B_7H_7$, which has been shown by X-ray investigation to consist of a distorted bicapped square antiprism with carbon in one apex; the metal and carbon atoms are in nonapical locations, but are separated to create a four-sided open face. In this manner both the platinum and the carbon atom achieve low coordination, which presumably provides the driving force for distortion of the polyhedron.[219]

A few 12-vertex platinum metallocarboranes have been described, and all of those known contain a single $C_2B_9H_{11}^{2-}$ ligand. Attempts to prepare the $(1,2-C_2B_9H_{11})_2Pt^q$ system ($q = 2$-, 1-, or 0) from several different starting reagents were unsuccessful, but cyclooctadienedichloroplatinum(II) and the $1,2-C_2B_9H_{11}^{2-}$ ion react to form $3,1,2-(1,5-C_8H_{12})PtC_2B_9H_{11}$, a stable yellow solid.[212] The same product is reportedly obtained from the reaction of $Tl^+[TlC_2B_9H_{11}]^-$ with $(1,5-C_8H_{12})PtCl_2$.[191] The reaction[69,71,191] of *closo*-2,3-$(CH_3)_2$-2,3-$C_2B_9H_9$ with platinum organometallics yields $1,7-(CH_3)_2$-3,1,7-$[(C_6H_5)(CH_3)_2P]_2PtC_2B_9H_9$ and $1,7-(CH_3)_2$-3,1,7-$[(C_2H_5)_3P]_2PtC_2B_9H_9$, the former species having been structurally confirmed in an X-ray study.[218] (A numbering system different from that used here was employed in Refs. 69, 71, 191, and 218).

The treatment of $(cis-PR_3)_2PtCl_2$ reagents with $[R'R''C_2B_9H_9]^{2-}$ anions generated *in situ* by degradation of *closo*-$R'R''C_2B_{10}H_{10}$ carboranes, yields platinacarboranes of the type $3,1,2-(PR_3)_2PtC_2B_9H_{11}$ where R, R', and R'' are small hydrocarbon groups.[130] The compounds formed are listed in Table I.

The preparation of mixed-metal cobalt–platinum and nickel–platinum metallocarboranes was described in Section IV.F.4.

At this writing, all of the known palladacarborane species are icosahedral 12-vertex systems. The reaction of a large excess of $1,2-C_2B_9H_{11}^{2-}$ with palladium salts generates a diamagnetic $(1,2-C_2B_9H_{11})_2Pd^{2-}$ species which is oxidized by iodine to the air-stable, diamagnetic $(1,2-C_2B_9H_{11})_2Pd$. The reaction of the latter material with the $(1,2-C_2B_9H_{11})_2Pd^{2-}$ ion produces a formal palladium(III) species, $(1,2-C_2B_9H_{11})_2Pd^-$, containing one unpaired electron.

The red-brown $(1,2-C_2B_9H_{11})_2Pd^{2-}$ ion, which is analogous to the series A nickel species discussed above, undergoes air oxidation at room temperature to yield the yellow $(1,7-C_2B_9H_{11})_2Pd^{IV}$ species (series C), evidently bypassing the series-B compound. In fact, no series-B species of palladium has been isolated.[212]

The neutral $(1,7\text{-}C_2B_9H_{11})_2Pd^{IV}$ exhibits a chemistry similar to that of the corresponding nickel compound, forming adducts with soft Lewis bases and undergoing degradation to palladium metal in the presence of hydroxide ion. Although no detailed X-ray study of a palladium carborane has been published, the bis(carboranyl)palladium species are assumed from spectral and chemical similarities[92,212] to be isostructural with the corresponding nickel complexes; i.e., $(1,2\text{-}C_2B_9H_{11})_2Pd^{2-}$ would be of the slipped, centro-symmetric (series-A) type, while $(1,2\text{-}C_2B_9H_{11})_2Pd$ would be nonslipped and eclipsed (Series C).

Treatment of the $1,2\text{-}C_2B_9H_{11}^{2-}$ ion with π-tetraphenylcyclobutadienyl-palladium(II) chloride yields $3,1,2\text{-}[(C_6H_5)_4C_4]PdC_2B_9H_{11}$.[100,214]

The C,C'-dimethyl derivative of this material, prepared from $(CH_3)_2$-$C_2B_9H_9{}^{2-}$, has been structurally characterized by X-ray diffraction (Ref. 100, footnote 28) and contains a dimethylcyclobutadiene ring symmetrically π-bonded to the metal. These compounds are analogous to the known $[(\eta\text{-}C_6H_5)_4C_4]Pd[\eta\text{-}C_5H_5]$ sandwich species, and are apparently the only cyclobutadienyl metallocarboranes to have been prepared. A $1,2\text{-}(CH_3)_2\text{-}3,1,$ $2\text{-}(C_4H_9NC)_2PdC_2B_9H_9$ icosahedral system has been prepared from the direct insertion of $Pd(C_4H_9NC)_2$ into the 11-vertex $closo$-carborane $2,3\text{-}$ $(CH_3)_2\text{-}2,3\text{-}C_2B_9H_9$ at room temperature or below in toluene.[71,191]

Experimental

Preparation of $[\pi\text{-}(C_6H_5)_4C_4]Pd[1,2\text{-}C_2B_9H_9(CH_3)_2]$.[100] A THF solution containing 3.94 mmol of $Na_2[1,2\text{-}C_2B_9H_9(CH_3)_2]$ was added slowly under nitrogen to a stirred suspension of 2.10 g (3.94 mmol) of $[\pi\text{-}C_4(C_6H_5)_4]PdCl_2$ dimer in 100 ml THF. The reaction mixture was stirred under nitrogen at room temperature for 3 hr and filtered, the solvent removed *in vacuo,* and the residue extracted with benzene. The benzene extract was chromatographed on a silica-gel column. Elution with 60:40 benzene–hexane gave a red material as the major fraction. The yield of red $[\pi\text{-}C_4(C_6H_5)_4]Pd[1,2\text{-}C_2B_9H_9(CH_3)_2]$, after crystallization from methylene chloride–hexane, was 0.16 g (0.254 mmol, 6.5%).

J. Copper Compounds

The only copper metallocarboranes that have been described are the $(1,2\text{-}C_2B_9H_{11})_2Cu^{2-}$ and $(1,2\text{-}C_2B_9H_{11})_2Cu^-$ ions, in which the metal has a formal oxidation state of $+2$ and $+3$, respectively. The copper(II) species is produced by treatment of $1,2\text{-}C_2B_9H_{12}^-$ salts with Cu^{2+} ion in cold concentrated NaOH solution.[9,100] The $(1,2\text{-}C_2B_9H_{11})_2Cu^{2-}$ ion has been structurally characterized from an X-ray study of its tetraethylammonium salt, and is found to have a slipped structure analogous to those of other icosahedral metallocarboranes of d^8 and d^9 metals[227] (see Section IV.H.3). This deep-blue ion is paramagnetic with one unpaired electron and is stable over indefinite periods in cold aqueous basic solution.[211] However, in diethyl ether, aceto-nitrile, or acetone solution it is oxidized by air to $(1,2\text{-}C_2B_9H_{11})_2Cu^-$, a red diamagnetic copper(III) species. This complex is also of the slipped-structure type.[228] Cyclic voltammetry measurements also indicate the existence of a copper(I) species, $Cu(1,2\text{-}C_2B_9H_{11})_2^{3-}$, although this ion has not been isolated. (A B-substituted pyridine derivative of a copper(I) species, $(C_5H_5N\text{-}1,2\text{-}C_2B_9H_{10})_2Cu^-$, has been reported without supporting data.[9]) It is significant that the copper(III) metallocarborane is the most stable in the series,[211] underlining the ability of the $1,2\text{-}C_2B_9H_{11}^{2-}$ ligand to stabilize unusually high metal-oxidation states; copper(III) compounds are quite rare, and even "cupricene," bis(cyclopentadienyl)copper(II), has not been isolated.

K. Silver and Gold Compounds

The only report of a silver metallocarborane in the literature is a state-ment, unaccompanied by data, that the reaction of $C_2B_9H_{12}^-$ with $AgNO_3$ in aqueous base generates the $(C_2B_9H_{11})_2Ag^{3-}$ ion which can be precipitated as the black tetraethylammonium salt.[9] The known gold species are limited to the icosahedral $(1,2\text{-}C_2B_9H_{11})_2Au^{2-}$ and $(1,2\text{-}C_2B_9H_{11})_2Au^-$ ions con-taining formal gold(II) and (III), respectively. The former complex is ob-tained by treatment of $C_2B_9H_{11}^{2-}$ with gold(III) chloride, and the blue-green paramagnetic $(1,2\text{-}C_2B_9H_{11})_2Au^{2-}$ species can be reversibly oxidized to the red diamagnetic gold(III) ion.[211] As expected from the formal d^9 electron configuration on the metal (see above), the $(1,2\text{-}C_2B_9H_{11})_2Au^-$ system is slipped[228]: The corresponding gold(I) system has not been isolated, but as

$$2\ C_2B_9H_{11}^{2-} + AuCl_3 \longrightarrow (1,2\text{-}C_2B_9H_{11})_2Au^{2-} \underset{Na/Hg}{\overset{H^+,\ H_2O_2}{\rightleftharpoons}} (1,2\text{-}C_2B_9H_{11})_2Au^-$$

in the case of copper, its existence is indicated from cyclic voltammetry studies.[211]

V. MONOCARBON TRANSITION-METAL METALLOCARBORANES

A. Introduction

With the exception of the $CoCB_7$ and $CoNiCB_7$ cages, obtained by base degradation of the 13-atom $(\eta\text{-}C_5H_5)CoC_2B_{10}H_{12}$ system (Section IV.F.3), and the phospha- and arsametallocarboranes (Section VI), all other known monocarbon metallocarboranes are icosahedral MCB_{10} cage systems. Although these icosahedral species are isoelectronic analogs of the well-known complexes of $C_2B_9H_{11}^{2-}$, there is no known synthetic method for their interconversion, and the preparative routes to the two classes are entirely different.

B. Synthesis via Monocarbon Carborane Anions

Metallocarboranes containing the formal $CB_{10}H_{11}^{3-}$ ligand (an analog of $C_2B_9H_{11}^{2-}$) are obtained[108,121] by reactions of metal reagents with salts of the open-cage anions $CB_{10}H_{11}^{3-}$ or $CB_{10}H_{13}^{-}$, the latter species having a hydrogen bridge on the open face analogous to $C_2B_9H_{12}^{-}$. Complexes of the analogous $NH_3CB_{10}H_{10}^{2-}$ ligand (formally related to $CB_{10}H_{11}^{3-}$ by replacement of H^- with NH_3) are prepared from the neutral compound NH_3CB_{10}-H_{12}, itself an analog of $CB_{10}H_{13}^{-}$. All of these monocarbon anions are derived from C-aminocarboranes which in turn are prepared from decaborane(14) cyano,[119,120,122] or alkyl isocyanide[109,122] derivatives.

$$B_{10}H_{14} + 2\ NaCN \xrightarrow[-HCN]{H_2O} Na_2B_{10}H_{13}CN \xrightarrow[\text{(ion exchange)}]{H^+} B_{10}H_{12}CNH_3$$

$$B_{10}H_{14} + RNC \longrightarrow B_{10}H_{12}CNH_2R$$
$$(R = CH_3,\ C_2H_5,\ n\text{-}C_3H_7,\ t\text{-}C_4H_9)$$

Other reported routes[184] to $B_{10}H_{12}CNH_3$ include the reaction of the $B_{10}H_{13}CN^{2-}$ ion with trimethylsilyl or trimethyltin chloride under basic conditions; the treatment of this same ion with alkyl iodides yields N-trialkyl derivatives, $B_{10}H_{12}CNR_3$. The $B_{10}H_{12}CNR_3$ aminocarboranes ($R = H$ or alkyl) react directly with metal ions to form *closo*-metallocarboranes; alternatively, they can be converted to N-methyl derivatives and then deaminated with sodium or sodium hydride to produce the $CB_{10}H_{13}^{-}$ ion.[110,119,120] If the deamination is carried out in dry refluxing THF, an adduct of the $CB_{10}H_{11}^{3-}$ salt is formed, which on hydrolysis yields CB_{10}-H_{13}^{-} [213]:

$$B_{10}H_{12}CNH_3 \xrightarrow[NaOH]{(CH_3)_2SO_4} B_{10}H_{12}CN(CH_3)_3 \xrightarrow[THF]{Na}$$

$$Na_3CB_{10}H_{11}(C_4H_8O)_2 \xrightarrow{H_2O} CB_{10}H_{13}^{-}$$

The *nido*-carborane species $CB_{10}H_{11}^{3-}$, $CB_{10}H_{13}^{-}$, and $B_{10}H_{12}CNH_3$ and their substituted derivatives, form metallocarboranes on treatment with suitable reagents.[108,119,121,123,168,181]

A recent report[17] describes the synthesis of several nickel-group 12-vertex monocarbon metallocarboranes by direct metal insertion into the *closo*-carborane species $CB_{10}H_{11}^{-}$ (not to be confused with the open-cage anion $CB_{10}H_{11}^{3-}$ described above) and *closo*-$(CH_3)_3NCB_{10}H_{10}$. Treatment with organometallic reagents such as $Ni(cod)(t-C_4H_9NC)_2$ (cod = cyclo-octadiene), $Pd(t-C_4H_9NC)_2$, and $Pt(trans$-stilbene$)[(C_2H_5)_3P]_2$ affords species of the type $3,1-L_2MCB_{10}H_{11}$ and $1-(CH_3)_3N-3,1-L_2MCB_{10}H_{10}$ (M = Ni or Pd, L = t-C_4H_9NC; M = Pt, L = $(C_2H_5)_3P$). An X-ray investigation has shown that the palladium compound is a distorted icosahedron with a long (2.60 Å) Pd–C bond.[17]

The syntheses in Scheme 3 are illustrative; a complete list of the reported compounds is given in Table III.

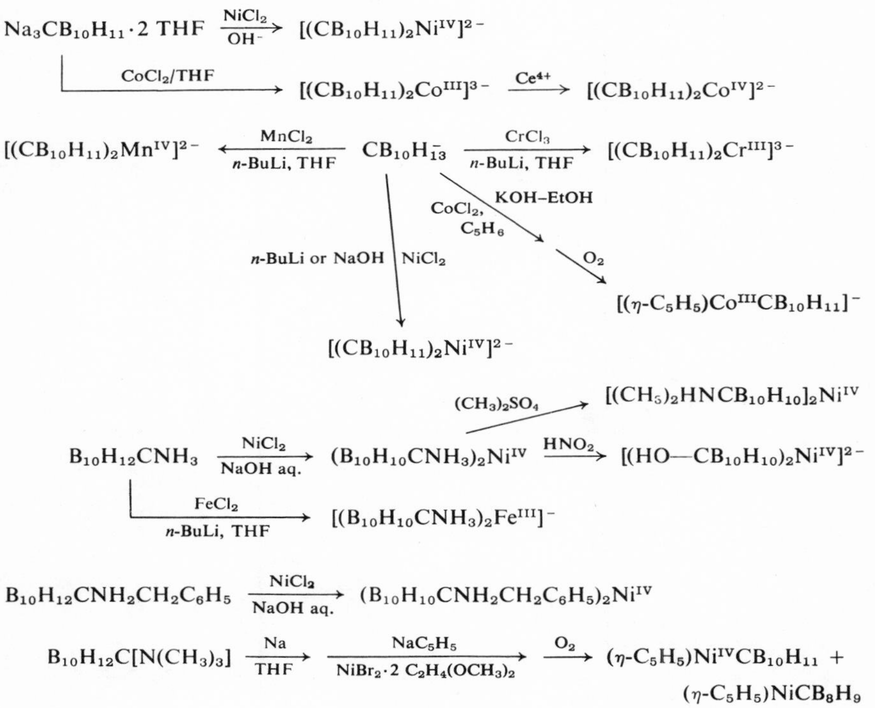

$$Na_3CB_{10}H_{11} \cdot 2\ THF \xrightarrow[OH^-]{NiCl_2} [(CB_{10}H_{11})_2Ni^{IV}]^{2-}$$

$$\xrightarrow{CoCl_2/THF} [(CB_{10}H_{11})_2Co^{III}]^{3-} \xrightarrow{Ce^{4+}} [(CB_{10}H_{11})_2Co^{IV}]^{2-}$$

$$[(CB_{10}H_{11})_2Mn^{IV}]^{2-} \xleftarrow[n\text{-BuLi, THF}]{MnCl_2} CB_{10}H_{13}^{-} \xrightarrow[n\text{-BuLi, THF}]{CrCl_3} [(CB_{10}H_{11})_2Cr^{III}]^{3-}$$

$$CB_{10}H_{13}^{-} \xrightarrow[C_5H_6]{CoCl_2,} \xrightarrow[O_2]{KOH-EtOH} [(\eta\text{-}C_5H_5)Co^{III}CB_{10}H_{11}]^{-}$$

n-BuLi or NaOH | NiCl₂

$$[(CB_{10}H_{11})_2Ni^{IV}]^{2-}$$

$$B_{10}H_{12}CNH_3 \xrightarrow[NaOH\ aq.]{NiCl_2} (B_{10}H_{10}CNH_3)_2Ni^{IV} \xrightarrow{(CH_3)_2SO_4} [(CH_3)_2HNCB_{10}H_{10}]_2Ni^{IV}$$

$$(B_{10}H_{10}CNH_3)_2Ni^{IV} \xrightarrow{HNO_2} [(HO{-}CB_{10}H_{10})_2Ni^{IV}]^{2-}$$

$$\xrightarrow[n\text{-BuLi, THF}]{FeCl_2} [(B_{10}H_{10}CNH_3)_2Fe^{III}]^{-}$$

$$B_{10}H_{12}CNH_2CH_2C_6H_5 \xrightarrow[NaOH\ aq.]{NiCl_2} (B_{10}H_{10}CNH_2CH_2C_6H_5)_2Ni^{IV}$$

$$B_{10}H_{12}C[N(CH_3)_3] \xrightarrow[THF]{Na} \xrightarrow[NiBr_2 \cdot 2\ C_2H_4(OCH_3)_2]{NaC_5H_5} \xrightarrow{O_2} (\eta\text{-}C_5H_5)Ni^{IV}CB_{10}H_{11} + (\eta\text{-}C_5H_5)NiCB_8H_9$$

SCHEME 3

Experimental

Preparation of Cs_2 $[(CB_{10}H_{11})_2Ni^{IV}]$.[123] To a 250-ml, three-necked-flask fitted with a reflux condenser, magnetic stirring bar, and air bubbler, was added a solution of 10.84 g of NaOH in 100 ml of water, followed by 5.87 g (22.1 mmol) of solid $CsB_{10}H_{12}CH$. An oil bath, maintained at 40°, was placed around the flask. Stirring was begun, and 3.2 g (13.5 mmol) of

TABLE III

Monocarbon Transition-Metal Metallocarboranes

Compound[a]	Color	Mp, °C	Other data[b]	References
Chromium and molybdenum				
12-Atom cages				
Cs$_3$[(CB$_{10}$H$_{11}$)$_2$Cr]·H$_2$O	Red		IR, E	119, 121
[(C$_4$H$_9$)$_4$N]$_2$[B$_{10}$H$_{10}$COMoCO(CO)$_3$]	Yellow		B, IR, X	213
[(CH$_3$)$_4$N][(B$_{10}$H$_{10}$COH)Mo(CO)$_4$]			B, IR	213
Manganese				
12-Atom cages				
[(CH$_3$)$_3$NH]$_2$[(CB$_{10}$H$_{11}$)$_2$Mn]	Black		E, IR	119, 121
[(CH$_3$)$_4$N]$_2$[(H$_2$NCB$_{10}$H$_{10}$)$_2$Mn]			E	119
[(CO)$_3$MnCB$_{10}$H$_{11}$]$^{2-}$				108
Iron				
12-Atom cages				
Cs$_3$[(CB$_{10}$H$_{11}$)$_2$Fe]·H$_2$O	Black		IR, E, EC	119
[(CH$_3$)$_4$N]$_3$[(CB$_{10}$H$_{11}$)$_2$Fe]	Red			108
[(CH$_3$)$_4$N][(H$_3$NCB$_{10}$H$_{10}$)$_2$Fe]	Green		E	119, 121
Cobalt				
9-Atom cages				
Cs[1,5-(C$_5$H$_5$)CoCB$_7$H$_8$]	Black		B, H, E	39, 41
			X	15
[(CH$_3$)$_4$N][1,5-(C$_5$H$_5$)CoCB$_7$H$_8$]	Black		B, H, IR, E, EC	39
[(CH$_3$)$_4$N][5-CH$_3$-1,5-(C$_5$H$_5$)CoCB$_7$H$_7$]	Black		B, H, IR, EC	39, 41
1,5-(C$_5$H$_5$)Co^{IV}CB$_7$H$_8$	Very dark green		IR, MS	39
5-CH$_3$-1,5-(C$_5$H$_5$)Co^{IV}CB$_7$H$_7$			IR, MS	39
10-Atom cages				
1,6,10-(C$_5$H$_5$)$_2$Ni^{IV}Co^{III}CB$_7$H$_8$	Green	171	B, H, MS, IR, E, EC	178, 180
2,4,10-(C$_5$H$_5$)$_2$Ni^{IV}Co^{III}CB$_7$H$_8$	Dark Green	199	B, H, MS, IR, E, EC	178, 180
1,2,10-(C$_5$H$_5$)$_2$Ni^{IV}Co^{III}CB$_7$H$_8$	Red-black	201	B, H, MS, IR, E, EC	180
2,3,10-(C$_5$H$_5$)$_2$Ni^{IV}Co^{III}CB$_7$H$_8$	Brown	273	B, H, MS, IR, E, EC	180
12-Atom cages				
[(CH$_3$)$_4$N][(C$_5$H$_5$)Co^{III}(CB$_{10}$H$_{11}$)]	Yellow		B, H, IR, E	168
[(CH$_3$)$_4$N]$_3$[(CB$_{10}$H$_{11}$)$_2$Co^{III}]	Yellow			108, 121
Cs$_3$[(CB$_{10}$H$_{11}$)$_2$Co^{III}]·H$_2$O	Yellow		IR, E	119

Compound	Color	M.p.	Methods	Ref.
[(C₆H₃₄...)₂[(CB₁₀H₁₁)₂Co^III)]	Black		H, IR, E	113
Cs₃[(CB₁₀H₃,₅Cl₇,₅)₂Co^III]·H₂O^c	Orange		E	119
[(CH₃)₄N]₂[(H₃NCB₁₀H₁₀)Co^III(H₂NCB₁₀H₁₀)]	Yellow		E	119, 121
[(CH₃)₄N][(H₃NCB₁₀H₁₀)₂Co^III]	Orange		E	119
[(CH₃)₄N][(CB₁₀H₁₀NH₂C₂H₅)₂Co^III]	Orange			108
(C₁₀H₈)Co^IIICB₁₀H₁₁		250–260 (dec.)	B, H, IR, MS, E, EC	180
Nickel				
10-Atom cages^d				
1,10-(C₅H₅)NiCB₈H₉	Yellow	133–135	B, H, IR, MS, E, EC	181
12-Atom cages				
Cs₂[(CB₁₀H₁₁)₂Ni^IV]	Yellow-orange^e		IR, E	108, 121
(H₃NCB₁₀H₁₀)₂Ni^IV	Orange		E	119
			X-ray photoelectron	165
[(CH₃)₂NHCB₁₀H₁₀]₂Ni^IV	Orange		E	119, 121
(C₃H₇NH₂CB₁₀H₁₀)₂Ni^IV	Orange			108
[C₃H₇(CH₃)NHCB₁₀H₁₀]₂Ni^IV	Orange			108
[(C₆H₅)CH₂NH₂CB₁₀H₁₀]₂Ni^IV	Orange		E	119
[(CH₃)₄N]₂[(HOCB₁₀H₁₀)₂Ni^IV]	Yellow-orange		E	119, 121
3,1-(C₅H₅)Ni^IVCB₁₀H₁₁	Orange		B, H, IR, E, MS	168
7,1-(C₅H₅)Ni^IVCB₁₀H₁₁	Yellow-orange		B, H, IR, E, MS	168
12,1-(C₅H₅)Ni^IVCB₁₀H₁₁	Yellow		B, H, IR, E, MS	168
[(CH₃)₄N][3,1-(t-C₄H₉NC)₂NiCB₁₀H₁₁]	Orange		H, IR	17
1-(CH₃)₃N-3,1-(t-C₄H₉)₂NiCB₁₀H₁₀	Orange		B, H, IR	17
Cs[((CH₃)₂NCB₁₀H₁₀)Ni((CH₃)₂NHCB₁₀H₁₀)]	Yellow		B, H, IR	119
Palladium				
[(CH₃)₄N][3,1-(t-C₄H₉NC)₂PdCB₁₀H₁₁]	Yellow		H, IR	17
1-(CH₃)₃N-3,1-(t-C₄H₉NC)₂PdCB₁₀H₁₀	Yellow	255–260 (dec.)	B, H, IR, X	17
1-(CH₃)₃N-3,1-[(C₆H₅)₂PCH₂]₂PdCB₁₀H₁₀	Yellow	294–296		17
Platinum				
[(CH₃)₄N][3,1-[(C₂H₅)₃P]₂PtCB₁₀H₁₁]	Yellow		H, IR	17
1-(CH₃)₃N-3,1-[(C₂H₅)₃P]₂PtCB₁₀H₁₀	Yellow		B, H, IR, ³¹P	17
Copper				
12-Atom cages				
[(CB₁₀H₁₁)₂Cu^III]³⁻				108

^a See footnote a in Table I. ^b See footnote b in Table I. ^c Mixture. ^d See also mixed cobalt–nickel species, above.
^e May be dark in color due presence of Ni(III) salt.
Note: See Table V for germanium monocarbon compounds.

$NiCl_2 \cdot 6H_2O$ in 37 ml of water was added slowly to the flask from a dropping funnel. Air was blown through the bubbler into the reaction mixture and up through the condenser, in order to reduce absorption of CO_2 formed in the next step. After a 20-hr period, the products were transferred to a 500-ml Erlenmeyer flask and neutralized by dropwise addition of concentrated hydrochloric acid. Foaming occurred during this latter step. A saturated solution of cesium chloride was added to the ice-cold reaction mixture until precipitation was complete. The crude $Cs_2[(CB_{10}H_{11})_2Ni^{IV}]$ was isolated by filtration and crystallization from hot water. Second and third fractions were obtained by reducing the volume of solution. The yield from three fractions was 5.00 g (78%) of an orange, crystalline, air-stable solid.

Preparation of $[(CB_{10}H_{11})_2Co^{III}]^{3-}$.[119] Three grams (15.5 mmol) of $[(CH_3)_3NH][CB_{10}H_{13}]$[119] was dissolved in 75 ml of 50% aqueous sodium hydroxide. This solution was added to a solution of $CoCl_2 \cdot 6H_2O$ (12 g, 50 mmol) in 25 ml of water and the resulting mixture was heated 10 min on a steam bath with occasional agitation. Then it was diluted with 100 ml of water and filtered through Celite to obtain a dark-yellow filtrate. The addition of 30 ml of 50% aqueous cesium hydroxide solution to this filtrate precipitated a dingy yellow solid which was recrystallized from water. The yield of $Cs_3[(CB_{10}H_{11})_2Co] \cdot H_2O$ was 3.8 g (66%). The analytical sample was recrystallized a second time.

Preparation of $[(CB_{10}H_{11})_2Co^{IV}]^{2-}$.[119] A solution of $(NH_4)_4Ce(SO_4)_4 \cdot 2H_2O$ (8 g, 12.6 mmol) in water (50 ml) and concentrated sulfuric acid (2 ml) was added to $Cs_3[(CB_{10}H_{11})_2Co] \cdot H_2O$ (3.9 g 5.3 mmol) in water (100 ml). Immediately a blue-green color appeared. The solution was filtered. Tetramethylammonium hydroxide was added but not in sufficient amount to basify the solution. A blue precipitate of $[(CH_3)_4N]_2[(CB_{10}H_{11})_2Co]$ formed. This was removed by filtration; the filtrate (yellow) was treated with more $(NH_4)_4 Ce(SO_4)_4 \cdot 2H_2O$ to precipitate additional product. After dissolving the crude product in acetonitrile, the solution was filtered and concentrated slowly at room temperature. Black crystals of $[(CH_3)_4N]_2[(CB_{10}H_{11})_2Co]$ separated and were dried at room temperature *in vacuo*.

The formation of high metal-oxidation states is a notable feature of these reactions,[119] as shown by the formation of Mn(IV) and Ni(IV) species. Even more unusual is the oxidation of the yellow-orange diamagnetic $[(CB_{10}H_{11})_2Co^{III}]^{3-}$ ion to a black paramagnetic $[(CB_{10}H_{11})_2Co^{IV}]^{2-}$ complex. The stabilization of high metal-oxidation states by the CB_{10} ligands is attributable to the bonding of the metal to a CB_4 face; in contrast, complexes of $C_2B_9H_{11}^{2-}$ ligands in which the metal is bonded to a C_2B_3 face have noticeably less ability to stabilize high oxidation states on the metal.[119]

Presumably, the larger boron orbitals interact more effectively with the metal than do the carbon orbitals, so that withdrawal of electron density from the metal is more effective as carbon is replaced by boron in the bonding face. This trend is evident in the studies of Hawthorne *et al.*[118] (Section IV.F.5.f) who found that the B_5-bonded $C_2B_9H_{11}CoC_5H_5$ isomers are more thermally stable than the B_4C- and B_3C_2- bonded isomers. In line with these observations, the $C_5H_5^-$ ion is far less effective than carborane ligands in stabilizing high metal-oxidation states.

The anionic species $[(\eta\text{-}C_5H_5)Co^{III}CB_{10}H_{11}]^-$ mentioned above has recently been shown to react with sodium naphthalide, forming a neutral η^6-naphthalene complex $(\eta\text{-}C_{10}H_8)Co^{III}CB_{10}H_{11}$.[180]

C. Synthesis via Carbonyl Insertion into $B_{10}H_{13}^-$

A different synthetic route to MCB_{10} metallocarboranes utilizes the insertion of a carbonyl group together with a metal atom into a B_{10} cage system, via the reaction of the $B_{10}H_{13}^-$ ion with hexacarbonyls of chromium, molybdenum, or tungsten.[213]

$$NaB_{10}H_{13} \xrightarrow[\text{THF, } h\nu]{M(CO)_6} (B_{10}H_{10}COH)M(CO)_4^- \xrightarrow{NaH} [B_{10}H_{10}\overline{COMCO}(CO)_3]^{2-}$$

$$\text{I, 50\%} \qquad\qquad \text{II, 100\%}$$

$$M = Cr, Mo, W$$

The type I species found initially have been spectroscopically characterized as *closo*-metallocarboranes containing a metal with four terminal CO groups, and the type II products are *closo*-cages containing a bridging CO_2 group between the metal and the lone framework carbon atom. The structure of the molybdenum species of type II has been established in an X-ray study and is shown in Figure 37. In both I and II, a carbonyl group has been fused into the metalloboron cage system to generate a monometallocarborane. The conversion of I to II is proposed[213] to occur via intramolecular attack of an alkoxide ion, which is produced by the action of sodium hydride upon a CO ligand on the metal.

The I → II conversion is reversible, as shown by the quantitative production of I on treatment of II with air-free aqueous acid. In this reaction the C–O ether linkage is cleaved and the cage-bond oxygen is protonated, leaving a C–OH group and four terminal CO groups on the metal. Both the type I and type II species react irreversibly with aqueous base to remove the cage carbon atom and form air-sensitive metalloboranes having the formula $B_{10}H_{12}M(CO)_4$, which apparently have a *nido* (open-cage) structure[213]:

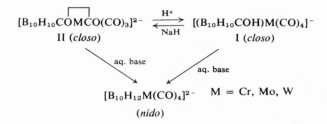

This reaction is the only reported instance in which a metallocarborane is converted to a metalloborane. It is related to the well-known degradation of polyhedral carboranes and metallocarboranes to *nido* species by base attack, but differs in that in all other cases a boron atom, rather than carbon, is eliminated from the cage (in one instance, a B_3C unit is eliminated from a 13-vertex cobaltacarborane system: see Section IV.F.3). No rationale of this distinction has been offered, and a full report on these reactions has not yet appeared.

VI. TRANSITION-METAL PHOSPHA- AND ARSAMETALLOCARBORANES

A. Synthetic Routes

The icosahedral cages $1,2\text{-}PCB_{10}H_{11}$, $1,2\text{-}AsCB_{10}H_{11}$, and $1,2\text{-}SbCB_{10}H_{11}$ are isoelectronic analogs of $C_2B_{10}H_{12}$, and are *formally* derived from the latter by replacement of a CH group with a "bare" phosphorus or arsenic atom. [The actual synthesis, however, involves the reaction of PCl_3,[127,201] $AsCl_3$,[199,201] or SbI_3[199] with $Na_3CB_{10}H_{11}(C_4H_8O)_2$]. The phospha-, arsa-, and stibacarboranes undergo thermal rearrangement to the 1,7 isomers in which the carbon and the heteroatom are nonadjacent[127,200,241]; the arsenic system can be further isomerized to $1,12\text{-}AsCB_{10}H_{11}$,[199,200,241] a sequence analogous to the conversion of $1,2\text{-}C_2B_{10}H_{12}$ to its 1,7 and 1,12 isomers. The similarity to $C_2B_{10}H_{12}$ extends to the reaction of 1,2- or $1,7\text{-}PCB_{10}H_{11}$ with mild Lewis bases such as piperidine, which removes a boron atom and forms the 1,2- or $1,7\text{-}PCB_9H_{11}^-$ ion. These ions are analogous to the $C_2B_9H_{12}^-$ species and are believed to have an icosahedral-fragment structure with a bridge proton on the open PCB_3 face.[201] Treatment of either isomeric PCB_9-H_{11}^- ion with triethylamine or sodium hydride removes the bridging hydrogen to form 1,2- or $1,7\text{-}PCB_9H_{10}^{2-}$, which are analogs of 1,2- and $1,7$-$C_2B_9H_{11}^{2-}$ and like the latter ions are capable of accepting a transition-metal atom in the open face.[202] The arsenic species 1,2- and $1,7\text{-}AsCB_9H_{10}^{2-}$ are prepared in similar fashion and are also reported to undergo metal insertion

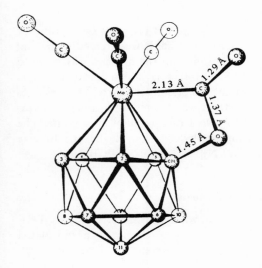

Figure 37. Structure of the $B_{10}H_{10}$-COMoCO(CO)$_3^{2-}$ ion.[213] (From *J. Amer. Chem. Soc.*, **92**, 3473 (1970). Copyright by the American Chemical Society.)

although no data are given.[200] Thus, the preparation and chemistry of the metallocarboranes derived from the phospha- and arsacarboranes generally parallels that based on 1,2- and 1,7-$C_2B_{10}H_{12}$. However, as one would expect from the presence of group V heteroatoms in the cage, there are significant differences between the MPCB$_9$-type systems and their MC$_2$B$_{10}$ analogs which were discussed in Section IV. At this writing, all known metallocarboranes containing phosphorus or arsenic in the cage are 12-atom frameworks having icosahedral geometry.

B. Manganese Compounds

The 1,7-PCB$_9$H$_{10}^{2-}$ ion (numbered 7,9-PCB$_9$H$_{10}^{2-}$ in Ref. 128) reacts with BrMn(CO)$_5$ in refluxing THF to yield the [2,1,7-(CO)$_3$MnPCB$_9$H$_{10}$]$^-$ ion, at the same time releasing two mole equivalents of carbon monoxide. Alternatively, the neutral P-methyl derivative, 1-CH$_3$-2,1,7-(CO)$_3$MnPCB$_9$-H$_{10}$, can be prepared from the 1,7-CH$_3$PCB$_9$H$_{10}^-$ ion in similar fashion (the latter ion is obtained by methylation of 1,7-PCB$_9$H$_{10}^{2-}$ at the phosphorus atom with methyl iodide[128,202]). Irradiation of a benzene solution of the P-methyl species in the presence of triphenylphosphine is reported to effect the replacement of one carbonyl group by a $(C_6H_5)_3P$ moiety.[242]

1-CH$_3$-2,1,7-(CO)$_3$MnPCB$_9$H$_{10}$ + $(C_6H_5)_3$P $\xrightarrow{h\nu}$

1-CH$_3$-2,1,7-(CO)$_2$[(C$_6$H$_5$)$_3$P]MnPCB$_9$H$_{10}$ + CO

Experimental

Preparation of $[(CH_3)_4N][2,1,7-(CO)_3MnPCB_9H_{10}]$.[128] One gram of $[(CH_3)_4N][1,7-PCB_9H_{11}]$ (4.5 mmol) in 25 ml of THF was treated with 3 ml of a 1.6-N solution of butyllithium in hexane and the mixture was refluxed briefly. Then $BrMn(CO)_5$, 1.32 g (4.8 mmol), was added and reflux was continued for 24 hr. The solvent was removed under vacuum and the residues were extracted with methanol. The extract was treated with aqueous tetramethylammonium chloride solution to give a bright-yellow precipitate. The product was recrystallized twice from methylene chloride–hexane to give 0.08 g (5% yield) of light yellow $[(CH_3)_4N][2,1,7-(CO)_3MnPCB_9H_{10}]$.

C. Iron Compounds

The 1,2- and $1,7-PCB_9H_{10}^{2-}$ ions react with anhydrous $FeCl_2$ in THF to give the corresponding red $(PCB_9H_{10})_2Fe^{2-}$ ions which have been isolated as the tetramethylammonium salts.[128,202] Treatment of these ions with methyl iodide generates the neutral bis(P-methyl) $(CH_3PCB_9H_{10})_2Fe$ compounds; the same species can also be prepared directly from the $CH_3PCB_9H_{10}^{-}$ ion.[128] The $(1,7-CH_3PCB_9H_{10})_2Fe$ complex exists in two isomeric forms which have been identified as a *dd,ll*-racemate and a *dl*-meso form, respectively. An X-ray study of a cocrystallite containing both isomers has been reported,[202] and the structure is depicted schematically in Figure 38.

The $(1,7-CH_3PCB_9H_{10})_2Fe$ system can be demethylated by reaction with sodium hydride, yielding the $[(1,7-CH_3PCB_9H_{10})Fe(1,7-PCB_9H_{10})]^{-}$ ion. Unlike the $(C_2B_9H_{11})_2Fe^{2-}$ complexes, the $(PCB_9H_{10})_2Fe^{2-}$ ions are not airoxidized in aqueous solution, but oxidation of $(1,7-PCB_9H_{10})_2Fe^{2-}$ to the green paramagnetic species $(1,7-PCB_9H_{10})_2Fe^{-}$ containing formal iron(III) is accomplished by treatment with ceric ion.[128]

The $(1,7-PCB_9H_{10})_2Fe^{2-}$ ion and its arsenic analog, $(1,7-AsCB_9H_{10})_2-Fe^{2-}$, undergo photolytic reactions with carbonyls of chromium, molybdenum, and tungsten, giving σ-bonded metal carbonyl derivatives in which the external $M(CO)_5$ ligands are bonded to phosphorus[4]:

$$(1,7-MCB_9H_{10})_2Fe^{2-} + W(CO)_6 \xrightarrow{h\nu} \{1,7-[(CO)_5W]MCB_9H_{10}\}_2Fe^{2-}$$
$$M = P, As$$

Unfortunately, the characterization data of the unsubstituted arsenic system $(1,7-AsCB_9H_{10})_2Fe^{2-}$ do not seem to have been published, nor those of any other arsacarborane transition metal complexes other than the bis(chromium pentacarbonyl)-substituted iron species[4] and the cyclopentadienyl cobalt arsacarborane described below (see Table IV). These compounds are unusual in that the arsacarborane cages function simultaneously as σ- and π-bonded

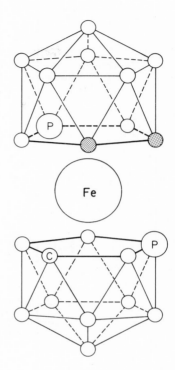

Figure 38. Structure of $(1,7\text{-}CH_3PCB_9H_{10})_2Fe$,[202] with the P-methyl groups omitted. Shaded atoms represent disordered boron and carbon atom locations. (From *J. Amer. Chem. Soc.*, **90**, 4489 (1968). Copyright by the American Chemical Society.)

ligands to transition metal atoms. All are stable at room temperature under nitrogen, although the arsenic species are less stable than the phosphorus complexes.

Few cyclopentadienyl transition metal metallophosphacarboranes have been prepared, but $1\text{-}CH_3\text{-}2,1,7\text{-}(\eta\text{-}C_5H_5)FePCB_9H_{10}$ is obtainable via the reaction of $1,7\text{-}CH_3PCB_9H_{10}^-$ ion, $FeCl_2$, and $C_5H_5^-$, a process analogous to the formation of cyclopentadienyl metallocarboranes from $C_2B_9H_{11}^{2-}$.[128]

Experimental

Preparation of $[(CH_3)_4N]_2[(1,2\text{-}PCB_9H_{10})_2Fe]$.[128] A solution of $(C_5H_{10}\text{-}NH_2)$ $(1,2\text{-}PCB_9H_{11})$ (0.50 g, 2.1 mmol) in 50 ml of THF was treated with excess sodium hydride at reflux for 1 hr. The supernatant liquid was transferred via syringe to a flask containing a slurry of anhydrous ferrous chloride (0.21 g, 1.6 mmol) in THF at reflux and reflux was continued for 12 hr. The solvent was removed under vacuum after adding a small amount of 100–200-mesh silica gel to the mixture. The products were chromatographed on a silica gel column and a red fraction was eluted with acetone. This fraction

TABLE IV

Transition-Metal Metallophosphacarboranes and Metalloarsacarboranes

Compound[a]	Color	Mp, °C	Other data[b]	References
Manganese				
$[(CH_3)_4N][2,1,7-(CO)_3MnPCB_9H_{10}]$	Yellow		IR, E	128
$1-CH_3-2,1,7-(CO)_3MnPCB_9H_{10}$		99.5–100	H, IR	128
$2,1,7-[(C_6H_5)_3P](CO)_2MnPCB_9H_{10}$	Yellow	198–200	IR	242
Iron				
$1-CH_3-2,1,7-(C_5H_5)FePCB_9H_{10}$		165–167	H, IR, MS, E	128
$[(CH_3)_4N]_2[(1,2-PCB_9H_{10})_2Fe^{II}]$	Red	294–295	H, IR, E, EC	128, 202
$(1-CH_3-3,1,2-PCB_9H_{10})_2Fe^{II}$		325 (dec)	H, IR, E, EC	128, 202
$[(CH_3)_4N][(1,7-PCB_9H_{10})_2Fe^{II}]$			H, IR, E, EC	128, 202
$[(CH_3)_4N][1,7-PCB_9H_{10})_2Fe^{III}]$	Green		B, E, MAG, IR	128
$[(CH_3)_4N][(1,7-PCB_9H_{10})Fe^{II}(1-CH_3-1,7-PCB_9H_{10})]$	Orange-red	293–295	H, IR, EC	128
$(1-CH_3-1,7-PCB_9H_{10})_2Fe^{II}$, isomer I	Red	239–240	B, H, IR, MS, E, EC	128
			X	202
isomer II	Orange	233–234	B, H, IR, MS, E	128
			X	202
$[(CH_3)_4N]_2[(1-(CO)_5Cr-1,7-PCB_9H_{10})_2Fe]$			B, H, IR, C	4
$[(CH_3)_4N]_2[(1-(CO)_5Mo-1,7-PCB_9H_{10})_2Fe]$			B, H, IR	4

[(n-C$_4$H$_9$)$_4$N]$_2$[(1-(CO)$_5$W-1,7-PCB$_9$H$_{10}$)$_2$Fe]	Red	—	B, H, IR	4
[(n-C$_4$H$_9$)$_4$N]$_2$[(1-(CO)$_5$Cr-1,7-AsCB$_9$H$_{10}$)$_2$Fe]	Red	207–209	B, H, IR	4
Cobalt				
[(CH$_3$)$_4$N][(1,2-PCB$_9$H$_{10}$)$_2$Co^{III}]	Orange		H, IR, E	128
[(CH$_3$)$_4$N][(1,7-PCB$_9$H$_{10}$)$_2$Co^{III}]			H, IR, E	128, 202
(1,7-PCB$_9$H$_{10}$)Co^{III}(1-CH$_3$-1,7-PCB$_9$H$_{10}$)	Orange	181–183	H, IR, MS	128
(1-CH$_3$-1,2-PCB$_9$H$_{10}$)$_2$Co^{II}		276–278	H, IR, E, EC, MAG	128
(1-CH$_3$-1,7-PCB$_9$H$_{10}$)$_2$Co^{II} (isomer I)		278–279	H, IR, E, EC, MAG	128
(isomer II)		232–234	IR, E, MAG	128
3,1,2-(C$_5$H$_5$)CoAsCB$_9$H$_{10}$	Yellow	250–251	MS, X	199
Nickel				
2,1,7-(π-C$_3$H$_5$)Ni(CH$_3$PCB$_9$H$_{10}$)	Red	120–121	H, IR, MS, E	220
2,1,7-(π-CH$_3$C$_3$H$_4$)Ni(CH$_3$PCB$_9$H$_{10}$)	Red	123–124	H, IR, MS, E	220
2,1,7-[π-(C$_6$H$_5$)$_3$C$_3$]Ni(CH$_3$PCB$_9$H$_{10}$)	Red	202–203	H, IR, MS, E	220
2,1,7-(NO)Ni(CH$_3$PCB$_9$H$_{10}$)	Red	112–113	H, IR, MS, E	220
(1-CH$_3$-1,7-PCB$_9$H$_{10}$)$_2$Ni				202

[a] See footnote a in Table I. All species are twelve-atom cages.
[b] See footnote b in Table I.
Note: See Table V for germanium phosphacarboranes and germanium arsacarboranes.

was treated with aqueous tetramethylammonium chloride. The crude product was crystallized from acetone–methanol to give 0.44 g (83% yield) of $[(CH_3)_4N][(1,2\text{-}PCB_9H_{10})_2Fe]$, mp 294–295°.

D. Cobalt Compounds

The orange-red P-methylated species $(1,2\text{-}CH_3PCB_9H_{10})_2Co$ and $(1,7\text{-}CH_3PCB_9H_{10})_2Co$, containing formal cobalt(II), are obtained from the respective $CH_3PCB_9H_{10}^-$ anions and $CoCl_2$, but only in the 1,7-case have the stereoisomers been separated.[128,202] These neutral species are not readily oxidized in air, in contrast to the facile oxidation of cobaltocene, and indeed neither of the expected cobalt(III) cations, $(CH_3PCB_9H_{10})_2Co^+$, have been isolated, although cyclic voltammetry studies indicate that the 1,2 isomer should exist. The cobalt(III) parent species, $(1,2\text{-}PCB_9H_{10})_2Co^-$ and its 1,7-analog, are formed directly from the reaction of the isomeric $PCB_9H_{10}^{2-}$ ions with $CoCl_2$,[128,202] indicating again that the parent and methylated phosphacarborane ligands differ substantially in their ability to stabilize higher metal-oxidation states. The $(1,7\text{-}PCB_9H_{10})_2Co^-$ ion is readily methylated at the phosphorus atom by treatment with methyl iodide, giving neutral $(1,7\text{-}CH_3PCB_9H_{10})Co(1,7\text{-}PCB_9H_{10})$, a sublimable orange solid.

A cyclopentadienyl cobalt arsacarborane has been prepared by the reaction of $1,2\text{-}AsCB_9H_{11}^-$ with $C_5H_5^-$ and $CoCl_2$.[199] The expected sandwich structure of the product, $3,1,2\text{-}(\eta\text{-}C_5H_5)CoAsCB_9H_{10}$, has been confirmed in an X-ray study which established that the arsenic and carbon atoms occupy adjacent locations on the five-membered bonding face of the arsacarborane ligand.[199]

Experimental

Preparation of $3,1,2\text{-}(\eta\text{-}C_5H_5)CoAsCB_9H_{10}$.[199] A solution of 2.0 g (7 mmol) of $(C_5H_{10}NH_2)$ $(1,2\text{-}AsCB_9H_{11})$, 3 ml (36 mmol) of freshly cracked C_5H_6, 10 ml (0.1 mol) of triethylamine, and 2.7 g (20 mmol) of anhydrous cobalt(II) chloride in THF solution was refluxed overnight. The solvent was removed *in vacuo* and the solids were chromatographed on a silica gel column with benzene as eluent. Recrystallization from benzene–heptane gave 0.11 g of yellow crystalline product, mp 250–251°.

E. Nickel Compounds

The only reported nickel metallocarboranes containing a Group-V cage heteroatom are several species formed from the $1,7\text{-}CH_3PCB_9H_{10}^-$ icosahedral-fragment ligand, including $(CH_3PCB_9H_{10})_2Ni$[202] and several compounds of type $2,1,7\text{-}LNi(CH_3PCB_9H_{10})^{220}$ in which L = allyl, methallyl, triphenylcyclopropyl, or nitroso groups (Table IV). Each of these species is prepared directly from the $1,7\text{-}CH_3PCB_9H_{10}^-$ ion or neutral *nido*-$1,7\text{-}$ $CH_3PCB_9H_{11}$ by treatment with nickel reagents and the desired ligand. Since the icosahedra in these cage systems incorporate a d^8 metal atom and contain an excess of valence electrons beyond the $(2n + 2)$ required for a filled-shell electronic structure, a slip-distorted sandwich geometry analogous to those found in other metallocarboranes containing formal d^8 transition metals might be anticipated. An X-ray investigation of the bis(phosphacarbollyl) system would be of interest as a means of ascertaining whether the slip-distortion phenomenon found in certain $(C_2B_9H_{11})_2M^q$ species (Section IV.H.3) extends to phosphacarborane complexes as well. This is certainly to be expected on the basis of electronic considerations discussed in Section II and elsewhere throughout this review.

Experimental

Preparation of $2,1,7\text{-}(\pi\text{-}C_3H_5)NiP(CH_3)CB_9H_{10}.^{220}$ A THF solution of allylmagnesium chloride (ca. 12 mmol) was syringed into a flask containing 0.80 g (6 mmol) of anhydrous $NiCl_2$, maintained at $-15°$ in an ice-salt bath. A 0.346-g (2.1 mmol) sample of $1,7\text{-}CH_3PCB_9H_{11}$ in THF was added. The mixture was stirred for several hours during which time the temperature was raised from $-15°$ to room temperature. The reaction mixture was evaporated to dryness under vacuum and the residue was extracted with benzene. The extract was passed through a silica-gel column eluting with benzene. The main red band was crystallized from heptane or sublimed at $85°$ to give 0.216 g (34% yield) of product, mp 120–121°.

VII. METALLOCARBORANES OF THE MAIN-GROUP METALS

A. Introduction

The insertion of metal heteroatoms of Groups II, III, and IV into carborane frameworks has been accomplished via synthetic routes essentially analogous to the preparation of transition-metal metallocarboranes. However, with the exception of certain compounds of gallium, indium, and tin prepared in the author's laboratory, all of the known main-group metal metallocarboranes are 12-vertex icosahedral cage systems (Table V).

Heterocarboranes containing the Group-V elements phosphorus, arsenic, and antimony are known, but are discussed in this review only insofar as they also contain metal atoms of the transition or main groups.

TABLE V

Metallocarboranes of the Main-Group Metals

Compound[a]	Color	Mp, °C	Other data[b]	References
Beryllium				
12-Atom cages				
3-$(C_2H_5)_2O$-3,1,2-$BeC_2B_9H_{11}$	White	120–121	H	166, 167
3-$(CH_3)_3N$-3,1,2-$BeC_2B_9H_{11}$	White	221–223	B, H, MS	166, 167
Aluminum				
7-Atom cage, bridged				
μ-[$(CH_3)_2Al$]$C_2B_4H_7$	White		MS	131
12-Atom cages				
3-CH_3-3,1,2-$AlC_2B_9H_{11}$		97–100	B, H, IR	232
3-C_2H_5-3,1,2-$AlC_2B_9H_{11}$	White	97–99	B, H, IR, MS	232
			X	27
3-C_2H_5-3,1,2-$AlC_2B_9H_{11}$·2THF	White	119–120	IR	144, 232
2-C_2H_5-2,1,7-$AlC_2B_9H_{11}$	White	100–102	B, H, IR, MS	232
nido-μ-$(CH_3)_2Al$-1,2-$C_2B_9H_{12}$	White	120–122	B, H, IR, MS	232
			X	26, 29
nido-μ-$(C_2H_5)_2Al$-1,2-$C_2B_9H_{12}$	White	34.5–35.5	B, H, IR, MS	232
Gallium				
7-Atom cages				
1-CH_3-1,2,3-$GaC_2B_4H_6$	White	33.5–34.5	B, H, IR, MS, X	80, 81
μ-[$(CH_3)_2Ga$]$C_2B_4H_7$	White		B, MS, IR	131
12-Atom cages				
3-C_2H_5-3,1,2-$GaC_2B_9H_{11}$	White	114–115	B, H, IR, MS	232
2-C_2H_5-2,1,7-$GaC_2B_9H_{11}$	White	114–115	H	232
nido-μ-$(C_2H_5)_2Ga$-1,2-$C_2B_9H_{12}$	Clear oil		B, H, IR, MS	232
Indium				
7-Atom cages				
1-CH_3-1,2,3-$InC_2B_4H_6$	White		B, H, IR, MS	81

Compound		Color	Methods	Ref.
Thallium				
12-Atom cages				
Tl$^+$[3,1,2-TlC$_2$B$_9$H$_{11}$]		Pale Yellow		191
Tl$^+$[1-CH$_3$-3,1,2-TlC$_2$B$_9$H$_{10}$]		Yellow		191
Tl$^+$[1,2-(CH$_3$)$_2$-3,1,2-TlC$_2$B$_9$H$_9$]		Bright Yellow		191
Silicon Bridged				
7-Atom cages				
nido-μ-SiH$_3$-2,3-C$_2$B$_4$H$_7$		White	B, H, IR, MS	196
nido-μ-(CH$_3$)$_3$Si-2,3-C$_2$B$_4$H$_7$		White	B, H, IR, MS	196
nido-μ,μ'-SiH$_2$(2,3-C$_2$B$_4$H$_7$)$_2$		White	B, H, IR, MS	192
nido-μ-CH$_3$SiH$_2$-2,3-C$_2$B$_4$H$_7$		White	B, H, IR, MS	192
nido-μ,4-[(CH$_3$)$_3$Si]$_2$-2,3-C$_2$B$_4$H$_6$		White	B, H, IR, MS	192
nido-μ-(CH$_2$)$_4$SiCl-2,3-C$_2$B$_4$H$_7$		White	B, H, IR, MS	192
nido-μ-(CH$_3$)$_3$Si-2,3-(CH$_3$)$_2$-2,3-C$_2$B$_4$H$_5$		White	B, H, IR, MS	183, 209
nido-μ-(CH$_3$)$_2$ClSi-2,3-(CH$_3$)$_2$-2,3-C$_2$B$_4$H$_5$		White	B, H, IR, MS	183
Germanium				
7-Atom cages				
nido-μ-GeH$_3$-2,3-C$_2$B$_4$H$_7$		White	B, H, IR, MS	196
nido-μ-(CH$_3$)$_3$Ge-2,3-C$_2$B$_4$H$_7$		White	B, H, IR, MS	196
nido-μ,μ'-(CH$_3$)$_2$Ge[4-(CH$_3$)$_3$Si-2,3-C$_2$B$_4$H$_6$]$_2$		White	B, H, IR, MS	192
12-Atom cages, Dicarbon				
3,1,2-GeC$_2$B$_9$H$_{11}$	342	White	B, H, IR, MS	172, 204
1-CH$_3$-3,1,2-GeC$_2$B$_9$H$_{10}$		White	H	172
1,2-(CH$_3$)$_2$-3,1,2-GeC$_2$B$_9$H$_9$		White	H	172
2,1,7-GeC$_2$B$_9$H$_{11}$	>310	White	B, H, IR, MS	18
12-Atom cages, monocarbon				
1-CH$_3$-1,2-GeCB$_{10}$H$_{11}$	230, 231		B, H, MS, IR	200, 223a
[(CH$_3$)$_4$N][1,2-GeCB$_{10}$H$_{11}$]			B, H, IR, B	200
1-C$_2$H$_5$-1,2-GeCB$_{10}$H$_{11}$			H, B, IR	223a
[(CH$_3$)$_4$N][(CO)$_5$CrGeC$_2$B$_{10}$H$_{11}$]			B, IR	223a

(continued)

TABLE V (continued)

Compound[a]	Color	Mp, °C	Other Data[b]	References
[(CH₃)₄N][(CO)₅MoGeC₂B₁₀H₁₁]			B, IR	223a
[(CH₃)₄N][(CO)₅WGeC₂B₁₀H₁₁]			B, IR	223a
12-Atom cages, phospha- and arsacarboranes				
$3,1,2\text{-}GePCB_9H_{10}$	White	340–343 (dec)	B, H, IR, MS	5
$2,1,7\text{-}GePCB_9H_{10}$	White	410–415 (dec)	B, H, IR, MS	5
$GePCB_9H_{10}$[c]			B, H, IR, MS	5
$3,1,2\text{-}GeAsCB_9H_{10}$	White	385–390 (dec)	B, H, IR, MS	5
$2,1,7\text{-}GeAsCB_9H_{10}$	White	413–416 (dec)	B, H, IR, MS	5
Tin				
7-Atom cages				
$nido\text{-}\mu\text{-}(CH_3)_3Sn\text{-}2,3\text{-}C_2B_4H_7$	White		B, H, IR, MS	192
$1,2,3\text{-}SnC_2B_4H_6$			MS	230
$2,3\text{-}(CH_3)_2\text{-}1,2,3\text{-}SnC_2B_4H_4$			B, H, IR, MS	230
12-Atom cages				
$3,1,2\text{-}SnC_2B_9H_{11}$	White	210 (dec)	B, H, IR, MS, MB	170,172, 204
Lead				
7-Atom cages				
$nido\text{-}\mu\text{-}(CH_3)_3Pb\text{-}2,3\text{-}C_2B_4H_7$	Yellow		B, H, IR, MS	192
$1,2,3\text{-}PbC_2B_4H_6$	Yellow		B, H, IR, MS	230
$2,3\text{-}(CH_3)_2\text{-}1,2,3\text{-}PbC_2B_4H_4$			B, H, IR, MS	230
12-Atom cages				
$3,1,2\text{-}PbC_2B_9H_{11}$	Yellow		B, H, IR, MS	172, 204

[a] See footnote a in Table I.
[b] See footnote b in Table I.
[c] Isomer unknown.

B. Beryllium Compounds

The reaction of $nido$-1,2-$C_2B_9H_{13}$ with dimethyl- or diethylberyllium in ether produces an extremely air-sensitive white solid etherate of icosahedral $BeC_2B_9H_{11}$.[166,167]

$$1,2\text{-}C_2B_9H_{13} + BeR_2 \cdot 2[O(C_2H_5)_2] \xrightarrow{C_6H_6,\ (C_2H_5)_2O}$$
$$2\ RH + 3\text{-}[(C_2H_5)_2O]\text{-}3,1,2\text{-}BeC_2B_9H_{11} + (C_2H_5)_2O$$
$$R = CH_3,\ C_2H_5$$

The reaction of $C_2B_9H_{13}$ with diethylberyllium containing only a third of a mole equivalent of complexed diethyl ether proceeds differently, generating a polymer which is proposed to contain repeating $BeC_2B_9H_{11}$ units linked by B–H–B bridges.[166] The treatment of either the diethyl etherate monomer or the polymer with triethylamine produces a solid amine adduct which is considerably less air sensitive but is still degraded to 1,2-$C_2B_9H_{12}^-$ by ethanolic KOH. The ether and amine complexes can be regarded as analogs of a hypothetical $HBeC_2B_9H_{11}^-$ icosahedron, whose isoelectronic relationship to $C_2B_{10}H_{12}$ is seen by the formal replacement of a BH unit in the latter species by a BeH^- group; substitution of R_3N: or R_2O: for :H^- in $HBeC_2$-$B_9H_{11}^-$ yields the species described above.[166]

Experimental

Preparation of 3,1,2-$(CH_3)_3NBeC_2B_9H_{11}$.[166] The reaction vessel was a 100-ml, three-necked, round-bottomed flask, equipped with a pressure-equalized addition funnel, magnetic stirring bar, and reflux condenser. The round-bottom flask contained a solution of 6.44 g (48 mmol) of 1,2-$C_2B_9H_{13}$, dissolved in 50 ml of dry, oxygen-free benzene. The addition funnel was charged with 1.87 g (48 mmol) of dimethylberyllium, dissolved in 8 ml of dry diethyl ether. The solution in the flask was maintained at room temperature and the dimethylberyllium solution was added dropwise with stirring. Gas evolved immediately, but stopped when the addition of dimethylberyllium was completed. The evolved gas was identified as methane from its characteristic vpc (vapor-phase chromatography) retention time.

After the reaction was completed, a yellow oil layer had separated on the bottom of the flask. The upper colorless benzene layer was decanted, and the remaining oil was washed three times with 40 ml of dry benzene and dried under high vacuum at room temperature. The remaining semisolid was dissolved in 30 ml of dry dichloromethane and filtered, and dry trimethylamine was passed through the solution. A white material precipitated which was separated by filtration, washed with benzene, and dried under high vacuum. This white solid was recrystallized from a dichloromethane–hexane mixture by passing dry oxygen-free nitrogen over the surface of the solution. The white solid product (mp 221–223° dec) was obtained in a 51% overall yield. The product was less air-sensitive than the analogous etherate and was decomposed by water.

C. Aluminum, Gallium, Indium, and Thallium Compounds

1. General Comments

The vast scope of the chemistry of boron cage compounds, of which the area covered in this review represents only a fraction, suggests that the remaining elements in Group III of the Periodic Table might be similarly capable of forming electron-deficient molecular frameworks. As is well known, however, the chemistry of aluminum and the heavier Group-III elements differs greatly from that of boron, and boron-like cage compounds based on these elements are unknown at present. However, in 1968 Mikhailov and Potapova[144] prepared C_2H_5-$AlC_2B_9H_{11}$, and in 1969 Grimes and Rademaker[80,81] obtained $CH_3GaC_2B_4H_6$ and its indium analog, the metal in each case residing in an electron-deficient carborane cage. Hawthorne et al.[232,233] also have reported extensive studies of aluminum and gallium metallocarboranes. Many of the early structural assignments were later confirmed in X-ray studies, and other metallocarboranes of aluminum, gallium, indium, and thallium were later prepared as described below. In all of these compounds, the Group-III metal plays the role of a heteroatom bound into a boron cage; no molecular cage framework constructed of heavy Group-III metal atoms in the *absence* of boron has been prepared, and indeed, no metallocarborane having more than one Group-III metal per cage has yet been found. It should be pointed out, however, that these and other non-transition metal metallocarboranes have been far less studied than the transition metal metallocarboranes. The existence of polyhedral transition metal clusters, and of metallocarboranes having as many as three transition metal atoms, suggests that three-dimensional frameworks of aluminum or gallium atoms may yet be synthesized.

2. Seven-Atom Cages

The only known *closo*-carboranes of smaller than icosahedral size containing heteroatoms of the heavier Group-III elements are 1-CH_3-1,2,3-$GaC_2B_4H_6$ and 1-CH_3-1,2,3-$InC_2B_4H_6$, both compounds forming in vapor-phase insertions of the trimethylmetal into 2,3-$C_2B_4H_8$.[80,81]

The structure of the gallium species was proposed from spectroscopic data[80] to be a pentagonal bipyramid with the CH_3Ga group occupying one apex, and adjacent carbon atoms in the five-membered equatorial ring. This

$$C_2B_4H_8 + Ga(CH_3)_3 \xrightarrow{215°} CH_3GaC_2B_4H_6 + B(CH_3)_3 + \text{solids}$$

$$20\text{--}30\%$$

$$C_2B_4H_8 + In(CH_3)_3 \xrightarrow{95\text{--}110°} CH_3InC_2B_4H_6 + B(CH_3)_3 + \text{solids}$$

$$50\text{--}60\%$$

gross geometry (Figure 39) was confirmed in an X-ray investigation,[81] which however revealed two anomalies: (1) the gallium atom is shifted closer to the borons than to the carbon atoms in the equatorial ring, in a fashion suggestive of the slip-distortion observed in $(C_2B_9H_{11})_2M^q$ complexes in which M has more than six d electrons (Section IV.H.3), and (2) the CH_3–Ga axis is not perpendicular to the planar equatorial ring, but instead is tilted about 20° from the perpendicular. The first observation has been rationalized via a qualitative MO argument,[81] similar to that invoked by Warren and Hawthorne[211] for icosahedral metallocarboranes, and earlier by Mulliken[155] and Dewar[31] for benzene–Ag^+ and benzene–Cu^+ complexes. Essentially this approach assumes that the bonding of the CH_3Ga^+ group to the formal $C_2B_4H_6^{2-}$ ligand occurs primarily via an sp_z hybrid acceptor orbital on gallium, but that back bonding from the metal d orbitals to the carborane

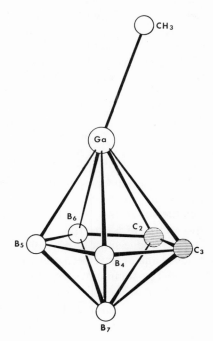

Figure 39. Structure of 1-CH_3-1,2,3-GaC_2-B_4H_6.[81] (From *J. Amer. Chem. Soc.*, **94**, 1865 (1972). Copyright by the American Chemical Society.)

ligand also occurs, and that this latter interaction requires a distortion from pseudo-C_{5v} symmetry in order to be effective.[81] The same point can be made by applying the $(2n + 2)$ rule discussed in Section II, in that the CH_3GaC_2-B_4H_6 system is formally analogous to the parent $C_2B_5H_7$ carborane which has the 16-electron bonding MO configuration required for a closed seven-atom polyhedron. An identical situation would occur in $CH_3GaC_2B_4H_6$ *if no d electron contribution to the bonding took place*, i.e., if all ten d electrons on gallium were in nonbonding orbitals, and in this event no distortion would be expected. The actual observation of apparent slip-distortion thus suggests some significant interaction of the d orbitals with the remainder of the cage framework. The tilting of the CH_3–Ga axis has been accounted for[81] as a consequence of the MO scheme previously referred to.

The indium analog, $CH_3InC_2B_4H_6$, is expected to exhibit similar distortions although no X-ray data are available. The corresponding aluminum species has not been reported, but a larger homolog, $1\text{-}C_2H_5\text{-}1\text{-}AlC_2B_9H_{11}$, described later in this section, has a nondistorted icosahedral structure. In light of the above discussion on the gallium species, it is tempting to conclude that d orbital participation in the aluminum species is not significant, but the comparison between the two molecules is complicated by the differences in cage size and geometry as well as in the metal atom. The extremely interesting questions of structure and bonding raised by these species clearly require detailed, quantitative MO treatment.

The chemistry of the small gallium and indium metallocarboranes has been examined to a limited extent.[81] Both compounds are degraded to 2,3-$C_2B_4H_8$ on pyrolysis, and each reacts easily with elemental bromine under Friedel–Crafts conditions at room temperature, destroying the cage structure.

$$CH_3GaC_2B_4H_6 + Br_2 \xrightarrow[\text{CS}_2,\ 25°]{\text{AlBr}_3} [(CH_3)_2GaBr]_2 + CH_3Br + HBr + \text{solids}$$
$$\sim 30\%$$
$$\xrightarrow[25\text{-}50°]{2\text{HCl}} 0.5(CH_3GaCl_2)_2 + C_2B_4H_8 + \text{solids}$$
$$40\text{-}53\%$$

The gallium compound is readily attacked by excess HCl to give $C_2B_4H_8$ as shown, but the reaction of $CH_3InC_2B_4H_6$ with HCl produces only nonvolatile solids and H_2. The gallium species is inert toward the electrophilic molecules $AlCl_3$ and BF_3.

A bridge-substituted gallium carborane and a probable aluminum-bridged analog have been prepared by insertion of the metal into the $C_2B_4H_7^-$ ion.[131] The aluminum species is extremely unstable, decomposing readily into $C_2B_4H_8$ at 25°.

$$C_2B_4H_7^- + (CH_3)_2MCl \longrightarrow \mu\text{-}(CH_3)_2MC_2B_4H_7 + Cl^-$$
$$(M = \text{Al or Ga})$$

Experimental

Preparation of $1\text{-}CH_3\text{-}1,2,3\text{-}GaC_2B_4H_6$.[81] Trimethylgallium (3.0 mmol) and 3.0 mmol of $2,3\text{-}C_2B_4H_8$ were sealed into a 100-ml Pyrex bulb equipped with a break-off tip and pyrolyzed at 215° for 24 hr, during which the interior of the bulb became coated with gray solids. The volatile contents were fractionated on the vacuum line through a $-45°$ trap for 1 hr. The material passing $-45°$ consisted of unreacted $C_2B_4H_8$ and trimethylboron, while the $-45°$ condensate was nearly pure crystalline $1\text{-}CH_3GaC_2B_4H_6$. Final purification was achieved by pumping on the gallacarborane for a few minutes at 0°, which removed the last traces of $C_2B_4H_8$. The yield of $1\text{-}CH_3GaC_2B_4H_6$ was 50 mg (20% yield based on $C_2B_4H_8$ consumed).

3. Twelve-Atom Cages

The icosahedral species $3\text{-}C_2H_5\text{-}3,1,2\text{-}AlC_2B_9H_{11}$ was originally prepared as a bis(tetrahydrofuran) adduct by Mikhailov and Potapova[144] by the reaction of $C_2B_9H_{11}^{2-}$ ion with $C_2H_5AlCl_2$ in THF at $-50°$. Although no structural data were presented, this compound was the first example of the incorporation of a Group-III element other than boron into an electron-deficient cage framework. This work was later confirmed by Hawthorne and co-workers,[232,233] and the structure shown in Figure 40 was confirmed in an X-ray investigation.[27] As was mentioned above, the geometry is symmetrical with approximately equal nearest-neighbor Al–C and Al–B distances, in contrast to the analogous seven-vertex system, $1\text{-}CH_3\text{-}1,2,3\text{-}GaC_2B_4H_6$ (Figure 39).

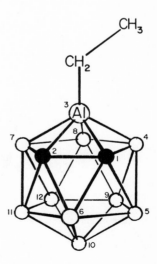

Figure 40. Structure of $3\text{-}C_2H_5\text{-}3,1,2\text{-}AlC_2B_9H_{11}$.[27,232] ●, CH; ○, BH. (From *J. Amer. Chem. Soc.*, 93, 5687 (1971). Copyright by the American Chemical Society.)

The reactions of aluminum and gallium trialkyls with *nido*-1,2-$C_2B_9H_{13}$ generate a novel type of *nido*-metallocarborane having the general formula μ-$R_2MC_2B_9H_{12}$ in which R = CH_3 or C_2H_5 and M = Al or Ga.[232,233] The unusual feature of these products is that they are fluxional at room temperature, the R_2M group evidently tautomerizing between equivalent B–H–M–H–B bridging locations on the edge of the open face (Figure 41), as indicated from low-temperature [11]B nmr studies.[232] An X-ray investigation[26] of the μ-dimethylaluminum compound at $-100°$ has confirmed the solid-state geometry indicated in Figure 41. Although the hydrogen atoms were not located, the rather long (2.30 Å) Al–B bond length has suggested[232] that these links are each bridged by a hydrogen atom as shown in Figure 41, similar to the binding in $Al(BH_4)_3$.

The bridged *nido* species—μ-$(C_2H_5)_2AlC_2B_9H_{12}$, μ-$(CH_3)_2AlC_2B_9H_{12}$, and μ-$(C_2H_5)_2GaC_2B_9H_{12}$—are converted on heating in dry benzene to the *closo*-metallocarboranes—3-C_2H_5-3,1,2-$AlC_2B_9H_{11}$, 3-CH_3-3,1,2-$AlC_2B_9H_{11}$, and 3-C_2H_5-3,1,2-$GaC_2B_9H_{11}$, respectively—with concomitant evolution of alkane.[232] These species are icosahedral (Figure 40), as confirmed by an X-ray investigation of the ethylaluminum compound,[27] and are analogs of 1,2-$C_2B_{10}H_{12}$. Not surprisingly, much of the chemistry of these compounds resembles that of 1,2-$C_2B_{10}H_{12}$; thus, $C_2H_5AlC_2B_9H_{11}$ and its gallium analog undergo thermal rearrangement, apparently to the 1,7 isomers, at elevated temperature. (The gallium species, however, decomposes extensively to give primarily *closo*-$C_2B_9H_{11}$ with only a 7% yield of 1,7-$C_2H_5GaC_2B_9H_{11}$.[232]) In neither case did higher temperatures produce any additional isomers.

All of the 1,2-*closo* and 1,2-*nido* species are readily hydrolyzed by moist air to the 1,2-$C_2B_9H_{12}^-$ ion, while 3-C_2H_5-3,1,7-$AlC_2B_9H_{11}$ gave

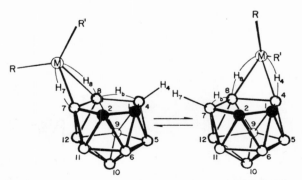

Figure 41. Proposed mechanism of tautomerism of μ-$R_2MC_2B_9H_{12}$ (M = Al or Ga).[232] ●, CH; ○, BH. (From *J. Amer. Chem. Soc.*, **93**, 5687 (1971). Copyright by the American Chemical Society.)

$1,7\text{-}C_2B_9H_{12}^-$. Hydrolysis by direct contact with water proceeds explosively.

Treatment of the *closo* species, $3\text{-}C_2H_5\text{-}3,1,2\text{-}AlC_2B_9H_{11}$, with HCl in benzene gave $nido\text{-}C_2B_9H_{13}$ quantitatively, a reaction analogous to that of $CH_3GaC_2B_4H_6$ with HCl to produce $nido\text{-}2,3\text{-}C_2B_4H_8$ as described earlier.[81] An interesting observation arising from this work is that both $nido\text{-}(C_2H_5)_2\text{-}AlC_2B_9H_{12}$ and $closo\text{-}3\text{-}C_2H_5\text{-}3,1,2\text{-}AlC_2B_9H_{11}$ react with $TiCl_4$ in *n*-heptane, yielding unidentified dark-red precipitates which on exposure to ethylene catalytically generate high-molecular-weight polyethylene.[232]

The only examples at present of thallium metallocarboranes are the series of yellow compounds having the empirical formula $Tl_2R^1R^2C_2B_9H_9$ ($R^1 = R^2 = H$; $R^1 = H$, $R^2 = CH_3$; and $R^1 = R^2 = CH_3$), which are prepared from thallium acetate and $nido\text{-}1,2\text{-}C_2B_9H_{12}^-$ ion or its C-alkyl derivatives in aqueous alkaline solution.[191] Cations such as $As(C_6H_5)_4^+$ easily displace one of the thallium atoms to give salts of the $TlR^1R^2C_2B_9H_9^-$ ion, so that the original compound is formulated as a Tl^+ salt. The facile replacement of the thallium in $TlR^1R^2C_2B_9H_9^-$ by transition-metal ions of iron, cobalt, platinum, and palladium is described in Section IV.

Experimental

Preparation of $\mu\text{-}(C_2H_5)_2AlC_2B_9H_{12}$.[222] Freshly sublimed $1,2\text{-}C_2B_9H_{13}$ (3.20 g, 23.8 mmol) was dissolved in 60 ml of dry benzene under nitrogen. This solution was stirred at room temperature while another solution, consisting of 2.72 g (23.8 mmol) of triethylaluminum in 50 ml of dry benzene, was slowly added. After stirring for 1 hr, 305 ml of ethane had been liberated. The solution was then warmed to 50° for 1 hr. Upon cooling to room temperature the total volume of ethane liberated was 490 ml (22.8 mmol). The benzene was then pumped off, leaving an essentially quantitative yield of colorless, microcrystalline crude product. Recrystallization from dry, cold *n*-hexane gave large colorless crystals of pure product, yield 4.09 g, 79.0%. The compound could be sublimed under high vacuum at 50°.

Preparation of $3\text{-}C_2H_5\text{-}3,1,2\text{-}AlC_2B_9H_{11}$.[232] A 5.00 g (22.9 mmol) sample of $\mu\text{-}(C_2H_5)_2AlC_2B_9H_{12}$ (see above) was dissolved in 20 ml of dry benzene under nitrogen and heated at 79° for 25 hr. During this time 485 ml (21.6 mmol) of ethane was released. The solution was then cooled to room temperature and the benzene pumped off, leaving an essentially quantitative yield of clear, microcrystalline crude product. This was then dissolved in a little dry benzene, filtered through a medium frit under nitrogen, and diluted with *n*-hexane. Upon cooling the solution, the product was obtained as clear, medium-size crystals, yield 3.11 g, 72.0%. The product sublimed in vacuum at 80°.

D. Germanium, Tin, and Lead Compounds

1. General Comments

Aside from the fact that silicon is not regarded as a metal for purposes of this review, no carboranes or boranes containing this element as a hetero-atom bound into the cage framework are presently known. (However, small boranes and carboranes containing bridging silyl groups linked to the cage via B–Si–B three-center bonds have been reported, e.g., μ-SiH$_3$-2,3-C$_2$B$_4$-H$_7$[192,196] and μ-SiH$_3$B$_5$H$_8$.[57,58]) The three heaviest Group-IV elements, in contrast, have been successfully incorporated into carborane cage systems and metallocarboranes of this type are described below.

2. Dicarbon-Cage Systems

Nearly all of the polyhedral Group-IV-heteroatom metallocarboranes are 12-vertex icosahedral species. The small *nido* cage system $C_2B_4H_7^-$ (obtained by bridge-deprotonation of *nido*-2,3-$C_2B_4H_8$) can accept a bridging MR_3 group in which M = Si, Ge, Sn, or Pb and R = H or alkyl[192,196]; the metal atom forms a B–M–B three-center bond on the edge of the open-cage pyramidal framework. The C,C'-dimethyl derivative of $C_2B_4H_7^-$ undergoes analogous reactions with silicon and germanium reagents.[183,209]

$$\text{Na}^+\text{C}_2\text{B}_4\text{H}_7^- + \text{R}_3\text{MX} \xrightarrow{\text{0–25°, THF}} \mu\text{-R}_3\text{MC}_2\text{B}_4\text{H}_7 + \text{NaX}$$

$$\text{M = Si, Ge, Sn, Pb; X = Br or Cl; R = H or alkyl}$$

Attempts to effect cage closure with incorporation of the metal into a poly-hedral metallocarborane system have led either to migration of the bridging MR_3 unit to a terminally bonded location, or to formation of *closo*-car-boranes with external MR_3 substituents.[192,196]

Recently, the compounds $SnC_2B_4H_6$ and $PbC_2B_4H_6$, which are proposed to be *closo* 7-vertex polyhedra analogous to $C_2B_5H_7$ with "bare" metal atoms in apical locations, have been prepared in the author's laboratory from the reaction of the $C_2B_4H_7^-$ ion with $PbCl_2$ or $SnCl_2$.[230] These species are apparently lower homologs of the $MC_2B_9H_{11}$ compounds (M = Ge, Sn, or Pb) previously prepared by Rudolph *et al.*, from the 1,2-$C_2B_9H_{11}^{2-}$ ion[172,204]:

$$\text{1,2-C}_2\text{B}_9\text{H}_{11}^{2-} + \text{MX}_2 \longrightarrow \text{3,1,2-MC}_2\text{B}_9\text{H}_{11} + 2\,\text{X}^-$$

X-ray data on these compounds are not available, but their gross icosahedral geometry is indicated from ^{1}H and ^{11}B nmr and mass spectral data, and from their direct analogy to the 1,2-$C_2B_{10}H_{12}$ system. The Ge, Sn, and Pb hetero-atoms each formally replace a BH group in the latter cage (e.g., $:\text{Ge} \equiv \text{BH}$) and thus presumably have an unshared electron pair in lieu of an attached hydrogen. It has been pointed out[172,204] that the $SnC_2B_9H_{11}$ species and its germanium and lead homologs are analogous to a hypothetical $C_3B_9H_{11}$ icosahedron which would be isoelectronic with $C_2B_{10}H_{12}$. While $C_3B_9H_{11}$ has not been found, a lower homolog, $closo$-$C_3B_5H_7$, has been reported[197] and characterized spectroscopically as an eight-vertex polyhedron analogous to $C_2B_6H_8$ and containing a "bare" carbon atom. The limited chemistry known for the $MC_2B_9H_{11}$ species is consistent with that of the Group-III-hetero-atom metallocarboranes discussed above; thus, all three compounds are degraded to 1,2-$C_2B_9H_{12}^-$ in methanolic KOH, and treatment of $SnC_2B_9H_{11}$ with HCl generates $nido$-1,2-$C_2B_9H_{13}$ and $SnCl_2$.[172] The germanium species undergoes cage isomerization at 600°, yielding the 2,1,7 isomer in which the cage carbon atoms are separated from each other but remain adjacent to the metal. The same 2,1,7 species, a volatile, sublimable solid, is obtained in the reaction of 7,9-$C_2B_9H_{11}^{2-}$ with germanium(II) iodide.[18]

$$3,1,2\text{-GeC}_2B_9H_{11} \xrightarrow{600°} 2,1,7\text{-GeC}_2B_9H_{11}$$

$$1,7\text{-}C_2B_9H_{11}^{2-} + GeI_2 \xrightarrow[-2\,I^-]{} \nearrow$$

Unexpectedly, the treatment of 1,7-$C_2B_9H_{11}^{2-}$ with $SnCl_2$ in an effort to prepare the unknown 2,1,7-$SnC_2B_9H_{11}$ directly, resulted instead in oxidative closure of the ion to $closo$-2,3-$C_2B_9H_{11}$:

$$1,7\text{-}C_2B_9H_{11}^{2-} + SnCl_2 \xrightarrow{C_6H_6} Sn^0 + 2,3\text{-}C_2B_9H_{11} + 2\,Cl^-$$

$$3,1,2\text{-SnC}_2B_9H_{11} \xrightarrow{\Delta} \nearrow$$

Recently analogous results have been noted in the oxidation of $C_2B_5H_7^{2-}$ with $SnCl_2$.[230] An attempt to effect the cage rearrangement of 3,1,2-SnC_2B_9-H_{11} to the 2,1,7 isomer gave 2,3-$C_2B_9H_{11}$ in 98% yield,[18] a reaction which parallels the thermal degradation of 3-C_2H_5-3,1,2-$GaC_2B_9H_{11}$ to 2,3-$C_2B_9H_{11}$ as described above. The behavior of the lead analog, 3,1,2-$PbC_2B_9H_{11}$, has not been described, although mass spectral data suggest that it will follow the pattern of the tin compound and decompose rather than isomerize on pyrolysis.[172]

A tin-119 Mössbauer study[170] of 3,1,2-$SnC_2B_9H_{11}$ supports the assignment of an Sn(II) valence state in this compound, with less directional character for the electron lone pair than in $Sn(\eta\text{-}C_5H_5)_2$.

Experimental

Preparation of μ-$(CH_3)_3SnC_2B_4H_7$.[192] A THF solution containing 3.20 mmol of $Na^+C_2B_4H_7^-$ was added to 3.88 mmol of $(CH_3)_3SnBr$ in THF at

$-196°$ *in vacuo*, after which the contents were warmed to $23°$ and stirred for 1 hr. The products were fractionated repeatedly through a $-12°$ trap, which retained only μ-$(CH_3)_3SnC_2B_4H_7$ (0.520 g, 66% yield). The mass spectrum exhibits a cutoff at m/e 229 (intensity 13% of base peak), assigned to $^{124}Sn^{12}C_4^{11}B_4^1H_{13}^+$ which is formed by loss of CH_3 from the parent ion. The cutoff group profile (base peak m/e 223) is consistent with the calculated intensities based on natural isotope distribution. The materials passing through $-12°$ consisted of THF, $C_2B_4H_8$, and $(CH_3)_3SnBr$.

Preparation of $3,1,2$-$SnC_2B_9H_{11}$.[172] In a 100-ml, three-necked flask, equipped with a condenser, nitrogen inlet, and magnetic stirrer, which had been flushed with nitrogen for 1 hr, were placed 1.0 g (5.16 mmol) of the trimethylammonium salt of $C_2B_9H_{12}^-$, 50 ml of benzene, and 5.7 mmol of NaH (0.47 g of 59% dispersion which had been washed twice with hexane). The mixture was brought to reflux and flushed with nitrogen until trimethylamine could no longer be detected in the exhaust stream (usually ca. 4 hr). Anhydrous $SnCl_2$ (1.07 g, 5.7 mmol) was then added. After 24 hr of reflux under nitrogen, the mixture was placed in a Soxhlet thimble and extracted with benzene for 24 hr. Evaporation of the extract gave the crude product (in ca. 75% yield) which could be purified further by recrystallization from dry benzene or sublimation *in vacuo* at $140°$ to give 0.84 g (3.45 mmol) of the product. The compound discolored at $210°$ and blackened at $265°$ (sealed tube).

3. Monocarbon-Cage Systems

In each of the species described above, a "bare" Group-IV heteroatom formally replaces a BH group in a carborane framework. A closely related series exists in which the Group-IV atom in a formal sense replaces carbon instead of boron, in which case the heteroatom has an attached hydrogen atom or other substituent in the neutral molecule. As with the phospha- and arsacarboranes containing a single cage carbon atom, the synthesis involves insertion into the $CB_{10}H_{11}^{3-}$ ion[200,223a]:

$$Na_3CB_{10}H_{11}(C_4H_8O)_2 \xrightarrow[\text{THF}]{CH_3GeCl_3} CH_3GeCB_{10}H_{11} \underset{CH_3I}{\overset{\text{piperidine}}{\rightleftharpoons}} GeCB_{10}H_{11}^-$$

The methyl group is removed in refluxing pyridine, yielding a $GeCB_{10}H_{11}^-$ species which is a presumably closed icosahedral analog of $C_2B_{10}H_{12}$; treatment with methyl iodide regenerates the neutral Ge-methylated compound.[200] The unsubstituted anion, which incorporates a "bare" germanium atom, reacts photolytically with chromium group hexacarbonyls to produce complexes containing transition-metal–germanium σ-bonds.[198,223a] In this

$$1,2\text{-}GeCB_{10}H_{11}^- + M(CO)_6 \xrightarrow[\text{THF}]{h\nu} [1\text{-}(CO)_5M\text{-}1,2\text{-}GeCB_{10}H_{11}]^-$$

$$M = Cr, \text{ Mo, or } W$$

type of reaction a carbonyl group is in effect displaced by a $GeCB_{10}H_{11}^-$ moiety which contains an unshared electron pair on germanium and can function as a two-electron donor.

A further extension of the same synthetic and structural principles involves the preparation of germaphospha- and germaarsacarboranes of general formula $GePCB_9H_{10}$ and $GeAsCB_9H_{10}$, which are novel in that they incorporate three different main-group heteroatoms (nonboron atoms) in the same cage framework. Beer and Todd[5] obtained these materials by treatment of 1,2- or $1,7\text{-}PCB_{10}H_{10}^{2-}$ salts, or their arsenic analogs, with germanium diiodide in refluxing benzene.

$$1,2\text{-}PCB_9H_{10}^{2-} + GeI_2 \longrightarrow 3,1,2\text{-}GePCB_9H_{10} + 2\ I^-$$

The proposed structure of the 3,1,2 species, shown in Figure 42, contains the three heteroatoms in a triangular group on the surface of the icosahedron. The carbon and all of the borons have attached hydrogen atoms, but the germanium and phosphorus (arsenic) atoms are "bare," so that these cages are isoelectronic with $C_2B_{10}H_{12}$ provided it is assumed that the d-shell electrons on germanium and arsenic are not significantly involved in the delocalized framework bonding (see, however, the discussions of CH_3GaC_2-B_4H_6 and $C_2H_5AlC_2B_9H_{11}$ above). The $2,1\text{-}7\text{-}GePCB_9H_{10}$ and $2,1,7\text{-}GeAs$-CB_9H_{10} isomers (somewhat confusingly numbered 1,2,7 in the original paper[5]) are formed by similar reactions of $1,7\text{-}PCB_9H_{10}^{2-}$ and $\text{-}AsCB_9H_{10}^{2-}$, and are more thermally stable than the 3,1,2 isomers. This observation is understandable both from the analogy with 1,2- and $1,7\text{-}C_2B_{10}H_{12}$ and also in terms of the presumably less polar charge distribution in the 2,1,7 species,[5] in which phosphorus and carbon atoms occupy nonadjacent 1- and 7-

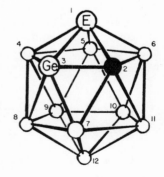

Figure 42. Proposed structure of $3,1,2\text{-}GePCB_9H_{10}$ and $3,1,2\text{-}GeAsCB_9H_{10}$.[5] ●, CH; ○, BH. E = P or As. (From *J. Organometal. Chem.*, **50**, 93 (1973). Copyright by the Elsevier Scientific Publishing Company.)

vertices in the icosahedron (see numbering in Figure 42). Efforts to prepare the corresponding tin and lead compounds have been unsuccessful.

The thermal rearrangement of $2,1,7$-$GePCB_9H_{10}$ at $512°$ gave three new isomers which have not been identified, the most abundant of which appears to be less polar and to have greater separation of the heteroatoms, than in the $2,1,7$ isomer.[5] This species is a good example of a cage system whose complexity, i.e., the large number of possible isomers, is such that nmr and other spectroscopic data cannot provide even a tentative structural characterization, and hence an X-ray investigation is mandatory. This situation is relatively new in carborane chemistry since normally the information obtainable from ^{11}B and ^1H nmr spectra in particular is sufficient to at least point to one or two isomers as highly probable. (In point of fact, nearly all X-ray studies of metallocarboranes have served to confirm the gross structures originally proposed from nmr data.) However, the recent trend toward synthesis of more complex cage systems such as those described here, indicates that X-ray diffraction will assume a role as a routine investigative tool in conjunction with synthesis, rather than as an ultimate procedure to be utilized on an occasional basis, as it has been up to the present.

Experimental

Preparation of $3,1,2$-$GePCB_9H_{10}$.[5] A solution of 0.700 g (3.28 mmol) of $[(CH_3)_3NH]^+[1,2$-$PCB_9H_{11}^-]$ in 25 ml of benzene was added to 0.158 g (6.59 mmol) of NaH (0.277 g of 57% oil dispersion which had been washed twice with hexane and once with benzene). The mixture was maintained at reflux for an additional 3 hr while a stream of argon swept over the system to remove the gaseous by-products formed (trimethylamine and hydrogen). Anhydrous GeI_2 was then added as a benzene slurry over a period of 1 hr and the resulting reaction mixture was heated at 70–$75°$ for 15 hr under an argon atmosphere. The yellow-orange residue obtained after removal of solvent *in vacuo*, was sublimed at $80°$ and 10^{-3} torr for 24 hr to give 0.500 g (68% yield) of product. An analytically pure sample was obtained by tlc. A benzene solution of the white sublimate was applied to a silica gel tlc plate (20 cm × 20 cm) and eluted with benzene. Sublimation of the extracted material yielded a pure crystalline product. The compound melted at 340–$343°$ with decomposition.

REFERENCES

1. R. G. Adler and M. F. Hawthorne, *J. Amer. Chem. Soc.*, **92**, 6174 (1970).
2. G. K. Barker, M. Green, J. L. Spencer, F. G. A. Stone, B. F. Taylor, and A. J. Welch, *J. Chem. Soc., Chem.Comm.*, 804 (1975).

3. D. C. Beer, V. R. Miller, L. G. Sneddon, R. N. Grimes, M. Mathew, and G. J. Palenik, *J. Amer. Chem. Soc.*, **95**, 3046 (1973).

4. D. C. Beer and L. J. Todd, *J. Organometal. Chem.*, **36**, 77 (1972).

5. D. C. Beer and L. J. Todd, *J. Organometal. Chem.*, **50**, 93 (1973).

6. P. Binger, *Tetrahedron Lett.*, 2675 (1966).

7. T. Birchall and I. Drummond, *Inorg. Chem.*, **10**, 399 (1971).

8. V. A. Brattsev and V. I. Stanko, *J. Organometal. Chem.*, **55**, 205 (1973).

9. V. A. Brattsev and V. I. Stanko, *Zh. Obshch. Khim.*, **38**, 2820 (1968).

10. J. P. Brennan, R. N. Grimes, R. Schaeffer, and L. G. Sneddon, *Inorg. Chem.*, **12**, 2266 (1973).

11. R. F. Bryan, W. M. Maxwell, and R. N. Grimes, submitted for publication.

12. K. P. Callahan, W. J. Evans, F. Y. Lo, C. E. Strouse, and M. F. Hawthorne, *J. Amer. Chem. Soc.*, **97**, 296 (1975).

13. K. P. Callahan and M. F. Hawthorne, *Pure Appl. Chem.*, **39**, 475 (1974).

14. K. P. Callahan, F. Y. Lo, C. E. Strouse, A. L. Sims, and M. F. Hawthorne, *Inorg. Chem.*, **13**, 2842 (1974).

15. K. P. Callahan, C. E. Strouse, A. L. Sims, and M. F. Hawthorne, *Inorg. Chem.*, **13**, 1393 (1974).

16. K. P. Callahan, C. E. Strouse, A. L. Sims, and M. F. Hawthorne, *Inorg. Chem.*, **13**, 1397 (1974).

17. W. E. Carroll, M. Green, F. G. A. Stone, and A. J. Welch, *J. Chem. Soc., Dalton Trans.*, 2263 (1975).

18. V. Chowdry, W. R. Pretzer, D. N. Rai, and R. W. Rudolph, *J. Amer. Chem. Soc.*, **95**, 4560 (1973).

19. M. R. Churchill and B. G. Deboer, *Inorg. Chem.*, **13**, 1411 (1974).

20. M. R. Churchill and B. G. Deboer, *J. Chem. Soc., Chem. Commun.*, 1326 (1972).

21. M. R. Churchill and K. Gold, *Inorg. Chem.*, **10**, 1928 (1971).

22. M. R. Churchill and K. Gold, *Inorg. Chem.*, **12**, 1157 (1973).

23. M. R. Churchill and K. Gold, *J. Amer. Chem. Soc.*, **92**, 1180 (1970).

24. M. R. Churchill and K. Gold, *J. Chem. Soc., Chem. Commun.*, 901 (1972).

25. M. R. Churchill, K. Gold, J. N. Francis, and M. F. Hawthorne, *J. Amer. Chem. Soc.*, **91**, 1222 (1969).

26. M. R. Churchill and A. H. Reis, Jr., *J. Chem. Soc., Dalton Trans.*, 1314 (1972).

27. M. R. Churchill and A. H. Reis, Jr., *J. Chem. Soc., Dalton Trans.*, 1317 (1972).

28. M. R. Churchill, A. H. Reis, Jr., J. N. Francis, and M. F. Hawthorne, *J. Amer. Chem. Soc.*, **92**, 4993 (1970).

29. M. R. Churchill, A. H. Reis, Jr., D. A. T. Young, G. R. Willey, and M. F. Hawthorne, *J. Chem. Soc., Sect. D*, 298 (1971).

30. B. G. DeBoer, A. Zalkin, and D. H. Templeton, *Inorg. Chem.*, **7**, 2288 (1968).

31. M. J. S. Dewar, *Bull. Soc. Chim. Fr.*, **18**, C79 (1951).

32. R. P. Dodge and V. P. Schomaker, *J. Organometal. Chem.*, **3**, 274 (1965).

33. E. Dubler, M. Textor, H-R. Oswald, and A. Salzer, *Angew. Chem. Intern. Ed.* **13**, 135 (1974).

34. G. B. Dunks and M. F. Hawthorne, *J. Amer. Chem. Soc.*, **92**, 7213 (1970).

35. G. B. Dunks, M. M. McKown, and M. F. Hawthorne, *J. Amer. Chem. Soc.*, **93**, 2541 (1971).

36. D. F. Dustin, G. B. Dunks, and M. F. Hawthorne, *J. Amer. Chem. Soc.*, **95**, 1109 (1973).

37. D. F. Dustin, W. J. Evans, and M. F. Hawthorne, *J. Chem. Soc., Chem. Commun.*, 805 (1973).

38. D. F. Dustin, W. J. Evans, C. J. Jones, R. J. Wiersema, H. Gong, S. Chan, and M. F. Hawthorne, *J. Amer. Chem. Soc.*, **96**, 3085 (1974).
39. D. F. Dustin and M. F. Hawthorne, *Inorg. Chem.*, **12**, 1380 (1973).
40. D. F. Dustin and M. F. Hawthorne, *J. Amer. Chem. Soc.*, **96**, 3462 (1974).
41. D. F. Dustin and M. F. Hawthorne, *J. Chem. Soc., Chem. Commun.*, 1329 (1972).
42. W. J. Evans, G. B. Dunks, and M. F. Hawthorne, *J. Amer. Chem. Soc.*, **95**, 4565 (1973).
43. W. J. Evans and M. F. Hawthorne, *Inorg. Chem.*, **13**, 869 (1974).
44. W. J. Evans and M. F. Hawthorne, *J. Amer. Chem. Soc.*, **93**, 3063 (1971).
45. W. J. Evans and M. F. Hawthorne, *J. Amer. Chem. Soc.*, **96**, 301 (1974).
46. W. J. Evans and M. F. Hawthorne, *J. Chem. Soc., Chem. Commun.*, 611 (1972).
47. W. J. Evans and M. F. Hawthorne, *J. Chem. Soc., Chem. Commun.*, 706 (1973).
48. W. J. Evans and M. F. Hawthorne, *J. Chem. Soc., Chem. Commun.*, 38 (1974).
49. W. J. Evans, C. J. Jones, B. Štíbr, R. A. Grey, and M. F. Hawthorne, *J. Amer. Chem. Soc.*, **96**, 7405 (1974).
50. W. J. Evans, C. J. Jones, B. Štíbr, and M. F. Hawthorne, *J. Organometal. Chem.*, **60**, C27 (1973).
51. G. Evrard, J. A. Ricci, Jr., I. Bernal, W. J. Evans, D. F. Dustin, and M. F. Hawthorne, *J. Chem. Soc., Chem. Commun.*, 234 (1974).
52. J. N. Francis and M. F. Hawthorne, *Inorg. Chem.*, **10**, 594 (1971).
53. J. N. Francis and M. F. Hawthorne, *Inorg. Chem.*, **10**, 863 (1971).
54. J. N. Francis and M. F. Hawthorne, *J. Amer. Chem. Soc.*, **90**, 1663 (1968).
55. J. N. Francis, C. J. Jones, and M. F. Hawthorne, *J. Amer. Chem. Soc.*, **94**, 4878 (1972).
56. D. A. Franz, V. R. Miller, and R. N. Grimes, *J. Amer. Chem. Soc.*, **94**, 412 (1972).
57. D. F. Gaines and T. V. Iorns, *J. Amer. Chem. Soc.*, **89**, 4249 (1967).
58. D. F. Gaines and T. V. Iorns, *J. Amer. Chem. Soc.*, **90**, 6617 (1968).
59. D. F. Gaines, J. W. Lott, and J. C. Calabrese, *J. Chem. Soc., Chem. Commun.*, 295 (1973).
60. P. M. Garrett, T. A. George, and M. F. Hawthorne, *Inorg. Chem.*, **8**, 2008 (1969).
61. T. A. George and M. F. Hawthorne, *Inorg. Chem.*, **8**, 1801 (1969).
62. T. A. George and M. F. Hawthorne, *J. Amer. Chem. Soc.*, **90**, 1661 (1968).
63. T. A. George and M. F. Hawthorne, *J. Amer. Chem. Soc.*, **91**, 5475 (1969).
64. E. D. German and M. E. Dyatkina, *Zh. Strukt. Khim.*, **7**, 866 (1966).
65. H. B. Gray, D. N. Hendrickson, and Y. S. Sohn, *Inorg. Chem.*, **10**, 1559 (1971).
66. B. M. Graybill and M. F. Hawthorne, *Inorg. Chem.*, **8**, 1799 (1969).
67. M. Green, J. Howard, J. L. Spencer, and F. G. A. Stone, *J. Chem. Soc., Chem. Commun.*, 153 (1974).
68. M. Green, J. Howard, J. L. Spencer, and F. G. A. Stone, *J. Chem. Soc., Dalton Trans.*, 2274 (1975).
69. M. Green, J. L. Spencer, F. G. A. Stone, and A. J. Welch, *J. Chem. Soc., Chem. Commun.*, 571 (1974).
70. M. Green, J. L. Spencer, F. G. A. Stone, and A. J. Welch, *J. Chem. Soc., Chem. Commun.*, 794 (1974).
71. M. Green, J. L. Spencer, F. G. A. Stone, and A. J. Welch, *J. Chem. Soc., Dalton Trans.*, 179 (1975).
72. P. T. Greene and R. F. Bryan, *Inorg. Chem.*, **9**, 1464 (1970).
73. N. N. Greenwood, C. G. Savory, R. N. Grimes, L. G. Sneddon, A. Davison, and S. S. Wreford, *J. Chem. Soc., Chem. Commun.*, 718 (1974).
74. R. N. Grimes, *Ann. N.Y. Acad. Sci.*, **239**, 180 (1974).
75. R. N. Grimes, *Carboranes*, Academic Press, New York, 1970.

76. R. N. Grimes, *J. Amer. Chem. Soc.*, **93**, 261 (1971).
77. R. N. Grimes, *Pure Appl. Chem.*, **39**, 455 (1974).
78. R. N. Grimes, Rev. Silicon, Germanium, Tin, and Lead Comp., in press.
79. R. N. Grimes, D. C. Beer, L. G. Sneddon, V. R. Miller, and R. Weiss, *Inorg. Chem.*, **13**, 1138 (1974).
80. R. N. Grimes and W. J. Rademaker, *J. Amer. Chem. Soc.*, **91**, 6498 (1969).
81. R. N. Grimes, W. J. Rademaker, M. L. Denniston, R. F. Bryan, and P. T. Greene, *J. Amer. Chem. Soc.*, **94**, 1865 (1972).
82. R. N. Grimes, A. Zalkin, and W. T. Robinson, *Inorg. Chem.*, **15**, 2274 (1976).
83. F. V. Hansen, R. G. Hazell, C. Hyatt and G. D. Stucky, *Acta Chem. Scand.*, **27**, 1210 (1973).
84. C. B. Harris, *Inorg. Chem.*, **7**, 1517 (1969).
85. D. T. Haworth, *Endeavour*, 16 (1972).
86. M. F. Hawthorne, *Accounts Chem. Res.*, **1**, 281 (1968).
87. M. F. Hawthorne, *Pure Appl. Chem.*, **29**, 547 (1972).
88. M. F. Hawthorne, *Pure Appl. Chem.*, **33**, 475 (1973).
89. M. F. Hawthorne and T. D. Andrews, *J. Amer. Chem. Soc.*, **87**, 2496 (1965).
90. M. F. Hawthorne and T. D. Andrews, *J. Chem. Soc., Chem. Commun.*, **19**, 443 (1965).
91. M. F. Hawthorne, T. D. Andrews, P. M. Garrett, F. P. Olsen, M. Reintjes, F. N. Tebbe, L. F. Warren, P. A. Wegner, and D. C. Young, *Inorg. Syn.*, **10**, 91 (1967).
92. M. F. Hawthorne and G. B. Dunks, *Science*, **178**, 462 (1972).
93. M. F. Hawthorne and T. A. George, *J. Amer. Chem. Soc.*, **89**, 7114 (1967).
94. M. F. Hawthorne, M. K. Kaloustian, and R. J. Wiersema, *J. Amer. Chem. Soc.*, **93**, 4912 (1971).
95. M. F. Hawthorne and R. L. Pilling, *J. Amer. Chem. Soc.*, **87**, 3987 (1965).
96. M. F. Hawthorne and A. D. Pitts, *J. Amer. Chem. Soc.*, **89**, 7115 (1967).
97. M. F. Hawthorne and H. W. Ruhle, *Inorg. Chem.*, **8**, 176 (1969).
98. M. F. Hawthorne, L. F. Warren, Jr., K. P. Callahan, and N. F. Travers, *J. Amer. Chem. Soc.*, **93**, 2407 (1971).
99. M. F. Hawthorne and P. A. Wegner, *J. Amer. Chem. Soc.*, **90**, 896 (1968).
100. M. F. Hawthorne, D. C. Young, T. D. Andrews, D. V. Howe, R. L. Pilling, A. D. Pitts, M. Reintjes, L. F. Warren, and P. A. Wegner, *J. Amer. Chem. Soc.*, **90**, 879 (1968).
101. M. F. Hawthorne, D. C. Young, and P. A. Wegner, *J. Amer. Chem. Soc.*, **87**, 1818 (1965).
102. R. H. Herber, *Inorg. Chem.*, **8**, 174 (1969).
103. E. L. Hoel and M. F. Hawthorne, *J. Amer. Chem. Soc.*, **96**, 4676 (1974).
104. E. L. Hoel, C. E. Strouse, and M. F. Hawthorne, *Inorg. Chem.*, **13**, 1388 (1974).
105. F. J. Hollander, D. H. Templeton, and A. Zalkin, *Inorg. Chem.*, **12**, 2262 (1973).
106. J. W. Howard and R. N. Grimes, *Inorg. Chem.*, **11**, 263 (1972).
107. J. W. Howard and R. N. Grimes, *J. Amer. Chem. Soc.*, **91**, 6499 (1969).
108. D. E. Hyatt, J. L. Little, J. T. Moran, F. R. Scholer, and L. J. Todd, *J. Amer. Chem. Soc.*, **89**, 3342 (1967).
109. D. E. Hyatt, D. A. Owen, and L. J. Todd, *Inorg. Chem.*, **5**, 1749 (1966).
110. D. E. Hyatt, F. R. Scholer, L. J. Todd, and J. L. Warner, *Inorg. Chem.*, **6**, 2229 (1967).
111. International Union of Pure and Applied Chemistry, *Pure Appl. Chem.*, **30**, 683 (1972).
112. C. J. Jones, W. J. Evans, and M. F. Hawthorne, *J. Chem. Soc., Chem. Commun.*, 543 (1973).

113. C. J. Jones, J. N. Francis, and M. F. Hawthorne, *J. Amer. Chem. Soc.*, **94**, 8391 (1972).
114. C. J. Jones, J. N. Francis, and M. F. Hawthorne, *J. Amer. Chem. Soc.*, **95**, 7633 (1973).
115. C. J. Jones, J. N. Francis, and M. F. Hawthorne, *J. Chem. Soc., Chem. Commun.*, 900 (1972).
116. C. J. Jones and M. F. Hawthorne, *Inorg. Chem.*, **12**, 608 (1973).
117. H. D. Kaesz, R. Bau, H. A. Beall, and W. N. Lipscomb, *J. Amer. Chem. Soc.*, **89**, 4218 (1967).
118. M. K. Kaloustian, R. J. Wiersema, and M. F. Hawthorne, *J. Amer. Chem. Soc.*, **94**, 6679 (1972).
119. W. H. Knoth, *Inorg. Chem.*, **10**, 598 (1971).
120. W. H. Knoth, *J. Amer. Chem. Soc.*, **89**, 1274 (1967).
121. W. H. Knoth, *J. Amer. Chem. Soc.*, **89**, 3342 (1967).
122. W. H. Knoth, J. L. Little, J. R. Lawrence, F. R. Scholer, and L. J. Todd, *Inorg. Syn.*, **11**, 33 (1968).
123. W. H. Knoth, J. L. Little, and L. J. Todd, *Inorg. Syn.*, **11**, 41 (1968).
124. R. N. Leyden and M. F. Hawthorne, *Inorg. Chem.*, **8**, 2018 (1975).
125. W. N. Lipscomb, *Boron Hydrides*, Benjamin, New York, 1963.
126. W. N. Lipscomb, *Science*, **153**, 373 (1966).
127. J. L. Little, J. T. Moran, and L. J. Todd, *J. Amer. Chem. Soc.*, **89**, 5495 (1967).
128. J. L. Little, P. S. Welcker, N. J. Loy, and L. J. Todd, *Inorg. Chem.*, **9**, 63 (1970).
129. F. Y. Lo, C. E. Strouse, K. P. Callahan, C. B. Knobler, and M. F. Hawthorne, *J. Amer. Chem. Soc.*, **97**, 428 (1975).
130. B. Longato, F. Morandini, and S. Bresadola, *Gazz. Chim. Ital.*, **104**, 805 (1974).
131. C. P. Magee, L. G. Sneddon, D. C. Beer, and R. N. Grimes, *J. Organometal. Chem.*, **86**, 159 (1975).
132. A. H. Maki and T. E. Berry, *J. Amer. Chem. Soc.*, **87**, 4437 (1965).
133. R. Mason, K. M. Thomas, and D. M. P. Mingos, *J. Amer. Chem. Soc.*, **95**, 3802 (1973).
134. D. S. Matteson and R. E. Grunzinger, Jr., *Inorg. Chem.*, **13**, 671 (1974).
135. W. M. Maxwell and R. N. Grimes, *J. Chem. Soc., Chem. Commun.*, 943 (1975).
136. W. M. Maxwell and R. N. Grimes, manuscript in preparation.
137. W. M. Maxwell and R. N. Grimes, to be submitted for publication.
138. W. M. Maxwell, V. R. Miller, and R. N. Grimes, *Inorg. Chem.*, **15**, 1343 (1976).
139. W. M. Maxwell, V. R. Miller, and R. N. Grimes, *J. Amer. Chem. Soc.*, **96**, 7116 (1974).
140. W. M. Maxwell, V. R. Miller, and R. N. Grimes, *J. Amer. Chem. Soc.*, **98**, 4818 (1976).
141. W. M. Maxwell, E. Sinn, and R. N. Grimes, *J. Amer. Chem. Soc.*, **98**, 3490 (1976).
142. W. M. Maxwell, E. Sinn, and R. N. Grimes, *J. Chem. Soc., Chem. Commun.*, 389 (1976).
143. G. D. Mercer, M. Tribo, and F. R. Scholer, *Inorg. Chem.*, **14**, 765 (1975).
144. B. M. Mikhailov and T. V. Potapova, *Izv. Akad. Nauk SSSR, Ser. Khim.*, 1153 (1968).
145. V. R. Miller and R. N. Grimes, *Inorg. Chem.*, **11**, 862 (1972).
146. V. R. Miller and R. N. Grimes, *J. Amer. Chem. Soc.*, **95**, 2830 (1973).
147. V. R. Miller and R. N. Grimes, *J. Amer. Chem. Soc.*, **95**, 5078 (1973).
148. V. R. Miller and R. N. Grimes, *J. Amer. Chem. Soc.*, **97**, 4213 (1975).
149. V. R. Miller and R. N. Grimes, *J. Amer. Chem. Soc.*, **98**, 1600 (1976).
150. V. R. Miller and R. N. Grimes, unpublished results.

151. V. R. Miller, L. G. Sneddon, D. C. Beer, and R. N. Grimes, *J. Amer. Chem. Soc.*, **96**, 3090 (1974).
152. V. R. Miller, R. Weiss, and R. N. Grimes, unpublished results.
153. D. M. P. Mingos, *Nature Phys. Sci.*, **236**, 99 (1972).
154. E. L. Muetterties and W. H. Knoth, *Polyhedral Boranes*, Marcel Dekker, New York, 1968.
155. R. S. Mulliken, *J. Amer. Chem. Soc.*, **74**, 811 (1952).
156. R. R. Olsen and R. N. Grimes, *J. Amer. Chem. Soc.*, **92**, 5072 (1970).
157. T. Onak, *Organometal. Chem.*, **1**, 104 (1972).
158. T. P. Onak and G. T. F. Wong, *J. Amer. Chem. Soc.*, **92**, 5226 (1970).
159. I. Pavlik, E. Maxova, and E. Vecernikova, *Z. Chem.*, **12**, 26 (1972).
160. T. E. Paxson and M. F. Hawthorne, *J. Amer. Chem. Soc.*, **96**, 4674 (1974).
161. T. E. Paxson, M. K. Kaloustian, G. M. Tom, R. J. Wiersema, and M. F. Hawthorne, *J. Amer. Chem. Soc.*, **94**, 4882 (1972).
162. J. R. Pladziewicz and J. H. Epsenson, *J. Amer. Chem. Soc.*, **95**, 56 (1973).
163. J. Plesek, S. Hermanek, and Z. Janousek, *Chem. and Ind.*, **108** (1974).
164. J. Plesek, B. Štíbr, and S. Hermanek, *Synth. Inorg. Metal-Org. Chem.*, **3**, 291 (1973).
165. L. O. Pont, A. R. Siedle, M. S. Lazarus, and W. L. Jolly, *Inorg. Chem.*, **13**, 483 (1974).
166. G. Popp and M. F. Hawthorne, *Inorg. Chem.*, **10**, 391 (1971).
167. G. Popp and M. F. Hawthorne, *J. Amer. Chem. Soc.*, **90**, 6553 (1968).
168. R. R. Rietz, D. F. Dustin, and M. F. Hawthorne, *Inorg. Chem.*, **13**, 1580 (1974).
169. W. T. Robinson and R. N. Grimes, *Inorg. Chem.*, **14**, 3056 (1975).
170. R. W. Rudolph and V. Chowdhry, *Inorg. Chem.*, **13**, 248 (1974).
171. R. W. Rudolph and W. R. Pretzer, *Inorg. Chem.*, **11**, 1974 (1972).
172. R. W. Rudolph, R. L. Voorhees, and R. E. Cochoy, *J. Amer. Chem. Soc.*, **92**, 3351 (1970).
173. H. W. Ruhle and M. F. Hawthorne, *Inorg. Chem.*, **7**, 2279 (1968).
174. D. St. Clair, A. Zalkin, and D. H. Templeton, *Inorg. Chem.*, **8**, 2080 (1969).
175. D. St. Clair, A. Zalkin, and D. H. Templeton, *Inorg. Chem.*, **10**, 2587 (1971).
176. D. St. Clair, A. Zalkin, and D. H. Templeton, *Inorg. Chem.*, **11**, 377 (1972).
177. D. St. Clair, A. Zalkin, and D. H. Templeton, *J. Amer. Chem. Soc.*, **92**, 1173 (1970).
178. C. G. Salentine and M. F. Hawthorne, *J. Chem. Soc., Chem. Commun.*, 560 (1973).
179. C. G. Salentine and M. F. Hawthorne, *J. Amer. Chem. Soc.*, **97**, 426 (1975).
180. C. G. Salentine and M. F. Hawthorne, *J. Amer. Chem. Soc.*, **97**, 6382 (1975).
181. C. G. Salentine, R. R. Rietz, and M. F. Hawthorne, *Inorg. Chem.*, **13**, 3025 (1974).
182. A. Salzer and H. Werner, *Angew. Chem. Internat. Ed.*, **11**, 930 (1972).
183. C. G. Savory and M. G. H. Wallbridge, *J. Chem. Soc., Dalton Trans.*, **8/9**, 918 (1972).
184. F. R. Scholer and L. J. Todd, *J. Organometal. Chem.*, **14**, 261 (1968).
185. A. R. Siedle, *J. Organometal. Chem.*, **90**, 249 (1975).
186. F. R. Scholer and L. J. Todd, *Prep. Inorg. React.*, **7**, 1 (1971).
187. A. R. Siedle, G. M. Bodner, and L. J. Todd, *J. Organometal. Chem.*, **33**, 137 (1971).
188. R. Snaith and K. Wade, *Int. Rev. Sci., Inorg. Chem. M.T.P.* (Med. Tech. Publ. Co.) *Ser. 1*, **1**, 139 (1972); *Ser. 2*, **1**, 95 (1975).
189. L. G. Sneddon, D. C. Beer, and R. N. Grimes, *J. Amer. Chem. Soc.*, **95**, 6623 (1973).
190. L. G. Sneddon and R. N. Grimes, *J. Amer. Chem. Soc.*, **94**, 7161 (1972).
191. J. L. Spencer, M. Green, and F. G. A. Stone, *J. Chem. Soc., Chem. Commun.*, 1178 (1972).
192. A. Tabereaux and R. N. Grimes, *Inorg. Chem.*, **12**, 792 (1973).
193. R. W. Taft, E. Price, I. R. Fox, I. C. Lewis, K. K. Andersen, and G. T. Davis, *J. Amer. Chem. Soc.*, **85**, 709 (1963).

194. R. W. Taft, E. Price, I. R. Fox, I. C. Lewis, K. K. Andersen, and G. T. Davis, *J. Amer. Chem. Soc.*, **85**, 3146 (1963).
195. F. N. Tebbe, P. M. Garrett, and M. F. Hawthorne, *J. Amer. Chem. Soc.*, **90**, 869 (1968).
196. M. L. Thompson and R. N. Grimes, *Inorg. Chem.*, **11**, 1925 (1972).
197. M. L. Thompson and R. N. Grimes, *J. Amer. Chem. Soc.*, **93**, 6677 (1971).
198. L. J. Todd, *Pure Appl. Chem.*, **30**, 587 (1972).
199. L. J. Todd, A. R. Burke, A. R. Garber, H. T. Silverstein, and B. N. Storhoff, *Inorg. Chem.*, **9**, 2175 (1970).
200. L. J. Todd, A. R. Burke, H. T. Silverstein, J. L. Little, and G. S. Wikholm, *J. Amer. Chem. Soc.*, **91**, 3376 (1969).
201. L. J. Todd, J. L. Little, and H. T. Silverstein, *Inorg. Chem.*, **8**, 1698 (1969).
202. L. J. Todd, I. C. Paul, J. L. Little, P. S. Welcker, and C. R. Peterson, *J. Amer. Chem. Soc.*, **90**, 4489 (1968).
203. D. Voet and W. N. Lipscomb, *Inorg. Chem.*, **6**, 113 (1967).
204. R. L. Voorhees and R. W. Rudolph, *J. Amer. Chem. Soc.*, **91**, 2173 (1969).
205. K. Wade, *Inorg. Nucl. Chem. Lett.*, **8**, 559 (1972).
206. K. Wade, *Inorg. Nucl. Chem. Lett.*, **8**, 563 (1972).
207. K. Wade, *Inorg. Nucl. Chem. Lett.*, **8**, 823 (1972).
208. K. Wade, *J. Chem. Soc.,Sect. D.*, 792 (1971).
209. M. G. H. Wallbridge and C. G. Savory, *J. Chem. Soc., Sect. D.*, 622 (1971).
210. L. F. Warren and M. F. Hawthorne, *J. Amer. Chem. Soc.*, **89**, 470 (1967).
211. L. F. Warren and M. F. Hawthorne, *J. Amer. Chem. Soc.*, **90**, 4823 (1968).
212. L. F. Warren and M. F. Hawthorne, *J. Amer. Chem. Soc.*, **92**, 1157 (1970).
213. P. A. Wegner, L. J. Guggenberger and E. L. Muetterties, *J. Amer. Chem. Soc.*, **92**, 3473 (1970).
214. P. A. Wegner and M. F. Hawthorne, *J. Chem. Soc., Chem. Commun.*, 861 (1966).
215. R. Weiss and R. F. Bryan, unpublished results.
216. R. Weiss and R. N. Grimes, *J. Organometal. Chem.*, **113**, 29 (1976).
217. R. Weiss and R. N. Grimes, unpublished results.
218. A. J. Welch, *J. Chem. Soc., Dalton Trans.*, 1473 (1975).
219. A. J. Welch, *J. Chem. Soc., Dalton Trans.*, 2270 (1975).
220. P. S. Welcker and L. J. Todd, *Inorg. Chem.*, **9**, 286 (1970).
221. H. Werner and A. Salzer, *Synth. Inorg. Metal-Organic Chem.*, **2**, 239 (1972).
222. R. J. Wiersema and M. F. Hawthorne, *J. Amer. Chem. Soc.*, **96**, 761 (1974).
223. R. A. Wiesboeck and M. F. Hawthorne, *J. Amer. Chem. Soc.*, **86**, 1642 (1964).
223a. G. A. Wikholm and L. J. Todd, *J. Organometal. Chem.*, **71**, 219 (1974).
224. R. E. Williams, *Inorg. Chem.*, **10**, 210 (1971).
225. R. E. Williams in *Progress in Boron Chemistry*, Vol. 2, Permagon Press, Oxford, 1970, Chapter 2, p. 37.
226. R. J. Wilson, L. F. Warren, and M. F. Hawthorne, *J. Amer. Chem. Soc.*, **91**, 758 (1969).
227. R. M. Wing, *J. Amer. Chem. Soc.*, **89**, 5599 (1967).
228. R. M. Wing, *J. Amer. Chem. Soc.*, **90**, 4828 (1968).
229. R. M. Wing, *J. Amer. Chem. Soc.*, **92**, 1187 (1970).
230. K-S. Wong and R. N. Grimes, manuscript in preparation.
231. D. A. T. Young, T. E. Paxson, and M. F. Hawthorne, *Inorg. Chem.*, **10**, 786 (1971).
232. D. A. T. Young, R. J. Wiersema, and M. F. Hawthorne, *J. Amer. Chem. Soc.*, **93**, 5687 (1971).
233. D. A. T. Young, G. R. Willey, M. F. Hawthorne, M. R. Churchill, and R. H. Reis, Jr., *J. Amer. Chem. Soc.*, **92**, 6663 (1970).

234. L. I. Zakharkin and R. Kh. Bikkineev, *Izv. Akad. Nauk SSSR, Ser. Khim.*, 2128 (1974).
235. L. I. Zakharkin and R. Kh. Bikkineev, *Izv. Akad. Nauk SSSR, Ser. Khim.*, 2377 (1974).
236. L. I. Zakharkin and R. Kh. Bikkineev, *Zh. Obshch. Khim.*, **44**, 2473 (1974).
237. L. I. Zakharkin and R. Kh. Bikkineev, *Zh. Obshch. Khim.*, **45**, 476 (1975).
238. L. I. Zakharkin and V. N. Kalinin, *Zh. Obshch. Khim.*, **42**, 714 (1972).
239. L. I. Zakharkin, V. N. Kalinin, and N. P. Levina, *Zh. Obshch. Khim.*, **44**, 2478 (1974).
240. L. I. Zakharkin, V. V. Kobak, A. Kovredov, and R. Kh. Bikkineev, *Izv. Akad. Nauk SSSR, Ser. Khim.*, 921 (1974).
241. L. I. Zakharkin and V. I. Kyskin, *Izv. Akad. Nauk SSSR, Ser. Khim.*, 2142 (1970).
242. L. I. Zakharkin and A. I. L'vov, *Zh. Obshch. Khim.*, **41**, 1880 (1971).
243. A. Zalkin, T. E. Hopkins, and D. H. Templeton, *Inorg. Chem.*, **5**, 1189 (1966).
244. A. Zalkin, T. E. Hopkins, and D. H. Templeton, *Inorg. Chem.*, **6**, 1911 (1967).
245. A. Zalkin, D. H. Templeton, and T. E. Hopkins, *J. Amer. Chem. Soc.*, **87**, 3988 (1965).

Chapter 3

Homogeneous Catalysis by Arene Group-VIB Tricarbonyls

MICHAEL F. FARONA

I. INTRODUCTION

Transition-metal complexes are particularly well suited for promoting certain organic reactions because they possess a combination of properties that makes them relatively unique as a class. First of all, they serve as a center for organic molecules to gather around, and once coordination takes place, a variety of processes can occur which are either directly or indirectly promoted by the metal. For example, a ligand migration might occur, there might be a transfer of electrons between the ligands or between the ligands and the metal, a hydrogen atom might be transferred via the metal, or *cis* ligands might react. After the new ligand is formed on the metal, if its donor properties are decreased with respect to the incoming ligands, the complex can break down, releasing the product; the initial metal site is now available to repeat the process.

In most cases, the transition metal must, at some time during the catalysis, be coordinatively unsaturated in order to provide a site of attachment for the incoming ligands. The coordinatively unsaturated state can come about by a variety of events. For example, a ligand in the coordination sphere might dissociate, the metal complex might be capable of accepting additional ligands and expand its coordination number (addition of ligand with no change in oxidation state of the metal), or the metal complex might be capable of undergoing an oxidative addition (addition of ligand accompanied by a change in oxidation state of the metal).

MICHAEL F. FARONA • Department of Chemistry, The University of Akron, Akron, Ohio.

Furthermore, the proper combination of coordination number and electronic configuration on the metal can cause the complex to be catalytically active. This combination suggests a comparison of certain transition-metal systems to reactive organic intermediates.[1] For example, a low-spin, five-coordinate, d^7 complex such as $[Co(CN)_5]^{3-}$ resembles a free radical, and indeed, reactions which parallel those of free radicals are observed for this complex, viz., abstraction and dimerization. A four-coordinate d^8 complex such as $IrCl(CO)(PPh_3)_2$ may be likened to a carbene, whereas a five-coordinate, low-spin d^6 complex may be compared to a carbonium ion and a five-coordinate d^8 system resembles a carbanion. By no means does the proper combination of electronic configuration and coordination number indicate that a particular molecule will be catalytically active; for example, a fairly large number of five-coordinate, low-spin cobalt(II) complexes are known, but only very few of them have been reported to be active hydrogenation catalysts. Other factors such as the bonding nature and steric properties of the coordinated ligands, reaction conditions, solvents, etc., must also be considered.

Following a brief, general review of the chemistry of the arene Group-VIB tricarbonyls (Section II), discussions of the areas of catalysis will be reviewed in succeeding sections.

The discussions of the areas of catalysis by arene Group-VIB tricarbonyls are arranged according to reaction type. The major areas where these catalysts have been shown to be active in homogeneous systems are Friedel–Crafts and related reactions, hydrogenation of olefins, isomerization of olefins, olefin metathesis, and reactions of acetylenes.

Whereas it is the intention of this review to present the chemistry promoted by arene Group-VIB tricarbonyls in homogeneous systems, other related catalysts derived from chromium, molybdenum, and tungsten hexacarbonyls, both homogeneous and heterogeneous, are also included. The choice of which related catalysts to include is purely arbitrary, but the catalytic systems are generally limited to mononuclear species which contain an $M(CO)_x$ group. Thus, catalysis by bis(arene) compounds is generally not considered in the following discussions. Also not included in this review are areas of catalysis where no specific example of the use of arene Group-VIB tricarbonyls has been reported but where extensive use of related (as defined above) catalysts has been published.

II. CHEMISTRY OF ARENE GROUP-VIB TRICARBONYLS

The molecules $ArM(CO)_3$ (Ar = arene; M = Cr, Mo, W) are *per se* coordinatively saturated. However, a variety of reactions have been exhibited by these molecules which indicate that the coordinatively unsaturated state can occur during the course of reaction; hence, it would be envisaged that these complexes might be catalytically active in certain organic reactions. This section is devoted to covering briefly certain aspects of the chemistry of $ArM(CO)_3$ molecules which are pertinent to the discussions that follow concerning their activity in catalytic systems.

The compounds $ArCr(CO)_3$ may be prepared by several different procedures. The most common method is by direct reaction of the appropriate aromatic compound with $Cr(CO)_6$ at elevated temperatures or under the influence of ultraviolet light, according to the following equation:

$$Ar + Cr(CO)_6 \longrightarrow ArCr(CO)_3 + 3CO$$

Often, the reaction is assisted by use of donor solvents such as diglyme (dimethyl ether of diethylene glycol). A very large variety of $ArCr(CO)_3$ complexes are known; both activating and deactivating substituents may be present on the coordinated ring. In cases where direct reaction is not successful, the final complex can be made by one of the following methods:

a. Ring exchange with $ArCr(CO)_3$.

$$Ar' + ArCr(CO)_3 \longrightarrow Ar'Cr(CO)_3 + Ar$$

b. Reaction of functional groups on coordinated arene.

$$(C_6H_5CH_2OH)Cr(CO)_3 \xrightarrow{\text{HCl}} (C_6H_5CH_2Cl)Cr(CO)_3$$

$$(C_6H_5COOCH_3)Cr(CO)_3 \xrightarrow{\text{H}_2\text{O}} (C_6H_5COOH)Cr(CO)_3$$

Examples of $ArM(CO)_3$ where M = Mo and W are much more limited than the corresponding Cr derivatives. It appears that π-arene complexes of molybdenum and tungsten are generally limited to benzene and its methylated derivatives. Some of the π-arene complexes can be made directly, and, particularly for tungsten, displacement of three acetonitrile molecules in the complex $(CH_3CN)_3W(CO)_3$ by several arenes has proved effective.

This area of preparative chemistry of $ArM(CO)_3$ has been reviewed thoroughly and the reader is referred to those sources for further information.[2–4]

$ArM(CO)_3$ molecules are known to undergo several different types of reactions:

1. Displacement of the arene ring by various donor molecules.
2. Ring exchange with other aromatic molecules.
3. Substitution on the coordinated arene.
4. Reactions of functional groups on the coordinated arene.
5. Substitution of CO.
6. Total substitution of the arene and CO by various ligands.
7. Addition and oxidative addition.

In the following section are examples of each type of reaction together with tables for completeness for the chemistry of $ArM(CO)_3$.

A. Displacement of the Arene Ring by Donor Molecules

In this process the reaction proceeds according to the following general equation:

$$ArM(CO)_3 + 3 L \longrightarrow L_3M(CO)_3 + Ar$$

Reactions of the above type have been observed for all three metals, and a variety of arenes have been utilized. In addition, the triolefin molecules, 1,3,5-cycloheptatriene and 1,3,5-cyclooctatriene, have also been used in place of arene groups. In general, the ligands (L) involved in the process have been molecules whose donor atom is a member of the Group-VA elements. Reactions are summarized in Table I.

In reactions of $ArM(CO)_3$ with tetralkylammonium halides and pseudo-halides, it was found that the arene ring was displaced with retention of the tricarbonylmetal group. However, dimeric anionic products (1) were obtained, as shown in Equation (1):

$$3 R_4NX + 2 ArM(CO)_3 \longrightarrow (R_4N)_3 \begin{bmatrix} OC & X & CO \\ OC-M-X-M-CO \\ OC & X & CO \end{bmatrix} + 2 Ar \quad (1)$$

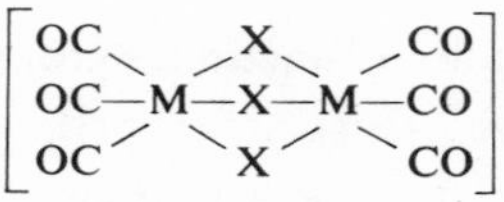

(1)

In this reaction Ar = toluene or mesitylene; M = Mo or W; X = F, Cl, Br, I, OH, SCN, and N_3.[16] In some cases, doubly bridged anions were formed, the proposed structure of which is shown in (2). For the doubly-bridged complexes, X = OH, Cl, I.[16]

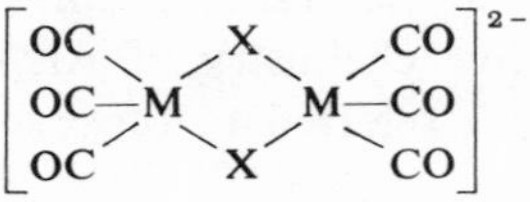

$$\begin{bmatrix} OC & X & CO \\ OC-M & M-CO \\ OC & X & CO \end{bmatrix}^{2-}$$

(2)

Attempts to replace the coordinated arene ring by tridentate ligands have generally been unfruitful. In one example, $(terpy)Mo(CO)_3$ (terpy = 2,2′,2″-terpyridyl) was isolated and characterized from the reaction of terpy with $(mesitylene)Mo(CO)_3$. The analogous chromium and tungsten compounds could not be prepared.[15]

King and co-workers have carried out a number of attempted displacements of 1,3,5-cycloheptatriene from $M(CO)_3$ groups with mono-, tri-, and tetradentate phosphines and arsines. With tris(dimethylamino)phosphine some $L_3M(CO)_3$ complexes were detected in solution but not isolated.[17] With the corresponding arsine ligand no $L_3M(CO)_3$ compounds were obtained.[17] The tridentate ligand $[(C_6H_5)_2PCH_2CH_2]_2PC_6H_5$ displaced 1,3,5-cyclo-heptatriene from $C_7H_8Mo(CO)_3$ to make an adduct with the $M(CO)_3$ group.[18] The potentially tetradentate ligands $(C_6H_5)_2PCH_2CH_2P(C_6H_5)CH_2CH_2P$-$(C_6H_5)CH_2CH_2P(C_6H_5)_2$ and $[(C_6H_5)_2PCH_2CH_2]_3P$ both displaced 1,3,5-cycloheptatriene from $C_7H_8Mo(CO)_3$ to make tridentate adducts of molybdenum tricarbonyl.[19]

Very important to the discussion of reactions catalyzed by $ArM(CO)_3$ are the kinetics and mechanistic studies which have been carried out on displacement of coordinated arene and 1,3,5-cycloheptatriene by various phosphorus-donor ligands on molybdenum and tungsten complexes.[5,11,12,14] These reactions have been found to be first order in $ArM(CO)_3$ and in phosphine. The mechanism shown in Equation (2) has been proposed for displacement of arene,[11,12,14] and that for displacement of C_7H_8 is similar.

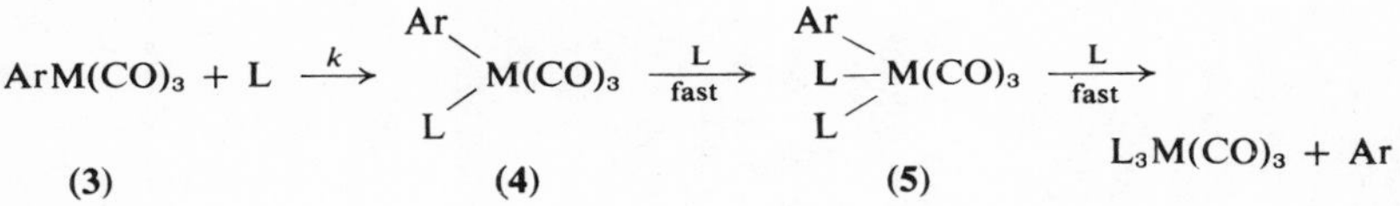

$$ArM(CO)_3 + L \xrightarrow{k} \underset{\underset{L}{(4)}}{\overset{\overset{Ar}{\diagdown}}{M(CO)_3}} \xrightarrow[\text{fast}]{L} \underset{\underset{L}{(5)}}{\overset{\overset{Ar}{\diagdown}}{L-M(CO)_3}} \xrightarrow[\text{fast}]{L} L_3M(CO)_3 + Ar \quad (2)$$

$$(3)$$

The starting complex (3) contains the arene ring as a six-electron donor, whereas the intermediates (4) and (5) coordinate the arene ring as four- and two-electron donors, respectively. Whereas the process from (4) to $L_3M(CO)_3$ is not precisely known, the kinetics of the reaction indicate that (4) is formed. In the above reaction of displacement of arene on molybdenum and tungsten complexes by Group-VA donor molecules, it was found that the reaction was catalyzed by certain oxygen-donor molecules such as trimethyl phosphate, dimethyl sulfoxide, and dimethylformamide.[14a]

B. Ring Exchange with Other Aromatic Molecules

This reaction is described by the general Equation (3):

$$ArM(CO)_3 + Ar' \longrightarrow Ar'M(CO)_3 + Ar \quad (3)$$

The Ar' entity may be chemically equivalent (^{14}C-labeled) or non-equivalent to Ar. The exchange reaction is normally carried out at elevated temperatures, but irradiation with ultraviolet light also induces the process to occur. Table II shows some examples of arene ring exchange.

Other aromatic ring-exchange reactions have been employed in the preparation of certain $Ar'M(CO)_3$ products, where $Ar' = C_6H_5OH$, C_6H_5-COOH, C_6H_5COOR, $C_6H_5NH_2$, $C_6H_5N(CH_3)_2$.[24,25] Where Ar' is diphenyl,

TABLE I

Displacement of Arene in ArM(CO)₃ by Donor Molecules

ArM(CO)$_3$[a]	Donor(L)	Reaction conditions	Comments on L$_3$M(CO)$_3$	Reference
(C$_7$H$_8$)Cr(CO)$_3$	P(OCH$_3$)$_3$	Hexane solvent, 4 hr	80% yield of *fac* isomer, colorless crystals	5
(tol)Cr(CO)$_3$	As(CH$_3$)$_2$(C$_6$H$_5$)	150–160°, 1 hr, ligand as solvent	41% yield, bright yellow crystals	6
(tol)Cr(CO)$_3$	P(C$_6$H$_5$)$_3$	160°, 2 hr, no solvent	52% yield	6
(tet)Cr(CO)$_3$	Pyridine	Reflux in pyridine 1 hr	42% yield	6
(mes)Cr(CO)$_3$	PF$_3$	140–150°, 15 hr	100% yield, colorless crystals which sublime (30°; 10 Torr)	7
(dur)Cr(CO)$_3$	P(OC$_6$H$_5$)$_3$	240°, 1.5 hr, ligand as solvent	30% yield, white crystals, light sensitive	8
(dur)Cr(CO)$_3$	P(OC$_4$H$_9$)$_3$	240°, 1.5 hr, ligand as solvent	Green liquid, 100% crude yield	9
(C$_6$H$_6$)Cr(CO)$_3$ or (dur)Cr(CO)$_3$	P(OC$_6$H$_5$)$_3$ As(C$_6$H$_5$)$_3$ Sb(C$_6$H$_5$)$_3$		White crystals Yellow crystals Yellow crystals	9
(C$_7$H$_8$)Cr(CO)$_3$	P(CH$_3$)(OCH$_3$)$_2$	Methylcyclohexane solvent, reflux 1 hr	Mixture of *fac* and *mer* isomers	10
(C$_7$H$_8$)Mo(CO)$_3$	P(OCH$_3$)$_3$	Room temp, 1 hr, ligand as solvent	90% Yield of *fac* isomer; colorless crystals	5
(mes)Mo(CO)$_3$	P(CH$_3$)(OCH$_3$)$_2$	Hexane solvent, reflux, 0.75 hr	*fac* Isomer, white crystals	10
(C$_6$H$_6$)Mo(CO)$_3$ (tol)Mo(CO)$_3$ (o-xyl)Mo(CO)$_3$ (m-xyl)Mo(CO)$_3$ (p-xyl)Mo(CO)$_3$ (mes)Mo(CO)$_3$ (TMB)Mo(CO)$_3$ (HMB)Mo(CO)$_3$ (N,N-DMA)Mo(CO)$_3$	P(OCH$_3$)$_3$	50° in either 1,2-Dichloroethane or methylcyclohexane	*fac*-Isomers obtained: kinetics and mechanistic studies	11

Complex	Ligand	Conditions	Product / Notes	Ref.
(mes)Mo(CO)₃ (p-xyl)Mo(CO)₃ (tol)Mo(CO)₃	PCl₃ PCl₂(C₆H₅) P(n-C₄H₉)₃	n-Heptane, CHCl₃, or Cl₂CH—CHCl₂, 25°	fac Isomers; kinetics and mechanistic studies	12
(1,3,5-COT)Mo(CO)₃	¹³CO	Cyclohexane solvent, room temperature, atmosphere of ¹³CO	¹³CO-enriched Mo(CO)₆	13
(tol)Mo(CO)₃	Morpholine	Benzene solution, room temperature	fac Isomer; white air-sensitive solid	14
(tol)Mo(CO)₃	Piperazine	Benzene solution, room temperature	fac Isomer; white solid	14
(mes)Mo(CO)₃	2,6-Dimethylpyrazine	Sealed-tube reaction, 60°, 3 hr	mer Isomer, red solid unstable in air and solution	15
(C₇H₈)W(CO)₃	P(OCH₃)₃	Ligand as solvent, room temperature, 1 hr	90% Yield of fac Isomer	5
(mes)W(CO)₃	P(CH₃)(OCH₃)₂	Hexane solvent, reflux 12 hr	fac Isomer, white crystals	10
(mes)W(CO)₃	P(CH₃)(OCH₃)₂	Methylcyclohexane	mer Isomer	10
(mes)W(CO)₃	2,6-Dimethylpyrazine	Sealed-tube reaction, 80°, 3 hr	mer Isomer, red solid, unstable in air and solution	15
(C₆H₆)W(CO)₃ (tol)W(CO)₃ (mes)W(CO)₃ (MBZ)W(CO)₃ (N,N-DMA)W(CO)₃ (anis)W(CO)₃	P(OCH₃)₃	1,2-Dichloroethane solvent, 50°	fac Isomers	14

TABLE II

Arene Ring-Exchange Reactions

$ArM(CO)_3{}^a$	Ar'	Conditions and comments	Reference
$(C_6H_6)Cr(CO)_3$	$^{14}C-C_6H_6$	80°, 12 hr; 0.9% exchange	20
		120°, 12 hr; 2.4% exchange	20
		160°, 12 hr; 22.5% exchange	20
$(tol)Cr(CO)_3$	$^{14}C-C_6H_5CH_3$	120°, 12 hr; 2.3% exchange	20
$(tol)Cr(CO)_3$	N,N-DMA	High temperature, 60% exchange	21
$(p\text{-}cym)W(CO)_3$	HMB	High temperature	22
$(p\text{-}cym)W(CO)_3$	C_6H_6	High temperature, no reaction	22
$(C_6H_6)Cr(CO)_3$	$^{14}C-C_6H_6$	Photochemical, $\lambda = 3660$ Å, 3 hr, 56.6% exchange	23
$(ClC_6H_5)Cr(CO)_3$	$^{14}C-C_6H_5Cl$	Same as with $^{14}C-C_6H_6$, 32.1% exchange	23
$(tol)Cr(CO)_3$	$^{14}C-C_6H_5CH_3$	Same as with $^{14}C-C_6H_6$, 17.4% exchange	23
$(tol)Mo(CO)_3$	$^{14}C-C_6H_5CH_3$	Same as with $^{14}C-C_6H_6$, 28.2% exchange	23
$(tol)W(CO)_3$	$^{14}C-C_6H_5CH_3$	Photochemical, $\lambda = 2540$ Å, 2.1% exchange	23

[a] Abbreviations: tol = toluene; p-cym = p-cymene; N,N-DMA = N,N-dimethylaniline; HMB = hexamethylbenzene.

benzpyrene, or stilbene, products of stoichiometry $Ar'[Cr(CO)_3]_2$ are obtained.[26]

Mechanistic studies of arene for arene exchange were carried out by Strohmeier and co-workers,[27-29] using ^{14}C-labeled aromatics as Ar'. The model for the reaction was [^{14}C]benzene with $(C_6H_6)Cr(CO)_3$.[27] The rate law for the reaction was found to be:

$$Rate = k_1[(C_6H_6)Cr(CO)_3]^2 + k_2[(C_6H_6)Cr(CO)_3][^{14}C—C_6H_6]$$

where $k_1 \gg k_2$. The two-term rate law suggests two competing reactions; the steps shown in Equations (4)–(8) were presented in accordance with the kinetics data.

$$2(C_6H_6)Cr(CO)_3 \xrightarrow[k_1]{slow} \quad (6) \tag{4}$$

$$(C_6H_6)Cr(CO)_3 + *C_6H_6 \xrightarrow[k_2]{slow} \quad (7) \tag{5}$$

$$(6) \xrightarrow{fast} C_6H_6 + (CO)_3Cr(C_6H_6) + Cr(CO)_3 \tag{6}$$
$$(inverted)$$

$$Cr(CO)_3 + *C_6H_6 \xrightarrow{fast} (*C_6H_6)Cr(CO)_3 \tag{7}$$

$$(7) \xrightarrow{fast} C_6H_6 + (CO)_3Cr(*C_6H_6) \tag{8}$$
$$(inverted)$$

Subsequent studies[28,29] on ^{14}C-labeled benzene derivatives such as toluene and chlorobenzene, and on the molybdenum and tungsten analogs showed a constant mechanism for all systems. In all cases, the k_1 step was about an order of magnitude greater than the k_2 step, and for corresponding systems the rate with respect to the metal decreased in the order Mo > W > Cr. The rate of exchange of arene decreased in the order $C_6H_5Cl > C_6H_6 > C_6H_5CH_3$.

C. Substitution on the Coordinated Arene

Two types of electrophilic substitution reactions have been observed on the coordinated arene in $ArCr(CO)_3$, namely, hydrogen–deuterium exchange and Friedel–Crafts acylation reactions. In H–D exchange reactions, a solution of $(C_6H_6)Cr(CO)_3$ in CH_3CH_2OD containing about 10% of sodium ethoxide was heated for 20 hr at 100°; nearly half of the hydrogen was replaced by deuterium.[30] With $(C_6H_5CH_3)Cr(CO)_3$, it was found that the tendency for the π-coordinated toluene to exchange hydrogen for deuterium was much greater than for free toluene.[30] Further work on the hydrogen–deuterium exchange in the tricarbonylchromium complexes of anisole,[31,32] ethyl benzoate,[32] and N,N-dimethylaniline[43] was also reported. It was found that for the recovered product 38% ortho, 5% meta, and 10% para hydrogens underwent exchange. The ethyl benzoate complex showed 60, 50, and 40% of the ortho, meta, and para hydrogens, respectively, had exchanged while, with benzoic acid, the percentage of ortho, meta, and para exchange was 50, 40, and 50, respectively. No exchange was observed on the ring of complexed N,N-dimethylaniline.

Several workers have shown that $(C_6H_6)Cr(CO)_3$ undergoes Friedel–Crafts acylation under rather mild conditions ($AlCl_3$ catalyst; refluxing carbon disulfide). Whereas Nicholls and Whiting[6] reported that no acylated product could be recovered, other groups[33,34] reported yields up to 88% for the reaction (9):

$$(C_6H_6)Cr(CO)_3 + CH_3COCl \xrightarrow{AlCl_3} (CH_3COC_6H_5)Cr(CO)_3 \qquad (9)$$

Herberich and Fischer[35] studied the directional properties of the methyl group in $(C_6H_5CH_3)Cr(CO)_3$ and compared the results of acylation with those of free toluene. For uncoordinated toluene, the ortho:meta:para distribution was 8:0:92, but π-bonded toluene gave a distribution of ortho:meta:para as 39:15:46.

Further acylation reactions of π-arene complexes were investigated by Jackson and Jennings.[36] They carried out acetylation reactions in dichloromethane solvent at 25° on the tricarbonylchromium complexes containing the π-arenes toluene, ethylbenzene, isopropylbenzene, t-butylbenzene, and anisole. The results were compared to those of the free arene, with the following distributions:

Arene	Free	π-Complexed
	o : m : p	o : m: p
Toluene	1.2:2 :96.8	43:17:40
Ethylbenzene	0.4:2.9:96.7	24:33:43
Isopropylbenzene	0.1:3.4:96.5	5:59:36
t-Butylbenzene	0 :4.3:95.7	0:87:13
Anisole	0.25:0 :99.75	77: 0:23

It should be noticed that the normally strongly *para*-directing substituent, *t*-butylbenzene, is strongly *meta*-directing when attached to a ring which is π-coordinated. The explanation for the effect is unlike that advanced earlier, and is based on ring–metal bonding in the transition state and the preferred conformations of the $Cr(CO)_3$ groups with respect to the size of the substituent on the ring. The conformational influence of the $Cr(CO)_3$ group was shown by the stereoselectivity exhibited in the acetylation of the tricarbonylchromium complex of *trans*-1,3-dimethylindane.

D. Reactions on Functional Groups on the Coordinated Arene

In this section are considered both nucleophilic attack on π-coordinated aromatic carbon and also reactions on functional groups on the ring. The $Cr(CO)_3$ group exerts a relatively strong electron-withdrawing effect on the ring. Supportive of this contention is the fact that pK measurements have shown that $(aniline)Cr(CO)_3$ is a weaker base than aniline itself, and (benzoic acid)- and (phenylacetic acid)$Cr(CO)_3$ are stronger acids than their respective uncomplexed forms.[6,37,38] Brown applied a molecular orbital treatment in order to elucidate the factors responsible for the enhanced reactivity of the complexed with respect to uncomplexed aromatic systems.[39] Assuming a Wheland-type transition state in which the carbon atom undergoing substitution is effectively removed from conjugation with the other five carbons and $Cr(CO)_3$ in conjugation, he predicted that nucleophilic substitution should occur more readily for the π-complexed than the free arene. Later work on the kinetics of replacement of halogen by methoxide ion supported the prediction of enhanced reactivity.[40]

Holmes *et al.*[41] also considered the problem of increased reactivity of the complexed ring, but took a different theoretical approach from that of Brown. They took into account the two main contributions to the bonding in $ArM(CO)_3$ molecules, namely, donation of π-electrons from the ring to the metal, and back donation from filled d orbitals of the metal to vacant π^* orbitals on the ring. The effect of the $M(CO)_3$ groups on the reactivity of the arene ring then depends on the relative importance of each of the two bonding factors. Table III lists a number of nucleophilic and functional reactions which are known to occur.

Several other studies have been carried out on functional groups on the coordinated ring. The complex $d,l(m\text{-}CH_3O\text{-}C_6H_4\text{-}CO_2CH_3)Cr(CO)_3$ was hydrolyzed to the racemic *m*-methoxybenzoic acid derivative, which was resolved to yield the pure enantiomers.[43]

An interesting application of functional group reactions on coordinated arene rings was carried out by Pittman and co-workers.[44] Treatment of $(C_6H_5CH_2OH)Cr(CO)_3$ with acrylyl chloride gave $(C_6H_5CH_2OCOCH\!\!=\!\!$

TABLE III
Nucleophilic and Functional Group Reactions

Reaction	Reference
$(C_6H_5Cl)Cr(CO)_3 + OCH_3^- \rightarrow (C_6H_5OCH_3)Cr(CO)_3$	6, 40
$(C_6H_5F)Cr(CO)_3 + OCH_3^- \rightarrow (C_6H_5OCH_3)Cr(CO)_3$	40
$(p\text{-}CH_3C_6H_4F)Cr(CO)_3 + OCH_3^- \rightarrow (p\text{-}CH_3C_6H_4OCH_3)Cr(CO)_3$	40
$(C_6H_5NH_2)Cr(CO)_3 + (CH_3CO)_2O \rightarrow (C_6H_5NHCOCH_3)Cr(CO)_3$	6
$(C_6H_5OH)Cr(CO)_3 + (CH_3CO)_2O \rightarrow (C_6H_5OCOCH_3)Cr(CO)_3$	6
$(C_6H_5COOCH_3)Cr(CO)_3 + H_2O \rightarrow (C_6H_5COOH)Cr(CO)_3$	6
$(C_6H_5COOCH_3)Cr(CO)_3 + LiAlH_4 \rightarrow (C_6H_5CH_2OH)Cr(CO)_3$	6
$(C_6H_5NH_2)Cr(CO)_3 + CH_3I(Na_2CO_3,CH_3OH) \rightarrow (C_6H_5NHCH_3)Cr(CO)_3$	6
$(C_6H_5CH_2OH)Cr(CO)_3 + HCl \rightarrow (C_6H_5CH_2Cl)Cr(CO)_3$	41
$[(C_6H_5)_2CO]Cr(CO)_3 + NaBH_4 \rightarrow [(C_6H_5)_2CHOH]Cr(CO)_3$	41
$[(C_6H_5)_2CO]Cr(CO)_3 + C_6H_5MgBr \rightarrow [(C_6H_5)_3COH]Cr(CO)_3$	41
$[(C_6H_5)_2CHOH]Cr(CO)_3 + HCl \rightarrow [(C_6H_5)_2CHCl]Cr(CO)_3$	41
$(C_6H_5CH_2Cl)Cr(CO)_3 + SCN^- \rightarrow (C_6H_5CH_2SCN)Cr(CO)_3$	42
$(C_6H_5CH_2SCN)Cr(CO)_3 \rightarrow (C_6H_5CH_2NCS)Cr(CO)_3$	42

$CH_2)Cr(CO)_3$. The latter compound was homopolymerized and copolymerized with styrene, methyl acrylate, and 2-ferrocenylethyl acrylate using free-radical initiation with azobis(isobutyronitrile).

Wittig reactions have also been carried out on functional groups on the coordinated arene ring.[45] Starting with $(C_6H_5CHO)Cr(CO)_3$, a variety of compounds of the type $(C_6H_5CH{=}CHR)Cr(CO)_3$ were prepared by reaction with the appropriate Wittig reagent. In this work R was phenyl, biphenyl, terphenyl, naphthyl, p-cyanophenyl, p-stilbene, and p-tolan.

In other studies it was found that protons benzylic to the coordinated arene ring undergo hydrogen–deuterium exchange fairly readily.[46] The rates of elimination of (2-phenylethyl)- and (2-phenylethyl p-toluenesulfonate)-$Cr(CO)_3$ were studied in side-chain reactions of $ArCr(CO)_3$ complexes.[47] The yields of complexed styrene were very high in both cases. Finally, the $Cr(CO)_3$ complexes of 3-phenylpropanoic and 3-phenylbutanoic acids were induced by phosphoric acid to cyclize yielding the indanone complexes, which were reduced to their corresponding carbinols.[48,49] The relative amounts of diastereoisomers obtained (65% exo, 35% $endo$) were influenced by the $Cr(CO)_3$ moiety of the molecules.

E. Substitution of CO

Whereas it would appear that the labile moiety of $ArM(CO)_3$ is the arene ring, and it has been shown that isotopic exchange of ^{14}CO for CO does not occur in $ArCr(CO)_3$ at elevated temperatures,[50] nonetheless, this exchange will occur under the influence of uv irradiation. The general mechanism is of the D type, and, once having produced the intermediate acceptor, $ArM(CO)_2$, the final product is formed by simple addition of a donor molecule[51]:

$$ArM(CO)_3 \xrightarrow{h\nu} ArM(CO)_2 + CO$$
$$ArM(CO)_2 + L \longrightarrow ArM(CO)_2L \tag{10}$$

Some examples of L are aniline, quinoline, acetonitrile, benzonitrile,[52] piperidine, pyridine, triphenylphosphine, dimethyl sulfoxide,[53] cyclohexyl isonitrile,[54] various sulfoxides,[55] ethylene,[56] and other olefins and acetylenes.[57] In all the above cases the metal was Cr, whereas several arenes were utilized.

Complexes containing molecular nitrogen have been prepared by photo-chemistry of $ArCr(CO)_3$. Thus, photochemical excitation of the molecules $ArCr(CO)_3$ (Ar = mesitylene and hexamethylbenzene) in N_2-saturated tetrahydrofuran (THF) produced the complexes $ArCr(CO)_2N_2$ and the dinuclear species $[ArCr(CO)_2]_2N_2$, where the two chromium atoms are bridged by the nitrogen molecule.[58,59] These represent the first N_2 complexes of chromium.

A novel compound was prepared upon irradiation of $(C_6H_6)Cr(CO)_3$ in the presence of triferrocenylphosphine. Substitution of CO was achieved and the compound $(C_6H_6)Cr(CO)_2P[(C_5H_4)Fe(C_5H_5)]_3$ was isolated.[60]

Substitution of CO in the complex (1,3,5-cycloheptatriene)$Cr(CO)_3$ occurred upon irradiation at 3660 Å to give the complexes $(C_7H_8)Cr(CO)_2L$, where L = triphenylphosphine and triphenyl phosphite.[61]

F. Total Substitution of Arene and CO

At high solution temperatures (160–220°) it is possible to substitute the arene and three CO groups in $(C_6H_6)Cr(CO)_3$ with pyridine-base chelates. The compounds $Cr(bipy)_3$, $Cr(phen)_3$, and $Cr(tripy)_2$ were prepared by Behrens and co-workers in this manner.[62] In these reactions, no change in oxidation state occurs in chromium.

Other reactions where total substitution was observed resulted in com-pounds which showed an increase in oxidation state of the metal. When $(C_6H_6)Cr(CO)_3$ dissolved in methanol was irradiated for 30 hr at 30°, the compound $Cr(OCH_3)_3$ was obtained. When a similar reaction in methanol–THF was carried out, the product $Cr(OCH_3)_2THF$ was isolated. The compounds $Cr(OC_2H_5)_3$ and $Cr(OC_6H_5)_3$ were prepared by similar means using the appropriate alcohol.[63]

Another reaction which resulted in total substitution without change in the oxidation state of the metal occurred when (tol)$Cr(CO)_3$ was treated with $NOPF_6$ in acetonitrile (tol = toluene). The expected product from this reaction would be the cationic complex $[(tol)Cr(CO)_2NO]^+$; however, the isolated product turned out to be $[Cr(NO)_2(CH_3CN)_4]^{2+}$ instead.[64]

G. Addition and Oxidative Addition

The complexes $ArCr(CO)_3$ and $ArCr(CO)_2P(C_6H_5)_3$ undergo protonation at the metal in trifluoroacetic or stronger acid medium. In an infrared spectral study it was found that the complexes $ArCr(CO)_2P(C_6H_5)_3$ (Ar = methyl benzoate, benzene, toluene, anisole, mesitylene, and *N,N*-dimethylaniline) protonated on chromium in CF_3COOH solution, and that the ease of protonation increased with respect to increasing electron-releasing substituents on the aromatic ring.[65] About a 100 cm^{-1} shift to higher frequencies occurred in the carbonyl stretching bands in the protonated, compared to the unprotonated, form. It was also found that under the same conditions, no protonation of $ArCr(CO)_3$ (Ar = same arenes as above) took place. However, protonation occurred at the metal in $ArCr(CO)_3$ complexes in the stronger acid medium ($BF_3 \cdot R_2O$—CF_3COOH) as determined by nmr spectrometric studies. In this case the arene ligands were benzene, toluene, anisole, and mesitylene. The basicity of the metal is enhanced with electron-releasing substituents on the aromatic ring, and, as implied from the infrared work and supported in this study, $ArCr(CO)_3P(C_6H_5)_3$ is more basic than the corresponding $ArCr(CO)_3$ complex.[66]

An nmr study of the protonation of mesitylene-, hexamethylbenzene-, and 1,3,5-trimethoxybenzenechromium tricarbonyl complexes in FSO_3H—SO_2 solutions was carried out, and it was concluded that protonation occurs at the metal exclusively and not on the ring or on the carbonyl oxygen.[67]

The complex $(C_6H_6)Cr(CO)_3$ was found to undergo an oxidative addition reaction with $HSiCl_3$ when irradiated with uv. The isolated product was identified as $(C_6H_6)Cr(CO)_2(H)(SiCl_3)$, and the general increase in the CO stretching frequencies of the product with respect to the starting material or $(C_6H_6)Cr(CO)_2P(C_6H_5)_5$ shows a decrease in electron density around the metal.[68]

H. Summary

It is apparent that $ArM(CO)_3$ molecules can undergo a variety of different types of reactions. The metal center can act as an electron acceptor (as in the reaction with donor ligands to replace the arene ring) and as an electron donor (as in the protonation and oxidative addition studies). The arene ring is labile in thermal reactions, whereas CO dissociates in photochemically induced processes. However, even though the arene ring can be replaced by a variety of monodentate ligands, the process occurs in an associative, rather than dissociative, mechanism. This point is very important in some of the catalytic reactions to be discussed below, i.e., that the arene ring can stay on the metal during the course of a reaction. At any one time, the arene ring does not necessarily function as a six-electron donor, but may, in fact, act as a four- or even two-electron donor. Whereas no complex of a Group-VIB carbonyl has been isolated which contains an arene ring coordinated in any form other than as a six-electron donor, a structural determination of $C_6H_6 \cdot CuAlCl_4$ showed that the benzene ring is attached to copper through only one π-bond.[69]

I. Experimental Procedures

Preparation of Tris(triphenylphosphine)tricarbonylchromium(O).[6] A mixture of 3.02 g (11.5 mmol) of triphenylphosphine and 0.467 g (2 mmol) of toluenetricarbonylchromium are slowly warmed to 160° under a dry nitrogen atmosphere. After maintaining the final temperature for 2 hr, the resulting deep-red solution is cooled and the volatile components removed under vacuum. The residue is recrystallized from chloroform yielding the pure product (0.97 g, 57%, mp 175–177°).

Preparation of Triphenylphosphinebenzenedicarbonylchromium(O).[52] A solution of 0.225 g (1.06 mmol) of benzenetricarbonylchromium and 0.536 g (2.02 mmol) of triphenylphosphine in 25 ml of benzene is irradiated with a high-pressure mercury lamp at ambient temperature under nitrogen. The process is continued until a quantitative amount of CO is evolved, at which time the solution is red in color. The solvent is removed on a rotary evaporator and the brown-red residue is washed first with heptane on a glass frit, then with concentrated HCl and water, and dried. The excess triphenylphosphine is sublimed from the crude product at 170° under high vacuum, the residue is taken up in benzene, filtered, and concentrated to 3 ml under vacuum. Addition of heptane affords an orange-yellow solid product, which is collected and dried (yield 0.42 g, 87%).

Preparation of Hydridotrichlorosilyl(benzene)dicarbonylchromium. A solution of 1.04 g (4.85 mmol) of benzenetricarbonylchromium in 10 ml of trichlorosilane (excess) and 40 ml of hexane is irradiated with a 100-W lamp (medium-pressure mercury arc) for 28 hr under nitrogen. The yellow, crystalline product precipitates from the solution during the irradiation process. The solid is collected and unreacted $(C_6H_6)Cr(CO)_3$ is removed by sublimation at 65° under high vacuum. The residue is taken up in a minimum amount of dichloromethane and filtered under nitrogen. Hexane is slowly added to the solution until it becomes turbid, whereupon the solution is cooled to $-78°$ which precipitates a fine yellow crystalline product. After a second recrystallization, the product is obtained in 90% yield.

III. FRIEDEL–CRAFTS AND RELATED REACTIONS

A. Reactions Promoted by the Catalysts

The most active catalysts, by far, in Friedel–Crafts reactions catalyzed by $ArM(CO)_3$ are those derived from molybdenum. Arenetricarbonylmolybdenum has been shown to be an effective catalyst in such reactions as alkylation, acylation, and sulfonylation of aromatic systems with organic halides,[70-72] alkylation of aromatic systems with olefins,[73] alkylation of olefins,[74] dehydrohalogenation, and polymerization.[71,72] Table IV shows some of the Friedel–Crafts reactions which have been reported.

B. Experimental Features

In the reactions carried out by White and Farona,[71] the aromatic system to be alkylated was normally used as the solvent. The catalyst added was either $Mo(CO)_6$ or $(toluene)Mo(CO)_3$. Where $Mo(CO)_6$ was added, an induction period was observed before the onset of reaction; the induction period is that time necessary to produce $ArMO(CO)_3$, which is the catalyst of the reaction (see below). In either case the catalyst concentration was always about 5×10^{-3} M. The reactions were carried out in open systems under nitrogen at elevated temperatures, usually at the reflux temperature of the solvent. Evolution of HCl gas was observed throughout the course of the reaction. Normally, at least 50% of the catalyst could be recovered at the end of the reaction.

In the alkylation reactions reported by Massie,[72,73] the processes were carried out in a glass-lined autoclave at elevated temperatures ($140°$). Pressures of up to 35 atm were attained by introduction of nitrogen gas. The concentration of catalyst employed in these reactions was about 5–10 times higher than those used in the reactions of White and Farona.

The conditions for alkylation of aromatic systems by olefins were similar to those applied in the autoclave reactions described above. It was found that cocatalysts such as iron filings, stainless steel turnings, iron carbonyls, or certain mineral acids enhanced the rates of these processes.

It is interesting to compare certain features of the results of alkylations in the two systems. In the open system, the catalyst was much more selective toward substitution in the *para* position on aromatic systems containing *ortho, para*-directing substituents. Often, products which were exclusively *para* substituted were obtained; occasionally there were small amounts of *ortho-*

substituted products (less than 5%) and there was never any *meta* substitution observed. In the closed system where the HCl is retained in the reaction medium, the distribution of products showed only a small preference for substitution in the *para* position; in fact, *meta* products predominated over *ortho* products. Similar results were found in the alkylation with olefins in a closed system.

C. Nature of the Catalyst

Clearly, the active catalyst (or catalyst precursor) in the Friedel–Crafts reactions is $ArMo(CO)_3$ and not $Mo(CO)_6$, even though the hexacarbonyl may be added instead of the arene derivative. Evidence for this contention comes from a variety of sources. First of all, $ArMo(CO)_3$ dehydrochlorinates *t*-butyl chloride and $Mo(CO)_6$ does not. Secondly, no Friedel–Crafts reaction of any type takes place under CO pressure with either the hexacarbonyl or arene derivative. In the latter case, the presence of excess carbon monoxide reversibly replaces the arene in $ArMO(CO)_3$ and converts it to $Mo(CO)_6$, thus rendering the catalyst inactive. Furthermore, where $Mo(CO)_6$ is added to the reaction, only $ArMo(CO)_3$ is recovered upon completion of the reaction. The hexacarbonyl may be added, however, since in the presence of the aromatic substrate, it is converted *in situ* to $ArMo(CO)_3$, which then promotes reaction.

D. Type of Reaction

The question arises as to the type of process promoted by $ArMo(CO)_3$ in Friedel–Crafts reactions, i.e., ionic or free radical. Although Friedel–Crafts reactions normally occur via carbonium-ion formation, other work with molybdenum hexacarbonyl shows that free radicals are produced with certain organic halides.[75,76]

The evidence overwhelmingly supports an ionic process in which carbonium ions are generated from the organic halides. From Table IV it can be seen that the polymerization of benzyl chloride is quantitative in the presence of the catalyst. When large amounts of free-radical traps such as galvinoxyl or 2,2-diphenyl-1-picrylhydrazyl were added to this reaction, the polymerization of benzyl chloride to polybenzyl proceeded unhindered yielding 100% polymer. Conversely, free-radical initiators such as azobis(isobutyronitrile), dibenzoyl peroxide, or cumene hydroperoxide did not induce the polymerization of benzyl chloride in the absence of $ArMo(CO)_3$.

In the reactions carried out by White and Farona, only *ortho* and *para* substitution were detected on aromatic systems containing *ortho,para*-directing substituents. Directional properties of substituents are not operative in free-radical reactions, hence, a substantial yield of *meta*-substituted products

TABLE IV

Friedel–Crafts Reactions Catalyzed by ArMo(CO)$_3$

Aromatic substrate	Organic halide or olefin	Comments	Reference
Alkylation of Aromatic Systems with Organic Halides			
Toluene	t-Butyl chloride	88% p-t-Butyltoluene	71
Toluene	Cyclohexyl chloride	84.5% Cyclohexyltoluene	71
Toluene	Benzyl chloride	100% Alkylation: 10% polybenzyl, 90% tolylphenylmethane	71
Toluene	1-Chloropropane	Reaction carried out in glass-lined Parr bomb; product exclusively p-cymene	71
t-Butylbenzene	1-Chloroheptane	Only secondary alkylates obtained	71
Toluene	Cyclohexyl fluoride	67.3% Cyclohexyltoluene	71
Toluene	Cyclohexyl bromide	23.4% Cyclohexyltoluene; extensive catalyst decomposition	71
Anisole	t-Butyl chloride	79% Alkylation	71
Phenol	t-Butyl chloride	96% Alkylation: 93% p-t-butylphenol, 3% 2,6-di-t-butylphenol	71
Benzene	2-Chloropropane	85% Conversion: 60% cumene, 20% p-diisopropylbenzene, 13% m-diisopropylbenzene, 4% 1,2,4-trisopropylbenzene, 3% 1,3,5-triisopropylbenzene	72
Benzene	Benzyl chloride	100% Diphenylmethane	72
Mesitylene	2-Chloropropane	60% 2,4,6-Trimethylcumene	72
Alkylation of Aromatic Systems with Olefins			
Benzene	Ethylene	Yield of ethylbenzene not reported	73

Benzene	Propylene	25% Conversion: 74% cumene, 1% o-diisopropylbenzene, 9.5% m-diisopropylbenzene, 15.5% p-diisopropylbenzene	73
Toluene	Propylene	25% Conversion: 75% cymene, 25% diisopropyltoluenes	73
Benzene	2-Butene	Yield of sec-butylbenzene not reported	73
Cumene	Propylene	Mixture of diisopropylbenzenes	73
Benzene	1-Octene	Yield of octylbenzene not reported	73
Acylation Reactions			
Toluene	Acetyl chloride	9% Yield of p-methylacetophenone	72
Toluene	Propionyl chloride	18% p-Acylation	72
Toluene	Benzoyl chloride	67% p-Methylbenzophenone	72
Anisole	Acetyl chloride	68% Yield: 90% p-methoxyacetophenone, 4% o-methoxyacetophenone	72
Anisole	Benzoyl chloride	70% p-Methoxybenzophenone	72
Sulfonylation Reactions			
Toluene	Tosyl chloride	43% 4,4'-Ditolylsulfone	72
Anisole	Tosyl chloride	25% 4-Methyl-4'-methoxydiphenylsulfone	72
Polymerization Reactions			
Benzyl chloride	—	100% Polybenzyl	72
Durene	p-Xylylene dichloride	98% Copolymer	72
Benzyl fluoride	—	100% Polybenzyl	72
Diphenyl ether	Benzene-1,3-disulfonyl chloride	32% Tan-colored copolymer	72

would be expected in the case of a free-radical reaction. Since no *meta*-substituted products were observed the type of reaction indicated must be ionic.

In classical cases of carbonium-ion behavior, it was found, for example, that in the reaction of toluene with *n*-chloropropane, only isopropyl derivatives were obtained. Alkylation with *n*-chloroheptane led only to secondary products. Apparently, carbonium ions are formed from the primary halides, and these rearrange to the more stable secondary ions before substitution takes place. This behavior is not expected for primary radicals; primary products would occur for radical reactions.

An interesting rearrangement of octachloro bicyclo[3.2.0]hepta-2,6-diene to octachlorocycloheptatriene, as promoted by $AlCl_3$, was speculated to proceed by abstraction of allylic chloride to give the intermediate heptachlorotropylium ion.[77] An attempt was made with (mesitylene)Mo(CO)$_3$ to induce the same reaction and trap the intermediate as $[(C_7Cl_7)Mo(CO)_3]^+$; this would be definitive proof for carbonium-ion generation by ArMo(CO)$_3$. However, whereas the perchlorotropylium complex was not isolated, the same rearrangement occurred with ArMo(CO)$_3$ as catalyst as with $AlCl_3$, presumably by the same mechanism.

E. Mechanism

In order to activate the organic halide, the molecule ArM(CO)$_3$ must, at some time during the reaction, become coordinatively unsaturated. The coordinatively unsaturated state can come about by several different processes, all of which are analogous to various aspects of the chemistry of ArM(CO)$_3$, discussed previously.

In a kinetic study by Korenz and Farona,[78] it was found that the length of induction time and the rate of reaction of alkylation of toluene by *t*-amyl chloride was dependent on the nature of the coordinated arene ring in ArMo-(CO)$_3$ (Ar = toluene, *m*-xylene, mesitylene, and 1,3,5-cycloheptatriene). The rate law for the reaction showed first-order dependence on the catalyst and alkyl chloride, and zero order in toluene (measured in a stoichiometric reaction in an inert solvent). The induction periods increased in the order 1,3,5-cycloheptatriene $<$ toluene $<$ *m*-xylene $<$ mesitylene, and ranged from 10 to 20 min. The rates of reactions decreased in order of arene: toluene $>$ *m*-xylene $>$ mesitylene $>$ 1,3,5-cycloheptatriene, with rate constants ranging from 0.23 to 0.03 mol^{-1} min^{-1}.

A mechanism was proposed which postulated that the arene ring remains attached to the metal during catalysis. Scheme 1 shows the proposed mechanism of activation of alkyl halide.

This mechanism accounts for the induction time as the period necessary to establish equilibrium with the arene ring as a four-electron donor (8), which

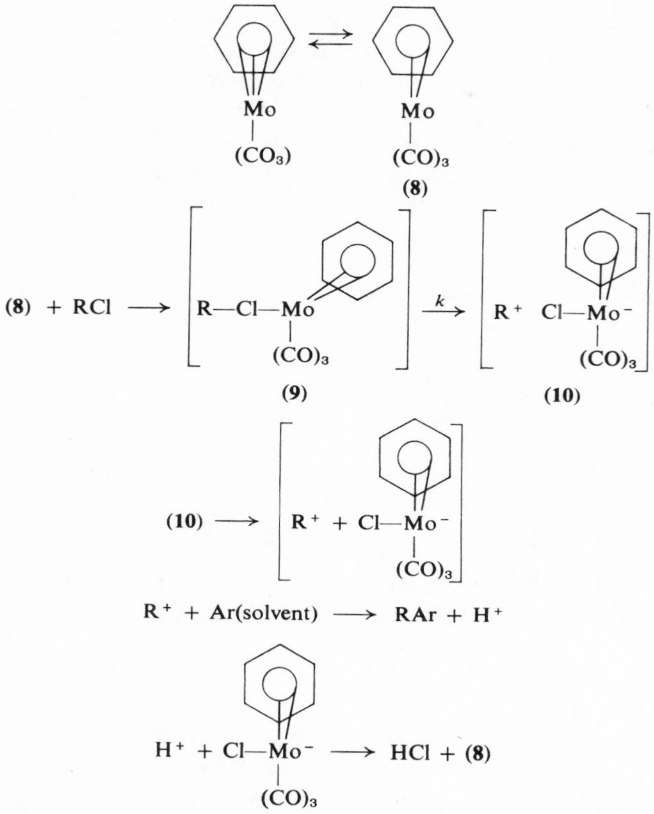

SCHEME 1

is the active catalytic species. The induction periods of the reactions would be influenced by the electron density of the coordinated ring, where the ring with the greatest amount of methylation should show the longest induction period, which was observed. By the same reasoning, the rates of reaction should also be influenced by the nature of the coordinated arene so that the active intermediate (8) would be present in greater effective concentration for toluene than for xylene than for mesitylene.

The possibility that the arene ring completely dissociates from $ArMo(CO)_3$ leaving $Mo(CO)_3$ as the active intermediate was rejected. Whereas a mechanism based on dissociation accounts for the observed induction time, it would be expected that all rates would be the same, irrespective of the arene ring, if the active intermediate is $Mo(CO)_3$.

Furthermore, in reactions of *t*-amyl chloride with toluene catalyzed by (mesitylene)$Mo(CO)_3$, only the starting catalyst was recovered at the end of the reaction, no $(tol)Mo(CO)_3$. This clearly shows that ring exchange does not occur, even in the presence of toluene as the solvent.

The ionic complex (10) was the favored form of the intermediate and was described as a tight ion pair. Also considered were the possibilities (11) and (12) for the intermediate, but the authors felt that more problems arose in explaining several chemical facts through these intermediates than by invoking structure (10).

           (11)   (12)

F. Related Reactions

The reaction of acyl halides with ethers has been investigated wherein a Group-VIb hexacarbonyl was used as the catalyst.[79] The products of the reaction are an alkyl halide and an ester:

$$ROR + R'COCl \xrightarrow{M(CO)_6} RCl + R'COOR \tag{11}$$

The order of catalytic effectiveness was found to be $Mo(CO)_6 > W(CO)_6 > Cr(CO)_6$, which is similar to that found for the Friedel–Crafts reaction promoted by $ArM(CO)_3$. Table V lists some of the reactions promoted by $Mo(CO)_6$.

The mechanism for the reaction appears to be like the Friedel–Crafts reactions discussed above, based on carbonium-ion generation. Evidence for this contention arises from the observation that the reactions are insensitive to air and to radical initiators. The mechanism shown in Equations (12)–(16) was postulated and accounts for (a) the reactivity pattern of the ether carbons (tertiary > secondary > primary); (b) the tendency to form elimination products (tertiary > secondary > primary); (c) effect of halogen; and (d) some stereochemical results.

$$\overset{O}{\overset{\|}{R}C}X + M(CO)_6 \longrightarrow \overset{O}{\overset{\|}{R}C^+} + M(CO)_5X^- + CO \tag{12}$$

$$\overset{O}{\overset{\|}{R}C^+} + R'OR'' \longrightarrow RCO^+\underset{R''}{\overset{R'}{\diagdown}} \tag{13}$$

$$RC{-}O^+\underset{R''}{\overset{R'}{\diagdown}} + M(CO)_5X^- \longrightarrow R'X + RCOR'' + M(CO)_5 \qquad (14)$$

Elimination products arise from an alternative pathway:

$$RC{-}O^+\underset{R''}{\overset{R'}{\diagdown}} \xrightarrow{\text{tertiary } R'} R^+ + RCOR'' \qquad (15)$$

$$R^+ \longrightarrow \text{elimination products} \qquad (16)$$

It was also observed that compounds such as $Mo(CO)_5[P(C_6H_5)_3]$, $Mo(CO)_5[As(C_6H_5)_3]$, and $Mo(CO)_4[P(C_6H_5)_3]_2$ were also effective catalysts for the reaction.

G. Experimental Procedures

Alkylation of Toluene with t-*Butyl Chloride.*[71] A solution of 160 ml of toluene, 13.9 g (0.15 mol) of *t*-butyl chloride and 0.02 g (0.08 mmol) of molyb-

TABLE V

Reaction of Acyclic and Cyclic Ethers with Acyl Halides Catalyzed by $Mo(CO)_6$[79]

Ether	Acyl halide	Products (yield %)
Acyclic		
2-Ethoxyoctane	CH_3COCl	$CH_3C(Cl)HC_6H_{13}$ (62), $CH_3CH{=}CHC_5H_{11}$ (31), $CH_3COOC_2H_5$
n-Butyl ether	C_6H_5COCl	C_4H_9Cl, $C_6H_5COOC_4H_9$ (73)
Ethyl triphenylmethyl ether	C_6H_5COCl	$(C_6H_5)_3CCl$, $C_6H_5COOC_2H_5$ (72)
Cyclic		
Tetrahydrofuran	CH_3COCl	4-Chlorobutyl acetate
	C_6H_5COBr	4-Bromobutyl benzoate (88)
2-Methyltetrahydrofuran	CH_3COCl	4-Chloropentyl acetate (80) 3-Pentenyl acetate
	C_6H_5COCl	4-Chloropentyl benzoate, 3-pentyl benzoate
7-Oxabicyclo[2.2.1]-heptane	CH_3COCl	*trans*-4-Chlorocyclohexyl acetate (55), 3-cyclohexenyl acetate (22)
7-Oxabicyclo[2.2.1]-heptane	C_6H_5COCl	*trans*-4-Chlorocyclohexyl benzoate (54), 3-cyclohexenyl benzoate
7-Oxabicyclo[2.2.1]-heptane	C_6H_5COBr	*trans*-4-Bromocyclohexyl benzoate (72), 3-cyclohexenyl benzoate (13)
7-Oxabicyclo[2.2.1]-heptane	C_6H_5COF	3-Cyclohexenyl benzoate (67)

denum hexacarbonyl [or (tol)Mo(CO)$_3$] is refluxed until the evolution of HCl ceases. After removal of excess toluene by distillation, the liquid product is obtained by vacuum distillation, bp 105° (57 torr). The yield of *p-t*-butyl-toluene is about 17 g (77%).

Acylation of Anisole with Acetyl Chloride.[71] A solution containing 7.8 g (0.1 mol) of acetyl chloride, 0.02 g (0.08 mmol) of molybdenum hexacarbonyl [or (tol)Mo(CO)$_3$] and 125 ml of anisole is heated at 100° for 36 hr. Isolation of the product by filtration, water washing, and removal of excess anisole on a rotary evaporator yields 10.2 g (68%) of a semi-solid. The residue is subjected to vacuum distillation; the first fraction, bp 118–120° (10 torr), weighs 0.4 g and the second fraction, bp 138–145° (10 torr), weighs 9.2 g. This corresponds to a product distribution of 4% *ortho*- and 90% *para*-methoxyacetophenone.

Sulfonylation of Toluene with p-*Toluenesulfonyl Chloride.*[71] A solution containing 160 ml of toluene, 3.8 g (0.02 mol) of tosyl chloride and 0.02 g (0.08 mmol) of molybdenum hexacarbonyl [or (tol)Mo(CO)$_3$] is refluxed for 36 hr. Slow evolution of HCl occurs accompanied by gradual decomposition of the catalyst, as is indicated by the dark color of the solution. The reaction solution is cooled, filtered, and washed successively with 50 ml of 5% Na$_2$CO$_3$ and two 100-ml portions of water. The toluene is removed in a rotary evaporator and the remaining light-tan crystals are recrystallized from 60% aqueous ethanol to give 2.1 g (43%) of 4,4'-ditolylsulfone (mp 157–159°).

Copolymerization of Durene with p-*Xylylene Dichloride.*[71] A solution of 100 ml of decalin, 7.3 g (41.5 mmol) of *p*-xylylenedichloride, 5.4 g (40.5 mmol) of durene and 0.02 g (0.08 mmol) of molybdenum hexacarbonyl [or (tol)Mo-(CO)$_3$] is heated at 110° for 3 hr. As HCl is evolved, the solution becomes more and more viscous. At the end of that time, the reaction mixture is poured into 800 ml of methanol from which the polymer precipitates as a fine, white powder. The white solid is collected by filtration and dried under vacuum. The yield is 9.6 g or 98% of theory. The polymer is essentially insoluble in all common organic solvents.

IV. HYDROGENATION OF OLEFINS

A. Reactions Promoted by ArM(CO)$_3$ and Related Catalysts

The largest area of research on catalysis by arene-Group-VIB tricarbonyls has been in hydrogenation of olefins. This is not surprising, since the catalyst system is very useful in promoting reductions of 1,3-dienes to internal 2-monoenes in very high yields and selectivities. Table VI presents some data on

TABLE VI

Hydrogenation Catalyzed by Group-VIB Carbonyls

Catalyst[a]	Substrate (% conversion)	Products (% distribution)	Reaction conditions	Reference
CpCr(CO)$_3$H	Isoprene (100)	2-Methyl-2-butene (95) 2-Methyl-1-butene (3) 3-Methyl-1-butene (2)	H$_2$ = 90 atm; 70°; 5 hr; benzene	80
Cr(CO)$_6$	2,3-Dimethyl-1,3-butadiene (30)	2,3-Dimethyl-2-butene (100)	Photochemical, 3600 Å; H$_2$ = 1 atm; 2.25 hr; decalin	81
(MBZ)Cr(CO)$_3$	2,3-Dimethyl-1,3-butadiene (100)	2,3-Dimethyl-2-butene (82) 2,3-Dimethyl-1-butene (18)	H$_2$ = 30 atm; 160°; pentane	82
CpCr(CO)$_3$H	cis-1,3-Pentadiene	cis-2-Pentene (12) trans-2-Pentene (88)	H$_2$ = 90 atm; 70°; 5 hr; benzene	80
(MBZ)Cr(CO)$_3$	2-Methyl-1,3-pentadiene (100)	2-Methyl-2-pentene (75) 2-Methyl-1-pentene (17) 4-Methyl-2-pentene (8)	H$_2$ = 30 atm; 160°; pentane	82
(MBZ)Cr(CO)$_3$	4-Methyl-1,3-pentadiene (100)	2-Methyl-2-pentene (66) 2-Methyl-1-pentene (14) 4-Methyl-2-pentene (14) 2-Methylpentane (6)	H = 30 atm; 160°; pentane	
CpCr(CO)$_3$H	4-Methyl-1,3-pentadiene	2-Methyl-2-pentene (78) 4-Methyl-2-pentene (22)	H$_2$ = 90 atm; 70°; 5 hr; benzene	80
CpMo(CO)$_3$H	4-Methyl-1,3-pentadiene (50)	2-Methyl-2-pentene (76) 4-Methyl-2-pentene (24)	Stoichiometric hydro- genation	83
CpW(CO)$_3$H	4-Methyl-1,3-pentadiene (50)	2-Methyl-2-pentene (72) 4-Methyl-2-pentene (28)	Stoichiometric hydrogenation	83

(continued)

[a] Abbreviations: Cp = π-cyclopentadienyl; MBZ = methyl benzoate; tol = toluene; mes = mesitylene; HMB = hexamethylbenzene; anis = anisole; C$_7$H$_8$ = 1,3,5-cycloheptatriene; 1,4-DPB = 1,4-diphenylbutadiene; 3-CMA = 3-carbomethoxyanisole; stil = stilbene; BZP = benzophenone; phen = phenanthrene; BHD = bicyclo[2.2.1]hepta-2,5-diene.

TABLE VI (*Continued*)

Catalyst	Substrate (% conversion)	Products (% distribution)	Reaction conditions	Reference
(MBZ)Cr(CO)$_3$	1,3-Hexadiene (100) (*cis* and *trans*)	1-Hexene (6) 2-Hexene (66 *cis*, 10 *trans*) 3-Hexene (18), *cis* and *trans*	$H_2 = 30$ atm; 160°; 6 hr; pentane	82
(MBZ)Cr(CO)$_3$	2,4-Hexadiene (100)	2-Hexene (10, *cis*) 3-Hexene (90, *cis*)	$H_2 = 30$ atm; 160°; 2 hr; pentane	82
CpCr(CO)$_3$H	2,4-Hexadiene (50)	*trans*-2- and -3-hexene (89) *cis*-2- and -3-hexene (11)	Stoichiometric hydrogenation	83
CpMo(CO)$_3$H	2,4-Hexadiene (50)	*trans*-2- and -3-hexene (79) *cis*-2 and -3-hexene (21)	Stoichiometric hydrogenation	83
(MBZ)Cr(CO)$_3$	1,4-Hexadiene (*cis* and *trans*)	*cis*-2-Hexene (40) 3-Hexene (60), *cis* and *trans*)	$H_2 = 30$ atm; 160°; 6 hr; pentane	82
(MBZ)Cr(CO)$_3$	1,4-Hexadiene (*cis* and *trans*) (77)	2-Hexene (22 *cis*, 25 *trans*) 3-Hexene (52 *cis* and *trans*)	$H_2 = 30$ atm; 175°; 6 hr; pentane	82
(MBZ)Cr(CO)$_3$	1,5-Hexadiene (2)	1-Hexene (50) 2-Hexene (50)	$H_2 = 30$ atm; 175°; 6 hr; pentane	82
(MBZ)Cr(CO)$_3$	2,5-Dimethyl-2,5-hexadiene (5)	2,5-Dimethyl-3-hexene (100)	$H_2 = 30$ atm; 160°; 6 hr; pentane	82
(MBZ)Cr(CO)$_3$	2,5-Dimethyl-2,5-hexadiene (45)	2,5-Dimethyl-3-hexene (100)	$H_2 = 30$ atm; 175°; 6 hr; pentane	82
Cr(CO)$_6$	1,3-Cyclohexadiene (100)	Cyclohexene (100)	Photochemical, 3600 Å; $H_2 = 1$ atm; decalin	81
(MBZ)Cr(CO)$_3$	1,3-Cyclohexadiene (100)	Cyclohexene (98) Cyclohexane (2)	$H_2 = 30$ atm; 160°; 6 hr; hexane	82
CpCr(CO)$_3$H	1,3-Cyclohexadiene (100)	Cyclohexene	$H_2 = $ atm; 70°; benzene	80
(MBZ)Cr(CO)$_3$	1,4-Cyclohexadiene (100)	Cyclohexene (93)	$H_2 = 30$ atm; 160°; 6 hr; hexane	82
(MBZ)Cr(CO)$_3$	1,3-Cyclooctadiene (100)	Cyclooctene (25)	$H_2 = 30$ atm; 160°; 6 hr; hexane	82

Catalyst	Substrate	Products	Conditions	Ref.
(MBZ)Cr(CO)₃	1,5-Cyclooctadiene (100)	1,3-Cyclooctadiene (64), 1,4-Cyclooctadiene (1), Cyclooctene (35)	H_2 = 30 atm; 160°; 6 hr; hexane	82
(MBZ)Cr(CO)₃	1,5-Cyclooctadiene (100)	1,3-Cyclooctadiene (28), 1,4-Cyclooctadiene (2), Cyclooctene (70)	H_2 = 30 atm; 170°; 6 hr; hexane	82
CpMo(CO)₃H/Al(i-Bu)₃ = (i-Bu)₂Al-Mo(CO)₃Cp	1,5-Cyclooctadiene (13)	Cyclooctene (30), Cyclooctane (70)	H_2 = 17 atm; 100°; 3 hr	84
(MBZ)Cr(CO)₃	trans-1,3,5-Hexatriene (80)	cis-3-Hexene (30), cis-2-Hexene (15), cis-1,4-Hexadiene (20), 2,4-Hexadiene (5), cis,cis and cis,trans, 1,3-Cyclohexadiene (3), Cyclohexene (1)	H_2 = 30 atm; 160°; 6 hr; pentane	85
(MBZ)Cr(CO)₃	cis-1,3,5-Hexatriene (100)	Cyclohexene (70), cis-1,4-Hexadiene (19), cis-3-Hexene (6), cis-2-Hexene (5)	H_2 = 30 atm; 160°; 1 hr; pentane	85
CpMo(CO)₃H	1,3,5-Hexatriene (50)	2,4-Hexadiene (trans,trans 66, trans,cis 31, cis,cis 3)	Stoichiometric hydrogenation	83
CpW(CO)₃H	1,3,5-Hexatriene (50)	2,4-Hexadiene (trans,trans 67, trans,cis 30, cis,cis 4)	Stoichiometric hydrogenation	83
Cr(CO)₆	1,3,7-Octatriene (99)	1,5-Octadiene (23), 1,6-Octadiene (56)	H_2 = 68 atm; 200°; cyclohexane	86
Cr(CO)₆	1,3,7-Octatriene (99)	1,5-Octadiene (13), 1,6-Octadiene (62)	H_2 = 88 atm; 200°; cylohexane	86
(MBZ)Cr(CO)₃	1,3,7-Octatriene (99)	1,5-Octadiene (34), 1,6-Octadiene (54)	H_2 = 82 atm; 150°; cyclohexane	86

(continued)

TABLE VI (*Continued*)

Catalyst	Substrate (% conversion)	Products (% distribution)	Reaction conditions	Reference
(MBZ)Cr(CO)$_3$	1,3,7-Octatriene (99)	1,5-Octadiene (21) 1,6-Octadiene (65)	H$_2$ = 88 atm; 150°; cyclohexane	86
(C$_6$H$_6$)Cr(CO)$_3$	1,3,7-Octatriene (95)	1,5-Octadiene (29) 1,6-Octadiene (40)	H$_2$ = 102 atm; 160°; cyclohexane	86
(C$_6$H$_6$)Cr(CO)$_3$	1,3,7-Octatriene (50)	1,5-Octadiene (22) 1,6-Octadiene (24)	H$_2$ = 88 atm; 140°; cyclohexane	86
Cr(CO)$_5$PPh$_3$	1,3,7-Octatriene (99)	1,5-Octadiene (14) 1,6-Octadiene (52)	H$_2$ = 95 atm; 210°; cyclohexane	86
Cr(CO)$_5$PPh$_3$	1,3,7-Octatriene (98)	1,5-Octadiene (17) 1,6-Octadiene (42)	H$_2$ = 88 atm: 225°; cyclohexane	86
CpMo(CO)$_3$H	2,4,6-Octatriene	3,5-Octadiene[b] (83) 2,4-Octadiene[b] (17)	Stoichiometric hydrogenation	83
CpCr(CO)$_3$H	1,3,5-Cycloheptatriene	Cycloheptene (66) 1,4-Cycloheptadiene (34)	H$_2$ = 90 atm; 70°; benzene	80
(MBZ)Cr(CO)$_3$	1,3,5-Cycloheptatriene (45)	1,3,5-Cycloheptatriene (55) 1,3-Cycloheptadiene (42) Cycloheptene (2)	H$_2$ = 90 atm; 70°; benzene	85
CpCr(CO)$_3$H	1,3,6-Cyclooctatriene (100)	1,5-Cyclooctadiene (74) 1,4-Cyclooctadiene (26)	H$_2$ = 90 atm; 70°; benzene	80
CpW(CO)$_3$H	1,3,5,7-Octatetraene (50)	2,4,6-Octatriene (89)	Stoichiometric hydrogenation	83
(C$_6$H$_6$)Cr(CO)$_3$	Methyl sorbate (100) (CH$_3$CH=CH—CH=CH-(COOCH$_3$)	Methyl 3-hexenoate (94) Methyl 2-hexenoate (4) Methyl hexanoate (2)	H$_2$ = 48 atm; 165°; 8 hr; cyclohexane	87
(C$_6$H$_6$)Cr(CO)$_3$	Methyl sorbate (100)	Methyl 3-hexenoate (92) Methyl 2-hexenoate (3) Methyl hexanoate (5)	H$_2$ = 48 atm; 165°; 0.5 hr; methylene chloride	87

Catalyst	Substrate	Products	Conditions	Ref.
$(C_6H_6)Cr(CO)_3$	Methyl sorbate (100)	Methyl 3-hexenoate (94); Methyl 2-hexenoate (4)	H_2 = 48 atm; 165°; 8 hr; cyclohexane	88
$(C_6H_6)Cr(CO)_3$	Methyl sorbate (100)	Methyl 3-hexenoate (94); Methyl 2-hexenoate (4); Methyl hexanoate (2)	H_2 = 30 atm; 175°; 4 hr; cyclohexane	89
$(tol)Cr(CO)_3$	Methyl sorbate (100)	Methyl 3-hexenoate (94); Methyl 2-hexenoate (5); Methyl hexanoate (1)	H_2 = 48 atm; 150°; 7 hr; cyclohexane	87
$(tol)Cr(CO)_3$	Methyl sorbate (100)	Methyl 3-hexenoate (95); Methyl 2-hexenoate (5)	H_2 = 30 atm; 175°; 4 hr; cyclohexane	89
$(mes)Cr(CO)_3$ (5 mole %)	Methyl sorbate (80)	Methyl 3-hexenoate (95); Methyl 2-hexenoate (5)	H_2 = 30 atm; 175°; 6 hr; cyclohexane	89
$(mes)Cr(CO)_3$ (5 mole %)	Methyl sorbate (1)	Methyl 3-hexenoate (100)	H_2 = 30 atm; 165°; 6 hr; cyclohexane	89
$(mes)Cr(CO)_3$ (10 mole %)	Methyl sorbate (63)	Methyl 2-hexenoate (95); Methyl 2-hexenoate (4); Methyl hexanoate (1)	H_2 = 30 atm; 165°; 6 hr; cyclohexane	89
$(mes)Cr(CO)_3$ (20 mole %)	Methyl sorbate (100)	Methyl 3-hexenoate (96); Methyl 2-hexenoate (3); Methyl hexanoate (1)	H_2 = 30 atm; 165°; 4.5 hr; cyclohexane	89
$(C_6H_5C_2H_5)Cr(CO)_3$	Methyl sorbate (95)	Methyl 3-hexenoate (95); Methyl 2-hexenoate (5)	H_2 = 48 atm; 150°; 7 hr; cyclohexane	87
$(HMB)Cr(CO)_3$	Methyl sorbate (92)	Methyl 3-hexenoate (91); Methyl 2-hexenoate (5); Methyl 4-hexenoate (1); Methyl hexanoate (3)	H_2 = 30 atm; 200°; 6 hr; cyclohexane	89
$(anis)Cr(CO)_3$	Methyl sorbate (12)	Methyl 3-hexenoate (100)	H_2 = 48 atm; 150°; 6 hr; cyclohexane	87
$(anis)Cr(CO)_3$	Methyl sorbate (88)	Methyl 3-hexenoate (97); Methyl 2-hexenoate (3)	H_2 = 48 atm; 150°; 6 hr; acetone	87

(continued)

[b] Probably.

TABLE VI (*Continued*)

Catalyst	Substrate (% conversion)	Products (% distribution)	Reaction conditions	Reference
$(C_7H_8)Cr(CO)_3$	Methyl sorbate (100)	Methyl 3-hexenoate (88) Methyl 2-hexenoate (3) Methyl hexanoate (9)	H_2 = 48 atm; 150°; 0.25 hr; cyclohexane	87
$(C_7H_8)Cr(CO)_3$	Methyl sorbate (100)	Methyl 3-hexenoate (98) Methyl 2-hexenoate (1) Methyl hexanoate (1)	H_2 = 30 atm; 120°; 1 hr; cyclohexane	89
$(C_6H_5Cl)Cr(CO)_3$	Methyl sorbate (100)	Methyl 3-hexenoate (96) Methyl hexanoate (4)	H_2 = 48 atm; 150°; 2 hr; cyclohexane	87
$(C_6H_5Cl)Cr(CO)_3$	Methyl sorbate (100)	Methyl 3-hexenoate (96) Methyl hexanoate (4)	H_2 = 30 atm; 150°; 2 hr; cyclohexane	89
$(MBZ)Cr(CO)_3$	Methyl sorbate (100)	Methyl 3-hexenoate (99) Methyl hexanoate (1)	H_2 = 48 atm; 150°; 2 hr; cyclohexane	87
$(MBZ)Cr(CO)_3$	Methyl sorbate (100)	Methyl 3-hexenoate (99) Methyl hexanoate (1)	H_2 = 48 atm; 165°; 2 hr; cyclohexane	88
$(MBZ)Cr(CO)_3$	Methyl sorbate (100)	Methyl 3-hexenoate (95) Methyl 2-hexenoate (4) Methyl hexanoate (1)	H_2 = 30 atm; 160°; 4 hr; cyclohexane	89
$(1,4\text{-}DPB)Cr(CO)_3$	Methyl sorbate (100)	Methyl 3-hexenoate (95) Methyl 2-hexenoate (2) Methyl hexanoate (3)	H_2 = 48 atm; 150°; 2 hr; cyclohexane	87
$(3\text{-}CMA)Cr(CO)_3$	Methyl sorbate (99)	Methyl 3-hexenoate (96) Methyl 2-hexenoate (2) Methyl hexanoate (1)	H_2 = 48 atm; 150°; 1 hr; cyclohexane	87
$(stil)Cr(CO)_3$	Methyl sorbate (98)	Methyl 3-hexenoate (95) Methyl 2-hexenoate (5)	H = 48 atm; 150°; 2.2 hr; cyclohexane	88
$(BZP)Cr(CO)_3$	Methyl sorbate (100)	Methyl 3-hexenoate (97) Methyl 2-hexenoate (2) Methyl hexanoate (1)	H_2 = 48 atm; 150°; 1.5 hr; cyclohexane	88

Catalyst	Substrate	Products	Conditions	Ref.
(phen)Cr(CO)$_3$	Methyl sorbate (100)	Methyl 3-hexenoate (97); Methyl 2-hexenoate (3)	H$_2$ = 48 atm; 150°; 0.3 hr; cyclohexane	88
(C$_7$H$_8$)Mo(CO)$_3$	Methyl sorbate (100)	Methyl 3-hexenoate (90); Methyl 2-hexenoate (10)	H$_2$ = 30 atm; 100°; 6 hr; cyclohexane	89
(mes)Mo(CO)$_3$	Methyl sorbate (100)	Methyl 3-hexenoate (86); Methyl 2-hexenoate (11); Methyl 4-hexenoate (2); Methyl hexanoate (1)	H$_2$ = 30 atm; 100°; 2 hr; cyclohexane	89
(mes)Mo(CO)$_3$	Methyl sorbate (100)	Methyl 3-hexenoate (68); Methyl 2-hexenoate (10); Methyl 4-hexenoate (19); Methyl hexanoate (3)	H$_2$ = 30 atm; 120°; 1 hr; cyclohexane	89
(mes)Mo(CO)$_3$	Methyl sorbate (100)	Methyl 3-hexenoate (28); Methyl 2-hexenoate (8); Methyl 4-hexenoate (55); Methyl hexanoate (9)	H$_2$ = 30 atm; 150°; 0.5 hr; cyclohexane	89
(BHD)Mo(CO)$_4$	Methyl sorbate (100)	Methyl 3-hexenoate (90); Methyl 2-hexenoate (10)	H$_2$ = 30 atm; 120°; 6 hr; cyclohexane	89
[CpMo(CO)$_3$]$_2$	Methyl sorbate (31)	Methyl 3-hexenoate (19); Methyl 2-hexenoate (25); Methyl 4-hexenoate (13); Methyl hexanoate (42)	H$_2$ = 30 atm; 180°; 6 hr; cyclohexane	89
(mes)W(CO)$_3$ (5 mole %)	Methyl sorbate (29)	Methyl 3-hexenoate (21); Methyl 2-hexenoate (59); Methyl 4-hexenoate (10); Methyl hexanoate (10)	H$_2$ = 30 atm; 165°; 6 hr; cyclohexane	89
(mes)W(CO)$_3$ (10 mole %)	Methyl sorbate (77)	Methyl 3-hexenoate (17); Methyl 2-hexenoate (56)	H$_2$ = 30 atm; 165°; 6 hr; cyclohexane	

(continued)

TABLE VI (Continued)

Catalyst	Substrate (% conversion)	Products (% distribution)	Reaction conditions	Reference
$(C_6H_6)Cr(CO)_3$	Methyl cis-9-trans-11-octadecadienoate (100)	Methyl 4-hexenoate (12), Methyl hexanoate (16), Mixture of mainly cis-9- and cis-10-octadecenoate	H_2 = 48 atm; 175°; cyclohexane	82, 87
$(MBZ)Cr(CO)_3$	Methyl cis-9-trans-11-octadecadienoate (98)	Mixture of mainly cis-9- and cis-10-octadecenoate	H_2 = 30 atm; 175°; 1.5 hr; cyclohexane	82
$(C_7H_8)Cr(CO)_3$	Methyl cis-9-trans-11-octadecadienoate (95)	Mixture of mainly cis-9- and cis-10-octadecenoate	H_2 = 30 atm; 125°; 3.5 hr; cyclohexane	82
$(C_6H_6)Cr(CO)_3$	Methyl trans-9-trans-11-octadecadienoate (100)	10-Octadecenoate (92), Other octadecenoates (8)	H_2 = 30 atm; 175°; 2 hr; cyclohexane	82
$(MBZ)Cr(CO)_3$	Methyl linoleate (95) (cis-9-cis-12-octadecadienoate)	9-Octadecenoate (24), 10-Octadecenoate (24), 11-Octadecenoate (23), 12-Octadecenoate (22), Other octadecenoates (7)	H_2 = 30 atm; 175°; 3 hr; cyclohexane	82
$(MBZ)Cr(CO)_3$	Alkali-conjugated Methyl linoleate (99)	9-Octadecenoate (21), 10-Octadecenoate (22), 11-Octadecenoate (27), 12-Octadecenoate (26), Other octadecenoates (4)	H_2 = 30 atm; 165°; 4 hr; cyclohexane	82

the types of hydrogenation reactions carried out using $ArM(CO)_3$ and related compounds as catalysts. The table is arranged according to substrate in order to make comparisons of the effectiveness of the catalysis more convenient for the reader.

In addition to the very large number of 1,3-dienes subjected to hydrogenation shown in Table VI, $ArCr(CO)_3$ has also been used as an effective catalyst in the hydrogenation of unsaturated fatty esters. Cais *et al.*[87] observed that in a mixture of dienoic esters derived from dehydrated methyl ricinoleate (the mixture consisted primarily of methyl *cis*-9-*trans*-11-octadecadienoate and the nonconjugated methyl 9,12-octadecadienoate) the conjugated dienes were reduced to a mixture of *cis*-9- and *cis*-10-octadecadienoate, but the non-conjugated dienes were unaffected. However, Frankel and Little[89] showed that at higher temperatures both methyl linoleate (9,12-octadecadienoate) and methyl linolenate (9,12,15-octadecatrienoate) were reduced mainly to mono-enes and small amounts of conjugated dienes (from linolenate). Apparently, the reduction of these unsaturated fatty esters is preceded by conjugation[89] (See Section V on Isomerization of Olefins).

A two-step process for the selective hydrogenation of methyl linolenate in the presence of methyl linoleate was reported.[90] In this procedure, the linolenate was isomerized to the conjugated form using potassium *t*-butoxide as the catalyst. The conjugated linolenate was then selectively hydrogenated with (phenanthrene)$Cr(CO)_3$ to the corresponding diene under mild conditions; throughout the entire two-step process, the methyl linoleate was unaffected.[90]

An interesting application of the selectivity of the $ArCr(CO)_3$ catalysts in hydrogenation was reported by Frankel *et al.* They were able to produce simulated peanut oil by hydrogenating soybean oil, and simulated olive oil from soybeans and safflower oils. The simulated composition of safflower oil was produced by hydrogenation of tung oil, and the simulated structure of cocoa butter was produced by stereoselective hydrogenation of stearines from winterized cottonseed oil. The catalysts used for these conversions were (methyl benzoate)- and (ethylbenzene)$Cr(CO)_3$ and at lower temperatures (1,3,5-cycloheptatriene)$Cr(CO)_3$.[91]

Mixtures of α-methyl eleostearate (*cis*-9-*trans*-11,*trans*-13-octadeca-trienoate) and β-methyl eleostearate (*trans*-9,*trans*-11-*trans*-13-octadecatri-enoate) were subjected to hydrogenation using (methyl benzoate), (benzene), and (cycloheptatriene)$Cr(CO)_3$ as catalysts. With (methyl benzoate)$Cr(CO)_3$ the eleostearates were quantitatively reduced to dienes, which in turn were converted to monoenes. However, the major product was the diene when (benzene)- or (cycloheptatriene)$Cr(CO)_3$ was used as the catalyst. The major products were *cis,cis*- and *cis,trans*-1,4-dienes from α-methyl eleostearate and *cis,trans*-1,4-dienes from β-eleostearate.[92]

B. Features of Hydrogenation

The most important aspect of hydrogenation by $ArM(CO)_3$ compounds as catalysts is the extremely high selectivity for reduction of 1,3-dienes to mainly *cis*-monoenes. The most favorable stereochemistry for reduction of conjugated dienes is *trans,trans*, then *cis,trans*, with the least favorable being *cis,cis*. The relative rates of reduction of a mixture of 9,11-octadecadienoates were 1.0:8.0:25 for *cis,cis*, *cis,trans*, and *trans,trans*, respectively. Furthermore, the *trans*-9-*trans*-11-octadecadienoate yielded essentially only the 10-octadecenoate, whereas the corresponding *cis,trans*-diene gave a mixture of mainly the 9- and 10-octadecenoates. The 9-octadecenoate formation was explained on the basis of an isomerization of the 9,11-diene to an 8,10-diene, followed by 1,4-addition of hydrogen.[82] The isomerization of the 9,11-diene to an 8,10-diene, followed by 1,4-addition of hydrogen.[82] The isomerization of the 9,11- to the 8,10-diene was proposed to take place through a 1,5-hydrogen shift (see Scheme 2).

Various branched 1,3-dienes were investigated to determine the effects of substituents in different positions. Methyl substituents in the C-2 and C-3 positions of 1,3-dienes did not interfere with reduction; for example, 2,3-dimethyl-1,3-butadiene was reduced readily. However, methyl substituents on the C-1 and C-4 positions seriously hindered reduction; for example, 2,5-dimethyl-2,4-hexadiene was relatively unreactive, even at higher temperatures. Reduction of 4-methyl-1,3-pentadiene gave 2-methyl-2-pentene as the major product. The formation of that product was explained through a 1,5-hydrogen shift preceding hydrogenation,[82] as shown in Scheme 2.

The relative rates of hydrogenation of various conjugated and unconjugated dienes were determined using (methyl benzoate)$Cr(CO)_3$ as catalyst by observing the decrease in the concentration of the diene. The pseudo-first-order constants (hr^{-1}) were determined after an induction period of 30–60 min. For the acyclic hydrocarbon dienes, the constants decrease in the order 2,4-hexadiene (2.1), 4-methyl-1,3-pentadiene (1.1), 1,3-hexadiene (0.9), 2-methyl-1,3-pentadiene (0.69), 2,3-dimethylbutadiene (0.61), 1,4-hexadiene (0.22), 2,5-dimethyl-2,4-hexadiene (0.007). The rate constant for 1,3-cyclohexadiene was 0.50 hr^{-1} while that for 1,4-cyclohexadiene was 0.42 hr^{-1}.[82]

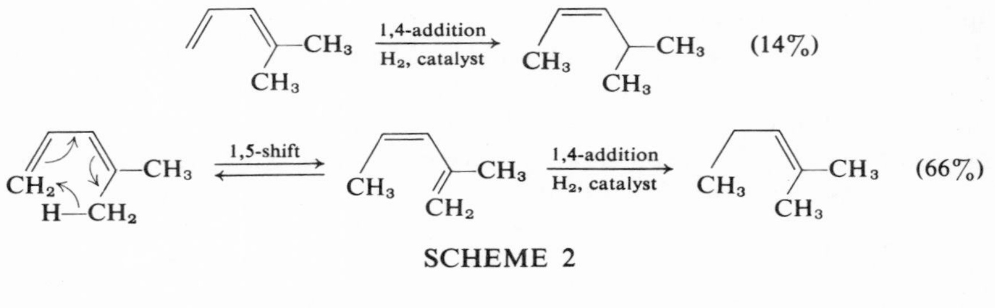

SCHEME 2

In the reduction of conjugated trienes such as α- and β-eleostearates, the products from β-eleostearate (a *trans*-9-*trans*-11-*trans*-13-triene) are the expected *trans*-9-*cis*-12- and *cis*-10-*trans*-13-octadecadienoates, depending on whether hydrogen addition is 9,12 or 11,14. The highest selectivity should have been with α-eleostearates (a *cis*-9-*trans*-11-*trans*-13-triene) since the most facile reductions occur with the *trans,trans*-1,3-dienes. However, whereas the expected *cis*-9-*cis*-12-octadecadienoate was obtained, the same products from reduction of β-eleostearate were also produced. This was explained on the basis of an isomerism of α- to β-eleostearate catalyzed by $ArCr(CO)_3$.[92] The results are shown in Scheme 3.

In the reduction of 1,3,5-cycloheptatriene, the expected product would be 1,4-cycloheptadiene; however, the major product obtained was 1,3-cyclo-heptadiene. Reduction studies using D_2 showed that the product 1,4-cyclo-heptatriene was initially formed from 1,4-addition, and this product isomerized to 1,3-cycloheptadiene by a 1,3-hydrogen shift.[85]

The rates and selectivity of hydrogenation of 1,3-dienes are dependent on the nature of the metal in the $ArM(CO)_3$ catalysts. For the complexes (mesity-lene)$M(CO)_3$, the relative order of reactivity was Mo > W > Cr in the reduction of methyl sorbate. However, the order of selectivity toward produc-tion of the methyl 3-hexenoate was Cr > Mo > W. (Mesitylene)$Mo(CO)_3$ was active at temperatures ranging from 100–150°, and complete conversion of sorbate was achieved in 0.5–2 hr. This contrasts with the corresponding chromium system, which is active at temperatures of 165–175° over 6-hr periods. Even then, for complete conversion of methyl sorbate, much higher catalyst concentrations are required for the chromium than for the molybdenum catalysts. Where quantitative conversions of methyl sorbate were achieved, the

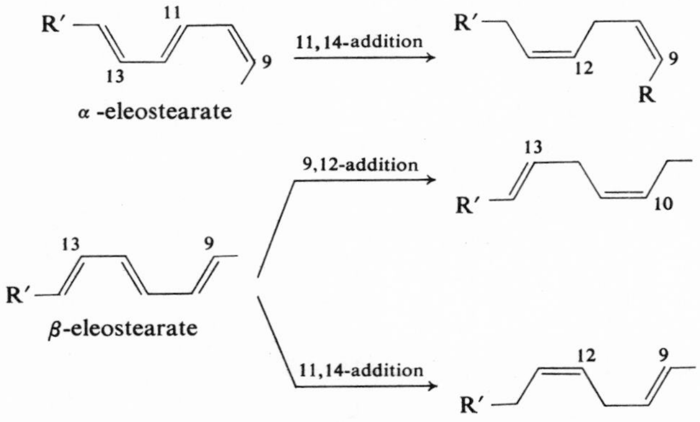

SCHEME 3

chromium catalyst was 96% selective toward production of the 3-hexenoate whereas (mesitylene)Mo(CO)$_3$ was selective toward the 3-hexenoate to about 88% at 100°. At higher temperatures the selectivity was considerably reduced. The selectivity promoted by (mesitylene)W(CO)$_3$ was very low in comparison with the other catalysts. Another feature which was observed in the comparison of the mesitylene complexes of the three metals in catalysis of methyl sorbate reduction was that no induction period was required for the molybdenum or tungsten species, whereas a significant induction period was observed for the chromium catalyst. This probably accounts for the greater reactivity of the molybdenum and tungsten compounds, which, unlike the chromium complex, decompose to other carbonyl species during hydrogenation.[89]

Whereas the selectivity was not affected significantly by the nature of the arene ring on ArCr(CO)$_3$, the relative activities of the catalysts were so dependent. In the reduction of methyl sorbate, it was found that the activities of the complexes decreased in the order of coordinated arene: 1,3,5-cycloheptatriene > C_6H_5Cl > $C_6H_5COOCH_3$ > C_6H_6 > $C_6H_5CH_3$ > $C_6H_3(CH_3)^3$ > $C_6(CH_3)_6$. This order was attributed to the stability of the arene–Cr bond, where electron-donating groups increase and electron-withdrawing or deactivating groups decrease the stability of bond.[82,89] However, the cycloheptatriene complexes were not effective at higher temperatures; they are apparently too labile and decompose completely after one hour of hydrogenation at 175°.[89]

Also affecting the efficiency of the catalyst is the presence of two Cr(CO)$_3$ moieties per molecule of complex, such as compounds (13) and (14):

(13)   (14)

The complexes containing two Cr(CO)$_3$ groups per molecule reduced methyl sorbate in 99% conversion yields to methyl 3-hexenoate in about 15–30 min.[87]

C. Mechanism

Cais and co-workers proposed a mechanism for hydrogenation by ArCr(CO)$_3$ as follows:

$$Cr + diene \underset{k_{-1}}{\overset{k_1}{\rightleftharpoons}} Cr \cdot diene$$

$$Cr \cdot diene + H_2 \underset{k_{-2}}{\overset{k_2}{\rightleftharpoons}} H_2 \cdot Cr \cdot diene$$

$$H_2 \cdot Cr \cdot diene \overset{k_3}{\longrightarrow} monoene + Cr$$

The mechanism was based on a comparison with studies on hydrogenation of olefins using dienetricarbonyliron as the catalyst. The observed induction period for all hydrogenation reactions involving $ArCr(CO)_3$ was explained as the time required for the formation of the Cr diene complex, which then becomes the active intermediate in the subsequent hydrogenation reaction.[88]

Frankel and co-workers[82,93,94] carried out experiments using deuterium in the reduction of methyl sorbate with (methyl benzoate)$Cr(CO)_3$ as the catalyst. The major product was methyl 2,5-dideuterio-cis-3-hexenoate, which showed that the primary process occurring in the reduction of conjugated dienes is 1,4-addition. The mechanism shown in Scheme 4 was presented to account for the reduction of methyl sorbate by $ArCr(CO)_3$.

According to Frankel et al. the mechanism accounts for the induction period as the time necessary to generate the active intermediate $[Cr(CO)_3]$, and this first step is also regarded as rate-determining. The proposed mech-

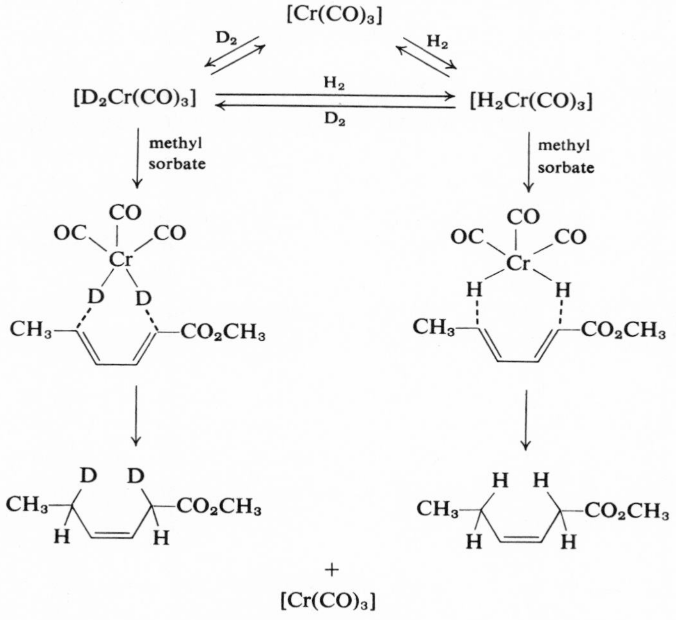

SCHEME 4

anism also accounts for the activity of the catalysts being dependent on the nature of the coordinated arene ring, i.e., the rate decreases in the order of arene in $ArCr(CO)_3$ as $C_6H_5Cl > C_6H_5COOCH_3 > C_6H_6 > C_6H_5CH_3 > C_6H_3(CH_3)_3 > C_6(CH_3)_6$. The electron-repelling substituents would be expected to decrease, whereas electron-withdrawing substituents would facilitate the ease of dissociation of $ArCr(CO)_3$.

Direct evidence for the process came from the observation of free methyl benzoate in solution (the possibility that free arene came from decomposition of the catalyst was discounted) and also that the addition of free benzene or other aromatic ligands to the reaction system strongly hindered hydrogenation. Both of these observations would be required by the proposed mechanism.

Other evidence in support of the proposed mechanism lies in examination of the products. The fact that the major product was the 2,5-dideuterio-monoene indicated that the $[D_2Cr(CO)_3]$ species was involved. In a partially reduced sample of methyl sorbate, the unreduced diene contained no deuterium, whereas the monoene contained two deuterium atoms per molecule. Using mixtures of H_2 and D_2, the amount of H_2–D_2 exchange was small, as evidenced in products containing only one D atom per molecule. In the absence of substrate, the H_2–D_2 exchange increased about twofold; this indicates that $ArCr(CO)_3$ alone is an effective catalyst for hydrogen activation and H_2–D_2 exchange. In the presence of the substrate, addition predominated over exchange.

The actual process of addition of hydrogen was implied, by analogy to hydrogenation by Rh(I) catalysts, to occur by a concerted mechanism.

However, Cais and Rejoan[95] have also studied various aspects of reduction of methyl sorbate by $ArCr(CO)_3$, such as the effect of various substituents on the coordinated arene ring, solvent effects, and others. Based on their studies, they proposed the mechanism for hydrogenation shown in Scheme 5.

The prime aspects of the proposed mechanism, which differ from that proposed by Frankel et al. are:

1. The arene ring remains attached to the metal throughout the course of catalysis.
2. Activation of the conjugated diene precedes activation of hydrogen.
3. Hydrogen addition to the coordinated diene occurs in stepwise, rather than concerted, fashion.

The essential features of the hydrogenation reaction were explained as follows: The observed induction time is the establishment of the equilibrium between $ArCr(CO)_3$ and (15) in which the arene ring functions as a four-

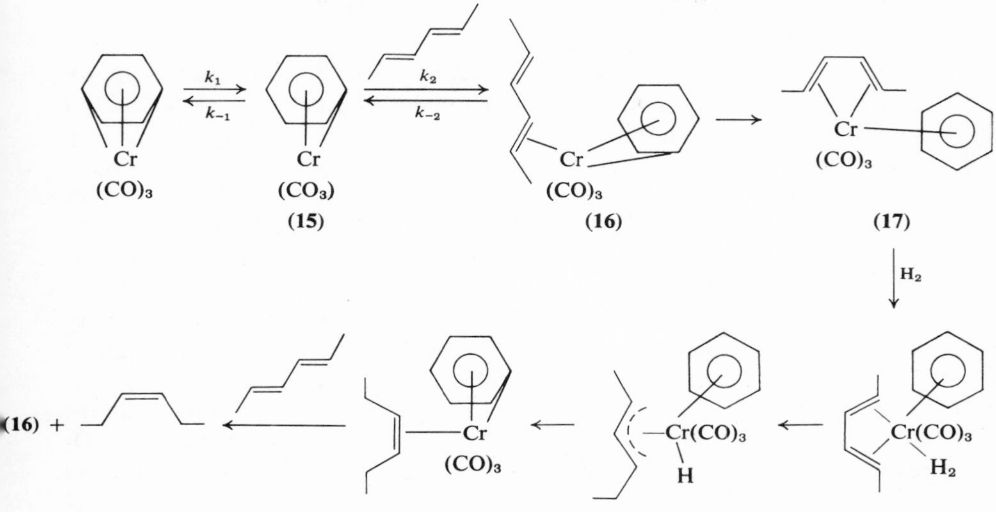

SCHEME 5

electron donor. The induction time will vary with substituents on the co-ordinated arene ring, electron-withdrawing substituents facilitating the partial dissociation of the ring. The partial dissociation would also be facilitated by nucleophilic solvents; this was borne out by experimental evidence. Aromatic solvents compete with the diene substrate and thus inhibit the reaction (15) → (16), and slow down the hydrogenation.

Moreover, a complete dissociation of arene as proposed by Frankel et al. resulting in the active species $Cr(CO)_3$ should only show differences in in-duction time with respect to varying substituents on the arene ring; once $Cr(CO)_3$ is produced, the rates of hydrogenation should all be the same, irrespective of the arene substituent on the original catalyst.

The transfer of hydrogen was proposed to proceed in a stepwise fashion creating the π-allylic intermediate (19), then addition of the second hydrogen atom completes the reduction to (20). A stepwise addition accounts for the small production of 1,2-addition products, e.g., methyl 2-hexenoate.

The free arene in solution detected by Frankel et al. and used in support of their proposal of complete arene dissociation was explained by Cais and Rejoan as arising from decomposition and/or disproportionation of the catalyst. For example, using (stilbene)$Cr(CO)_3$ as catalyst in the reduction of methyl sorbate, at 100% conversion to monoene, 76% of the catalyst was recovered unchanged, 18% was recovered in the form of metal-free arene, and the rest appeared in the form of a chromium-containing green, insoluble powder. Occasionally, $Cr(CO)_6$ could be recovered at the end of the reaction.

It was also observed that preheating the reactants before introduction of hydrogen significantly reduced the induction period for the onset of hydrogenation. Furthermore, it was found that preheating a mixture of the catalyst and substrate (no H_2) resulted in considerably higher initial rates of hydrogenation than when the catalyst and hydrogen were preheated (no substrate). Since the presence of hydrogen has a significantly smaller effect on the induction time relative to that of the diene, it would appear that these results are consistent with the proposed mechanism whereby coordination of diene precedes activation of hydrogen, i.e., (17) + $H_2 \rightarrow$ (18).

In a private communication[96] Frankel has raised certain objections to the mechanism proposed by Cais and Rejoan. The fact that the initial rate of hydrogenation is faster when the catalyst and substrate are preheated first than when the catalyst and hydrogen are preheated is also consistent with the observations of Frankel *et al.*[93,94] that the catalyst accelerates H_2–D_2 exchange only in the absence of diene. Furthermore, in the absence of H_2 and under conditions in which a metal-diene complex would be formed, mechanistic paths can be affected and new hydrogen-transfer reactions could result (see Section V, Isomerization of Olefins).

A second objection raised by Frankel to the mechanism proposed by Cais and Rejoan[95] lies in the stepwise addition of hydrogen via the π-allylic complex intermediate (19). Such stepwise addition of atomic hydrogen would be expected to yield significant amounts of monodeuterated products[93,94] and also produce reduced dienes in which positional and geometric isomerization of the double bond would be observed. These expectations are not supported by the facts.[93,94]

According to the mechanism of Cais and Rejoan,[95] stepwise addition on either end of the π-complexed allyl group would produce with D_2 a mixture of 2,4- and 3,5-dideuterio-3-hexenoate, if no isomerization occurs. However, established unequivocally by ^1H and ^2H nmr is the fact that the deuteration product of methyl sorbate is methyl 2,5-dideuterio-*cis*-3-hexenoate.[93,94] If π-allyl complexes are involved in the formation of isomers other than 3-hexenoate, as indicated by Cais and Rejoan, then the product distribution would be even more complicated and include: 2,4-dideuterio-*trans*-3-hexenoate, 3,5-dideuterio-*trans*-4-hexenoate, 2,4-dideuterio-*cis*-3-hexenoate, 3,5-dideuterio-*cis*-3-hexenoate, 2,2-dideuterio-*cis*-3-hexenoate, 5,5-dideuterio-*cis*-3-hexenoate, 3,3-dideuterio-*trans*-4-hexenoate, and 4,4-dideuterio-*trans*-2-hexenoate. By the same mechanism,[95] H–D scrambling would also be expected and result in hexenoate products with deuterium distributed on carbons 2–5. Such a complicated product distribution was not observed.[93,94]

It was also pointed out by Frankel in defense of his mechanism[96] that kinetics measurements and length of induction time are not reproducible.

Finally, it is difficult to explain why the formation of the 3-hexenoate is highly favored if one uses the mechanism proposed by Cais and Rejoan. According to the mechanism proposed by Frankel *et al.*, methyl *cis*-3-hexenoate is the product expected (and obtained) from 1,4-addition of hydrogen on a cisoid intermediate $H_2Cr(CO)_3$ complex of methyl sorbate.[93,94]

Obviously, the question on the mechanism of reduction of 1,3-dienes is far from settled.

Mechanistic details for the reduction of conjugated dienes by $[C_5H_5-(CO)_3M]_2$ (M = Cr, Mo, W) were not offered.[80,83] When the metal is chromium, the complex acts as a true catalyst; however, for the molybdenum and tungsten species, the reaction is stoichiometric in metal complex. The following description was proposed by Miyake and Kondo[80] for the reduction of 1,3-dienes to 2-monoenes, although 4-methyl-1,3-pentadiene gave a mixture of 2- and 3-monoenes.

$$[(C_5H_5)Cr(CO)_3]_2 + H_2 \rightleftharpoons 2(C_5H_5)Cr(CO)_3H$$

$$2(C_5H_5)Cr(CO)_3H + R-CH=CH-CH=CHR \longrightarrow$$
$$[(C_5H_5)Cr(CO)_3]_2 + RCH_2-CH=CH-CH_2R$$

In the case of chromium, the dinuclear complex reacts with molecular hydrogen to regenerate the mononuclear hydride species, but this conversion does not occur for the dinuclear molybdenum and tungsten species, presumably because of the stronger metal–metal bonds.[97]

No mechanistic details were presented for the photo-induced reduction of conjugated dienes by chromium hexacarbonyl.[81]

D. Experimental Procedures

Hydrogenation of Methyl Sorbate.[89,95] A solution of 0.24 g (19 mmol) of methyl 2,4-hexadienoate (methyl sorbate) and 0.27 g (1 mmol) of (methyl benzoate) chromium tricarbonyl in 90 ml of cyclohexane is placed in an autoclave and the system is flushed first with nitrogen, then with hydrogen. Hydrogen is then introduced to a pressure of 500 psig and the system is heated to 160° and maintained at that temperature for 4 hr. Solvent is removed from the final product on a rotary evaporator. The hydrogenated product is dissolved in 95% ethanol and nitrogen bubbled into the solution with magnetic stirring. Freshly ground $FeCl_3$ is then added (ca. 1 g) in three portions during 30 min. The solution is diluted with water and the product extracted with petroleum ether and dried over Na_2SO_4. The methyl sorbate is quantitatively converted, and the product mix is composed of 95% methyl 3-hexenoate, 4% methyl 2-hexenoate, and 1% methyl hexanoate.

V. ISOMERIZATION OF OLEFINS

A. Reactions Promoted by ArM(CO)$_3$ and Related Catalysts

Compounds derived from the Group-VIB carbonyls have been shown to be effective catalysts for positional and, to a lesser extent, geometrical isomerization of carbon–carbon double bonds. Among the positional changes promoted by the catalysts are isomerization of terminal to internal olefins, and conjugation of cyclic and acyclic polyenes. Table VII presents many of the isomerizations which have been reported using Group-VIB carbonyls as catalysts.

B. Discussion

In the case of (COD)W(CO)$_4$ as catalyst, the complex was originally tested for activity in an olefin disproportionation reaction; instead, it was found to be effective as an olefin isomerization catalyst. It was claimed[100] that not only terminal but also internal olefins could be isomerized, and that the product always contained the double bond in a more internal position. Other related catalysts were also claimed as effective in promoting double-bond migration, and these were norbornadiene and dicyclopentadiene tungsten tetracarbonyls, as well as 1,3,5-cycloheptatriene, cyclooctatetraene, 6-dimethylaminofulvene tungsten tricarbonyls, and 1,5-cyclooctadiene tungsten tetracarbonyl.

Using (methyl benzoate)Cr(CO)$_3$ as the catalyst it was found that 2- and 3-hexene (*cis* and *trans*) were not isomerized but 1-hexene was isomerized to a mixture of *cis* and *trans* 2- and 3-hexenes. Apparently an equilibrium strongly favoring 2- and 3-hexene is indicated.[82]

Isomerization of 1,3-hexadiene to 2,4-hexadiene resulted in a detectable equilibrium favoring the 2,4-isomer. Conjugation of 1,4-hexadiene with (methyl benzoate) Cr(CO)$_3$ gave a mixture of the 1,3- and 2,4-dienes, whereas 1,5-hexadiene was unaffected by the catalyst.[82]

In the case of cyclic dienes, 1,4-cyclohexadiene was readily isomerized to the conjugated diene, while 1,5-cyclooctadiene gave a mixture of 1,4- and 1,3-cyclooctadiene, with the latter being the major product.

An interesting disproportionation, which occurs as a side reaction, was observed in the isomerization of both 1,3- and 1,4-cyclohexadiene using the chromium tricarbonyl catalyst; small amounts of benzene and cyclohexene were detected in the product.

Methyl linoleate (methyl 9,12-octadecadienoate) was converted to conjugated isomers by chromium hexacarbonyl and methyl benzoate, or benzene and 1,3,5-cycloheptatriene chromium tricarbonyl. Investigation of the profile of the reaction with the (methyl benzoate) tricarbonyl catalyst showed that the *cis,trans* isomer was formed initially, followed by production of the *cis,cis* and *trans,trans* forms. The latter isomers arose from geometrical isomerization of the *cis,trans*-diene. While the major conjugated products were the 8,10- and 9,11-dienes, significant amounts of 7,9- and 10,12-isomers were also detected, along with smaller amounts of 5,7-, 6,8-, 11,13-, 12,14- and 13,15-dienes. Chromium hexacarbonyl also promoted conjugation to a mixture of isomers, similar to those obtained in the arene tricarbonyl-catalyzed reaction. Small amounts of conjugated triene and monoene, from disproportionation, were also detected in reactions promoted by both catalytic systems.

Unconjugated trienes, such as methyl linolenate (9,12,15-octadecatrienoate) and those contained in soybean, safflower, and linseed oils could also be isomerized to conjugated forms. In general, the major product obtained was a compound containing a conjugated diene, with one isolated double bond. These compounds were mostly mixtures of *trans,trans* and *cis,-trans* conjugation with isolated *trans* unsaturation. Significant amounts of conjugated triene were also detected.

C. Mechanism

Frankel[103] has studied the isomerization of 1,4-hexadiene and 1,4-cyclohexadiene. It was found that in a mixture of *trans*-1,4- and *cis*-1,4-hexadiene, the *trans*-diene was much more reactive than the *cis* form in yielding *cis*-2, *trans*-4-hexadiene as the major product using (methyl benzoate)Cr(CO)₃ as the catalyst. Minor products included *cis*-1,3- and *cis,cis*- and *trans,trans*-2,4-hexadienes.

The mechanism shown in Scheme 6 was proposed[103] to account for the formation of the major product. It also accounts for the production of 1,3-hexadiene, depending upon direction of readdition of the abstracted hydrogen atom. The observation that *trans*-1,3-hexadiene and *trans,trans*-2,4-hexadiene form very slowly in (methyl benzoate)Cr(CO)₃-catalyzed isomeriza-

$$(C_6H_5COOCH_3)Cr(CO)_3 \rightleftharpoons C_6H_5COOCH_3 + Cr(CO)_3$$

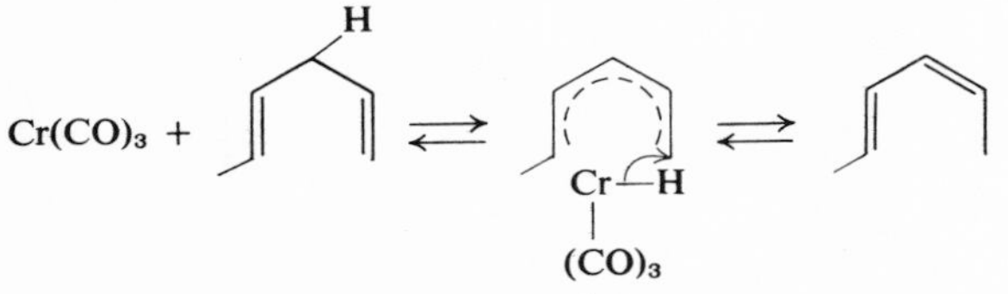

SCHEME 6

TABLE VII

Isomerization of Olefins by Group-VIB Carbonyls

Catalyst[a]	Olefin (% conversion)	Products (% selectivity)	Conditions	Reference
Mo(CO)$_6$[b]	1-Butene (19)	2-Butene	121°, 34 atm, 5 hr	98
	2-Butene (5)	1-Butene	230°, 34 atm, 5 hr	98
	1-Pentene (15)	2-Pentene	121°, 34 atm, 5 hr	98
	1-Hexene (15)	2-, 3- Hexenes	121°, 34 atm, 5 hr	98
(1,5-COD)W(CO)$_4$	1-Pentene (28)	2-Pentene (95)	115°, 15 min	99, 100
	1-Pentene (63)	2-Pentene (98)	215°, 15 min	99, 100
	1-Pentene (55)	2-Pentene (99)	265°, 15 min	99, 100
	2-Methyl-1-butene (62)	2-Methyl-2-butene (100)	115°, 15 min	99, 100
	2-Methyl-1-butene (72)	2-Methyl-2-butene (90)	175°, 15 min	99, 100
	2-Hexene	3-Hexene		100
	4-Methyl-1-pentene	4-Methyl-2-pentene		100
(MBZ)Cr(CO)$_3$	1-Hexene (23)	2- and 3-hexene, cis and trans	175°, 6 hr	82
M(CO)$_6$[c]	1-Dodecene (80)	2-Dodecene (100)	Reflux, 1.5 hr	101
(MBZ)Cr(CO)$_3$	1,3-Hexadiene (cis and trans) (21)	2,4-Hexadiene (88), 2- and 3-hexene (12)	175°, 3.4 atm N$_2$, 6 hr	82
(MBZ)Cr(CO)$_3$	2,4-Hexadiene (9) (48% cis,trans, 52% trans,trans)	1,3-Hexadiene (89), 2- and 3-hexene (11)	175°, 3.4 atm N$_2$, 6 hr	82
	1,4-Hexadiene (38) (28% cis, 72% trans)	cis,trans-2,4-Hexadiene (87), trans,trans-2,4-hexadiene (5), cis,cis,cis-2,4-hexadiene (8)	175°, 3.4 atm N$_2$, 6 hr	82
	1,5-Hexadiene	No isomerization	175°, 3.4 atm N$_2$, 6 hr	82
	4-Methyl-1,3-pentadiene (10)	2-Methyl-1,3- pentadiene (40)	175°, 3.4 atm N$_2$, 6 hr	82
	1,3-Cyclohexadiene (12)	1,4-Cyclohexadiene (25), cyclohexene (58), benzene (17)	160°, 3.4 atm N$_2$, 6 hr	82
(MBZ)Cr(CO)$_3$	1,4-Cyclohexadiene (63)	1,3-Cyclohexadiene (92), cyclohexene (6), benzene (2)	160°, 3.4 atm N$_2$, 6 hr	82

Catalyst	Substrate	Products	Conditions	Ref.
(MBZ)Cr(CO)₃	1,5-Cyclooctadiene (85)	1,4-Cyclooctadiene (16), 1,3-cyclooctadiene (84)	175°, 3.4 atm N₂, 6 hr	82
	1,3-Cyclooctadiene (2)	1,4-Cyclooctadiene (50), cyclooctene (50)	175°, 3.4 atm N₂, 6 hr	82
	Methyl linoleate (90) (9,12-octadecadienoate)	Conjugated diene (72) (*cis,trans*, 61%; *cis,cis*, 12%; *trans,trans*, 27%), conjugated triene (6), monoene (22)	175°, 3.4 atm He, 5 hr	102
(C₆H₆)Cr(CO)₃	Methyl linoleate (86)	Conjugated diene (74) (*cis,trans*, 73%; *cis,cis*, 11%; *trans,trans*, 16%), conjugated triene (6), monoene (20)	185°, 3.4 atm He, 6 hr	102
(C₇H₈)Cr(CO)₃	Methyl linoleate (85)	Conjugated diene (92) (*cis,trans*, 74%; *cis,cis*, 5%; *trans,trans*, 21%), conjugated triene (3), monoene (5)	185°, 3.4 atm, He, 5 hr	102
Cr(CO)₆	Methyl linoleate (93)	Conjugated diene (72) (*cis,trans*, 41%; *cis,cis*, 10%; *trans,trans*, 49%), conjugated triene (8), monoene (20)	195°, 3.4 atm, He, 2 hr	102
(MBZ)Cr(CO)₃	Methyl linolenate (50) (9,12,15-octadecatrienoate)	Triene with two conjugated double bonds and one isolated (62), conjugated triene (27), monoene (1), unconjugated diene (9)	185°, 3.4 atm N₂, 6 hr	102

a Abbreviations: 1,5-COD = 1,5-cyclooctadiene, MBZ = methyl benzoate, C₇H₈ = 1,3,5-cycloheptatriene.
b On activated alumina support.
c M = Cr, Mo.

tions of 1,4-hexadienes was noted as supportive evidence for the pentadienyl hydride complex intermediate (21). Steric interaction in the corresponding pentadienyl hydride intermediate from *cis*-1,4-hexadiene would account for the slow interconversion to *cis,cis*-2,4- and *cis*-1,3-hexadienes.[103]

The proposed mechanism of isomerization of 1,4-cyclohexadiene to the conjugated form is similar to that proposed above for the acyclic diene, and also accounts for the observed disproportionation of 1,4-cyclohexadiene to benzene and cyclohexene. The mechanism is shown in Scheme 7.

For the conjugation of methyl linoleate catalyzed by (methyl benzoate) $Cr(CO)_3$ or by chromium hexacarbonyl, the major products obtained were 8,10-, 9,11-, and 10,12-dienes, but several other dienes were also obtained. The proposed mechanism for production of 10,12- and 9,11-dienes from the 9,12-diene is shown in Scheme 8.

The active intermediate was proposed as $Cr(CO)_3$, which forms a π-allylic intermediate by hydrogen abstraction from C-11 of the *cis*-9,*cis*-12-diene, to give either (22) or (23). Depending upon the readdition of hydrogen, the *trans*-10-*cis*-12, or *cis*-9-*trans*-11-octadecadienoate can be obtained. The overall process of conjugation involves a 1,3-hydrogen shift. To account also for the formation of the 8,10- and 11,13-diene isomers, 1,5-hydrogen shifts of the conjugated products were invoked, shown in Scheme 9. It is apparent that a 1,5-hydrogen shift requires a *cis,trans* configuration.

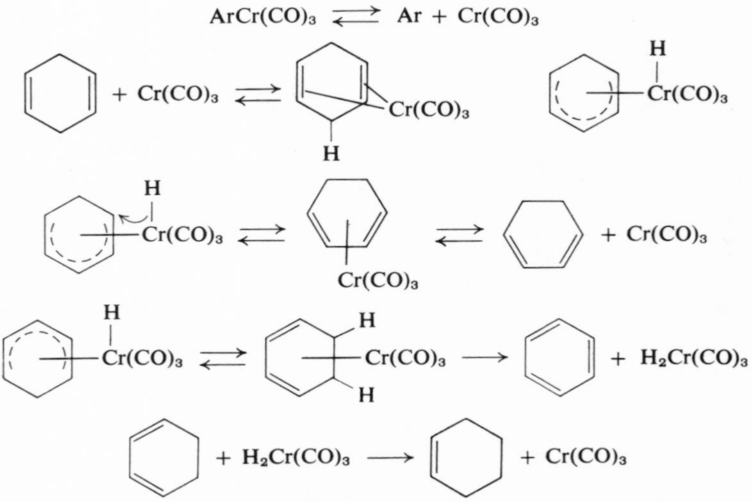

SCHEME 7

$$ArCr(CO)_3 \rightleftarrows Ar + Cr(CO)_3$$

$$\text{or } Cr(CO)_6 \rightleftarrows Cr(CO)_3 + 3CO$$

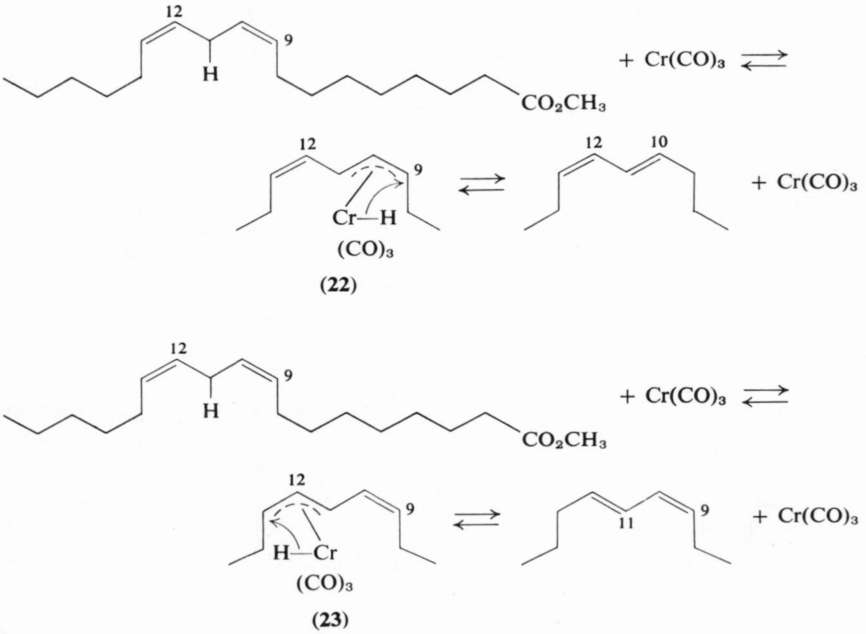

(22)

(23)

SCHEME 8

To explain the scattering of conjugated diene isomers throughout the fatty ester chain, a general scheme was proposed in which a number of 1,3- and 1,5-hydrogen shifts could take place. Thus:

9,12-diene $\underset{}{\overset{\text{1,3-shift}}{\rightleftarrows}}$ 9,11-diene

9,11-diene $\underset{}{\overset{\text{1,3-shift}}{\rightleftarrows}}$ 8,11-diene $\underset{}{\overset{\text{1,3-shift}}{\rightleftarrows}}$ 8,10-diene

9,11-diene $\underset{}{\overset{\text{1,5-shift}}{\rightleftarrows}}$ 8,10-diene $\underset{}{\overset{\text{1,3-shift}}{\rightleftarrows}}$ 8,11-diene

9,12-diene $\underset{}{\overset{\text{1,3-shift}}{\rightleftarrows}}$ 10,12-diene

10,12-diene $\underset{}{\overset{\text{1,5-shift}}{\rightleftarrows}}$ 11,13-diene $\underset{}{\overset{\text{1,3-shift}}{\rightleftarrows}}$ 10,13-diene

10,12-diene $\underset{}{\overset{\text{1,5-shift}}{\rightleftarrows}}$ 10,13-diene $\underset{}{\overset{\text{1,3-shift}}{\rightleftarrows}}$ 11,13-diene

Support for the above postulation of 1,3- and 1,5-hydrogen shifts to account for a variety of observed products comes from kinetics studies which

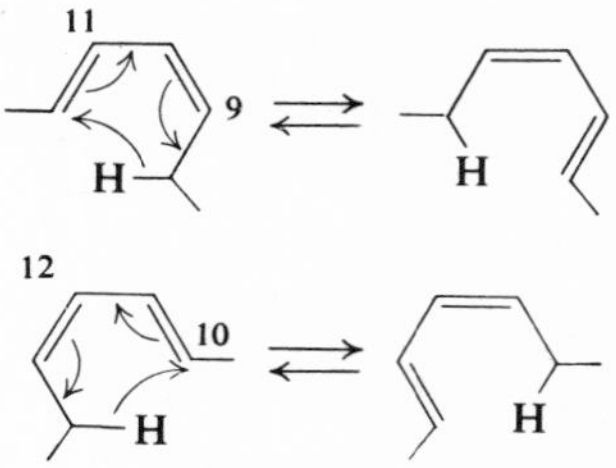

SCHEME 9

show that *cis,trans*-conjugated dienes are the initial products of conjugation. Apparently, all the other isomers arise from the initial products.

D. Experimental Procedures

Conversion of 1-Pentene to 2-Pentene.[99,100] The experiment is carried out in a reaction system composed of two stainless-steel pressure vessels connected by means of a section of stainless-steel tubing. One pressure vessel is employed as a reaction chamber (A) and the other as a mixing chamber (B). Capacity of each chamber is 30 ml. The catalyst system is 0.01 M solution of (COD)W-$(CO)_4$ in dry benzene. The 1-pentene is used as a 1.3 M solution in pentane. To chamber B is added 4.0 ml of the catalyst solution and 1.54 ml of the olefin solution; the mixture is thoroughly agitated. Chamber A is heated to 240° and the contents of chamber B are transferred to A by pressurizing to 460–680 psig with nitrogen. The reaction is continued for 15 min, whereupon analysis by vpc shows a 62% conversion of 1-pentene with 100% selectivity to 2-pentene.

VI. OLEFIN METATHESIS

One of the most interesting reactions of olefins promoted by the Group-VIB metal carbonyl systems and other catalysts is metathesis (also known as olefin disproportionation or dismutation). The general reaction may be exemplified as follows:

$$R_2C{=}CR_2 + R_2'C{=}CR_2' \rightleftarrows 2R_2C{=}CR_2'$$

Only one example has been reported wherein an arene Group-VIB tricarbonyl species was claimed to function as an olefin metathesis catalyst; however, reports where related chromium-group carbonyl compounds have been used effectively are extensive. Both homogeneous and heterogeneous systems have

been reported, and the active catalysts have generally been limited to molybdenum and tungsten carbonyl complexes.

A. Homogeneous Systems

The only example of the use of an arene tricarbonyl complex as an olefin
metathesis catalyst was reported by Lewandos and Pettit.[104] They used
(tol)W(CO)$_3$ as a catalyst in the disproportionation of 4-nonene to give 4-
octene and 5-decene in 28% conversion of the starting olefin. After about 20%
conversion, secondary disproportionation products were observed. The
catalyst also promoted a small amount of isomerization products of 4-nonene.
They mentioned that molybdenum and tungsten hexacarbonyls were also
active, but no details were given.

The study was carried out mainly to elucidate mechanistic aspects of the
reaction. It is notable that the catalysts reported in this work did not require
the use of Lewis acid cocatalysts in order to be active. With every other
homogeneous system reported to date, the use of a Lewis acid was necessary
for activity.

Herisson *et al.* have examined the metathesis of 4-methyl-2-pentene with
several tungsten compounds of varying oxidation states; among them was
W(CO)$_5$P(C$_6$H$_5$)$_3$ with AlCl$_3$ as the cocatalyst, the presence of which was
necessary for the tungsten compound to be active. With an Al/W ratio of 3.0
and an olefin/W ratio of 500, they observed 50% conversion of the starting
olefin at ambient temperatures in 48 hr. However, isomerization of the olefin
also occurred, and the product mix became more and more complicated with
increasing time of reaction. Also observed were by-products from oligo-
merization of the olefins.[105]

In another study, using the W(CO)$_5$P(C$_6$H$_5$)$_3$/C$_2$H$_5$AlCl$_2$ cocatalyst
system, Ramain and Trambouze found that in the complete absence of air,
the catalyst system was inactive toward the disproportionation of 2-pentene.
However, upon introduction of controlled amounts of oxygen, the system
became active. This was explained in terms of a reaction of O$_2$ with C$_2$H$_5$AlCl$_2$
forming an ethoxide complex which became the active cocatalyst. With this
system, the authors observed a 50% conversion of 2-pentene in less than 1 hr
at room temperature. They also determined a rate law for the reaction which
showed first-order dependence on the olefin, the tungsten complex, and the
aluminum cocatalyst. The activation energy for the disproportionation
reaction was determined to be 10.5 kcal/mol.[106]

Several patents have been issued on the use of various olefin dis-
proportionation catalyst systems. One patent[107] claimed that Mo(CO)$_6$,
Mo(CO)$_5$py, and Mo(CO)$_4$py$_2$ were active when used with promoters R$_4$NCl
(R = propyl, butyl) and activators (CH$_3$)$_3$Al$_2$Cl$_3$ or (CH$_3$)$_2$Al$_2$Cl$_4$. Both 1-

and 2-olefins were metathesized with the catalytic systems. An interesting disproportionation occurred with 1,7-octadiene, yielding cyclohexene and ethylene as products in 100% conversion. The process may be envisaged as an intramolecular metathesis:

$$
\begin{array}{l}
\text{CH}_2\text{—CH}_2\text{—CH}{=}\text{CH}_2 \\
\;|\qquad\quad\vdots\qquad\vdots \\
\text{CH}_2\text{—CH}_2\text{—CH}{=}\text{CH}_2
\end{array}
\longrightarrow
\bigcirc
\;+\;
\begin{array}{l}
\text{CH}_2 \\
\|\\
\text{CH}_2
\end{array}
\qquad (17)
$$

Another patent[108] claimed that the $Mo(CO)_6/C_2H_5AlCl_2$ and $Mo(CO)_5$-$P(C_6H_5)_3/C_2H_5AlCl_2$ systems were active toward the disproportionation of propylene with 2-pentene. The products of this reaction would be a mixture of 1- and 2-butenes as major products.

Bencze and Marko investigated noncarbonyl containing tungsten complexes as olefin metathesis catalysts and found that their activities increased considerably when used under an atmosphere of carbon monoxide.[109] Subsequent studies showed that tungsten carbonyl complexes (which were not identified) were formed, as evidenced by the presence of strong absorption bands in the 2100–1800 cm^{-1} (terminal CO stretching) region. Thus, using $Wpy_2Cl_4/C_2H_5AlCl_2$ as the catalyst system in the disproportionation of 2-pentene, they found that conversion of the olefin ranged from 7 to 22% under argon, but under carbon monoxide a conversion level of about 50% was attained within 15 min. Similarly, the catalyst system derived from $W[C_2H_4(PPh_2)_2]_2Cl_3$ and $C_2H_5AlCl_2$ transformed 90% of the initial 1-pentene to gaseous and liquid olefins when used under carbon monoxide. The liquid fraction contained about equal quantities of 2-pentenes, hexenes, heptenes, and octenes. Under argon, the same catalytic system showed mainly isomerizing activity; the major product obtained was 2-pentene. Only about 2% of the original olefin was converted to disproportionation products.

Some complexes containing metal–metal bonds have also been found to be active as olefin disproportionation catalysts. Ionic complexes of the type $A_n^+[(CO)_5M\text{-}M'(CO)_5]^{n-}$ (A = alkali metal or R_4N; M = Mo or W; M′ = Mo, W, Mn, Re; n = charge, 2 or 1).[110] These compounds require the synergistic use of a cocatalyst (a quaternary ammonium chloride) and an activator (alkylaluminum dichloride or alkylaluminum sesquichloride) in order to show activity. The highest activity was found for the Mo–Mo system, with the Mo–Re system only slightly less active. The catalyst systems were found to be effective in the disproportionation of several terminal and internal olefins, such as 1-pentene, 4-methyl-1-pentene, and 3-heptene. Conversions of around 60% for 1-pentene were found for the Mo–Mo system, and the intramolecular disproportionation of 1,7-octadiene gave cyclohexene and ethylene in greater than 99% selectivity.

Anionic carbene complexes of the Group-VIB metal carbonyls have been investigated as olefin metathesis catalysts.[111,112] Thus, compounds of the type $NR_4[M(CO)_5COR']$ (M = Mo, W; R' = CH_3, C_6H_5) with cocatalyst (activator) alkylaluminum dichloride or alkylaluminum sesquichloride provided active catalytic systems at ambient temperatures for the disproportionation reactions of terminal and internal olefins. The order of reactivity for the disproportionation of 1-pentene was Mo > W, and with respect to the carbene substituent, C_6H_5 > CH_3.

Interestingly, the nature of the cation showed an effect on the rate of disproportionation of 1-pentene; rates of the tetrabutylammonium compound were faster than tetramethylammonium or lithium salts.

Neutral carbenes of the type $W(CO)_5(COCH_3)C_2H_5$ and $W(CO)_5$-$[CN(CH_3)_2]CH_3$ were essentially inactive with CH_3AlCl_2 as cocatalyst. However, upon addition of tetrabutylammonium chloride to the system, similar activity to the anionic carbene complexes was observed.[111]

In the patent of Kroll and Doyle [112] on the anionic carbene complexes as olefin metathesis catalysts, a large variety of homo- and cross-disproportionation reactions of both terminal and internal olefins were reported.

The metathesis of cyclic olefins is particularly interesting. For example, the reaction of cyclopentene produces 1,6-cyclodecadiene, which can disproportionate further at each double bond with additional molecules of cyclopentene. The ring continues to grow to a larger and larger cyclic polyene, eventually yielding a polymer of high molecular weight.

The Group-VIB pentacarbonyl halide anions have also been reported to function as active metathesis catalysts in the presence of aluminum compounds as cocatalysts.[113] The most active system with respect to the aluminum cocatalysts were CH_3AlCl_2, i-$C_3H_7AlCl_2$, $(CH_3)_3Al_2Cl_3$, $(CH_3)_3Al_2Br_3$ and $C_2H_5AlCl_2$. Trialkylaluminum compounds as cocatalysts were not active, and dialkylaluminum chlorides showed only very feeble activity. Among the Group-VIB metals, the order of activity was Mo > W > Cr. The chloride derivatives were more active than the corresponding bromide systems. Surprisingly, the nature of the cation showed a significant effect on activity. In a study on the metathesis of 1-pentene by $Mo(CO)_5Cl^-/CH_3AlCl_2$, a variety of tetraalkylammonium counterions were employed. The most active systems were found with $(n$-$C_3H_7)_4N^+$, $(n$-$C_4H_9)_4N^+$, $(n$-$C_5H_{11})_4N^+$, and $(CH_3)(C_8H_{17})_3N^+$; the extent of metathesis was in the 79–91% range for these cations. The use of other counterions such as $(C_2H_5)_4N^+$, $(n$-$C_7H_{15})_4N^+$, $(CH_3)_3(C_6H_5)N^+$, and $(CH_3)(n$-$C_6H_{13})_3N^+$ gave metathesis reactions in a range of 10 to 53%.

The effect of the solvent was negligible. In the disproportionation of 1-pentene using $(C_4H_9)_4N[Mo(CO)_5Cl]/CH_3AlCl_2$, the conversion to 4-octene was essentially constant in such solvents as chlorobenzene, toluene, trichloroethylene, cyclohexane, pentane, benzene, or no solvent at all. There must be an excess of (at least 1.5:1) Al:Mo compounds in order for this catalytic system to show activity. Optimum temperature for the reactions appears to be about 60°.

Whereas both terminal and internal normal olefins appear to be very reactive toward disproportionation with the catalyst system, certain substituted olefins, particularly those where the substituent appears on the olefin carbons or one carbon removed from the double bond, are significantly less reactive.

A study on the catalytic system was made in an attempt to elucidate the nature of the active catalyst or catalyst precursor. The function of the cocatalyst was determined to be primarily that of a Lewis acid which removes chloride ion from $Mo(CO)_5Cl^-$:

$$CH_3AlCl_2 + (C_4H_9)_4N[Mo(CO)_5Cl] \longrightarrow (C_4H_9)_4N[CH_3AlCl_3] + Mo(CO)_5 \qquad (18)$$

The trichloromethylaluminate was isolated and identified in some reactions. Also isolated in some reactions was $Mo(CO)_6$, which indicates that the $Mo(CO)_5$ species undergoes some kind of disproportionation of its own. The other half of this $Mo(CO)_5$ disproportionation product appears, then, to be the active catalyst or precursor of it. It was speculated that perhaps the required excess of cocatalyst is necessary for the formation or stabilization of the active molybdenum species.

B. Heterogeneous Systems

The first example of an olefin disproportionation catalyst using a Group-VIB carbonyl compound in a heterogeneous system was reported by Banks and Bailey in 1964.[114] For these systems, activated alumina was impregnated with molybdenum or tungsten hexacarbonyls at 66°. Olefins such as propylene, 1-butene, 1-pentene, and 1-hexene were disproportionated in 10–60% conversions in a continuous-flow system under pressure using a fixed-bed reactor. A patent was issued on the same system.[98] Another patent used $Mo(CO)_6$ or $W(CO)_6$ on SiO_2-Al_2O_3 support for disproportionation.[115] Other catalyst supports of molybdenum and tungsten carbonyls have also been reported. These supports were high surface area, refractory materials such as the oxides of silicon, aluminum, thorium, and zirconium, and phosphates of aluminum, zirconium, titanium, and calcium.[116]

A novel disproportionation reaction of ethylene to propylene was reported by O'Neill and Rooney, using molybdenum hexacarbonyl on alumina support as the catalyst.[117] For the reaction

$$3\ CH_2{=}CH_2 \longrightarrow 2\ CH_3CH{=}CH_2$$

mechanistic details were not reported, but at least two possibilities were mentioned. Conversion of ethylene to two methylenes with subsequent addition to coordinated ethylene would result in a trimethylene intermediate. The trimethylene would then rearrange to propylene. A concerted reaction of three ethylenes to two propylene molecules could also not be discounted.

The nature of the active species of molybdenum on various supports was studied by infrared spectroscopy.[118,119] It was found that during the activation period at temperatures varying between 45 and 200°, changes in the CO stretching region occurred. This was interpreted as loss of CO to several subhexacarbonyl species, and upon further examination with respect to the activities of the various activated catalysts, it was concluded that none of the subhexacarbonyl species was active toward the disproportionation of propylene. It was further concluded that certain subhexacarbonyl species were the catalyst precursors, decomposing to metallic molybdenum. This in turn was oxidized to some higher oxidation state which became the active catalyst for disproportionation.

The support used was found to play a significant role in catalytic behavior, and the differences between the various catalysts were accounted for in terms of the surface hydroxyl groups on the supports. It was concluded that it is the role of the catalyst support to assist in complete decomposition of $Mo(CO)_6$ to form the active molybdenum species, so that catalysts on supports which hinder this decomposition by stabilizing molybdenum subhexacarbonyl intermediates are less active than those which do not. Thus, it was found that silica provides a more active support than alumina, and this system could be activated at lower temperatures than for example, alumina.

It was also found that treatment of $Mo(CO)_6/Al_2O_3$ with halogenated unsaturated molecules such as 1,1-dichloroethylene, 1,2-dichloroethylene, trichloroethylene, and 3,3,3-trifluoropropene resulted in a marked increase in activity of the catalyst.[120] Treatment with HCl or halogen compounds such as 1-chloropropane or chlorobenzene showed essentially no change in catalyst activity; hence it appears that the presence of a double bond is essential in the activating compound.

In a kinetics study using $Mo(CO)_6$ on alumina, second-order dependence was indicated for the disproportionation of propylene. It was proposed that reversible adsorption of propylene on the active catalyst led to $(C_3H_6)_2M$ (M = active catalyst), followed by the rate-determining rearrangement to $(C_2H_4)(C_4H_8)M$. This was thought to proceed through an activated complex resembling a substituted cyclobutane.

In a future volume of *Organometallic Reactions* a chapter will be devoted to the olefin metathesis reaction. Therefore, the section on this subject in this chapter is necessarily brief, and limited to the catalysts of Group-VIB carbonyls. Detailed mechanistic aspects of olefin metathesis, as well as the coverage of many other homogeneous and heterogeneous catalytic systems, will appear in the forthcoming chapter.

VII. REACTIONS OF ACETYLENES

A. Trimerization

It is apparent that the tendency for acetylenes to cyclotrimerize to benzene derivatives in the presence of a catalyst is overwhelming. There is a large number of transition metal catalysts which promote cyclotrimerization of alkynes;[121] among them are several arenetricarbonylchromium and -molybdenum complexes, as well as molybdenum and tungsten hexacarbonyls. Table VIII presents pertinent data.

All the reactions shown in Table VIII were carried out in the acetylene compound melt (no solvent) at 270°. At this rather high temperature, it would be expected that the active catalytic species would be $M(CO)_3$; for the arene derivatives, loss of the coordinated arene ring would be expected, and three CO molecules would dissociate from the hexacarbonyls. Coordination and activation of three acetylenic molecules in mutually *cis* positions would lead to the trimerized product.

TABLE VIII
Cyclotrimerization of Alkynes by Group-VIB Carbonyl Complexes

Catalyst[a]	Alkyne	Product (% yield)	Reference
$(C_7H_8)Cr(CO)_3$	Diphenylacetylene	Hexaphenylbenzene (59)	122
$(C_7H_8)Mo(CO)_3$	Diphenylacetylene	Hexaphenylbenzene (32)	122
$(DMB)Cr(CO)_3$	Diphenylacetylene	Hexaphenylbenzene (10)	122
$(DCB)Cr(CO)_3$	Diphenylacetylene	Hexaphenylbenzene (11)	122
$(C_6H_5Cl)Cr(CO)_3$	Diphenylacetylene	Hexaphenylbenzene (28)	122
$(C_6H_6)Cr(CO)_3$	Diphenylacetylene	Hexaphenylbenzene (43)	122
$(tol)Cr(CO)_3$	Diphenylacetylene	Hexaphenylbenzene (25)	122
$(naph)Cr(CO)_3$	Diphenylacetylene	Hexaphenylbenzene (22)	122
$(HMB)Cr(CO)_3$	Diphenylacetylene	Hexaphenylbenzene (27)	122
$Mo(CO)_6$	Diphenylacetylene	Hexaphenylbenzene (50)	123
$Mo(CO)_6$	$ClC_6H_4C\equiv CC_6H_4Cl$	$C_6(C_6H_4Cl)_6$ (40)	123
$W(CO)_6$	Diphenylacetylene	Hexaphenylbenzene (15)	123
$W(CO)_6$	$ClC_6H_4C\equiv CC_6H_4Cl$	$C_6(C_6H_4Cl)_6$ (25)	123

[a] Abbreviations: C_7H_8 = 1,3,5-cycloheptatriene; DMB = *p*-dimethoxybenzene; DCB = *p*-dichlorobenzene; tol = toluene; naph = naphthalene; MHB = hexamethylbenzene.

B. Polymerization

It would appear that in order to promote trimerization of acetylenes, the metal must have three vacant positions, mutually *cis*, for the coordination of the acetylene molecules. Using arene molybdenum tricarbonyls as catalysts for Friedel–Crafts reactions (at much lower temperatures than those employed for trimerization, see Section III), Farona *et al.* postulated that the arene ring remained coordinated to the metal during the course of the reactions.[71,78] The same contention was proposed by Cais and Rejoan in their mechanism of hydrogenation catalyzed by $ArCr(CO)_3$.[95] If the active form of the catalyst retains the arene ring, then the maximum number of acetylenes which could coordinate at one time is two. This would suggest that if the $ArM(CO)_3$ molecules were to act as homogeneous catalysts in acetylene reactions, aromatic trimers might not be formed. Indeed, upon application of this line of reasoning, it was found that $ArM(CO)_3$ molecules promoted linear polymerization reactions of acetylenes. Table IX reports polymerizations of alkynes catalyzed by $(tol)Mo(CO)_3$.

The concentration of catalyst in the reactions reported in Table IX was 3×10^{-2} M. The polymerization of phenylacetylene was studied in much more detail than the other alkynes, and it was pointed out that the number-

TABLE IX

Linear Polymerization Catalyzed by $(tol)Mo(CO)_3$[a]

Alkyne	Final product	Conditions	Mol wt	Reference
$C_6H_5C{\equiv}CH$	$\begin{array}{c} -C{=}CH- \\ \vert \\ C_6H_5 \end{array}_n$	Neat, 25° or benzene, reflux	12,000	125
$C_6H_{11}C{\equiv}CH$	$\begin{array}{c} -C{=}CH- \\ \vert \\ C_6H_{11} \end{array}_n$	n-Heptane, reflux	2,000	124
$C_4H_9C{\equiv}CH$	$\begin{array}{c} -C{=}CH- \\ \vert \\ C_4H_9 \end{array}_n$	n-Heptane, reflux	1,800	125
$C_2H_5C{\equiv}CH$	$\begin{array}{c} -C{=}CH- \\ \vert \\ C_2H_5 \end{array}_n$	n-Heptane, reflux	2,000	124
$C_6H_5C{\equiv}CCH_3$	$\begin{array}{c} -C{=}C- \\ \vert \quad \vert \\ C_6H_5 \ CH_3 \end{array}_n$	n-Heptane, reflux	1,500	125

[a] Reproduced by permission of John Wiley & Sons, New York.

average molecular weight of 12,000 is the largest reported to date for poly-(phenylacetylene).

Where the molybdenum catalyst was employed, the reaction was very fast, and only the final compound was obtained. However, when (mesitylene) M(CO)$_3$ (M = Cr, W) was used, the reaction being much slower, a mixture of two compounds could be separated and isolated. Whereas one of these was the final polymer, the other was identified as a ladder compound composed of fused cyclobutane rings. The proposed structure is shown in (24), and molecular-weight measurements indicated that the ladder compound was approximately an icosamer of phenylacetylene.

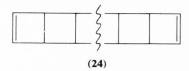

(24)

In the absence of the catalyst, the ladder compound was indefinitely stable in refluxing heptane; however, upon addition of small amounts of (tol)Mo(CO)$_3$, it was converted quantitatively to the final product, poly-(phenylacetylene). The presence of (24) was also detected spectroscopically in reactions promoted by molybdenum catalysts.

Apparently, phenylacetylene is not converted in concerted fashion to poly(phenylacetylene), but rather proceeds by way of (24). The ladder compound is formed most likely via a series of (2 + 2) cycloadditions, producing first a derivative of cyclobutadiene, then Dewar benzene, etc. The initiation is shown in Scheme 10 (reproduced by permission of John Wiley & Sons, New York). Support for the initiation process was obtained in the reaction of 2-butyne catalyzed by (tol)Mo(CO)$_3$. In this reaction, the formation of hexamethyl Dewar benzene was detected, which isomerized to hexamethylbenzene under the influence of the catalyst.

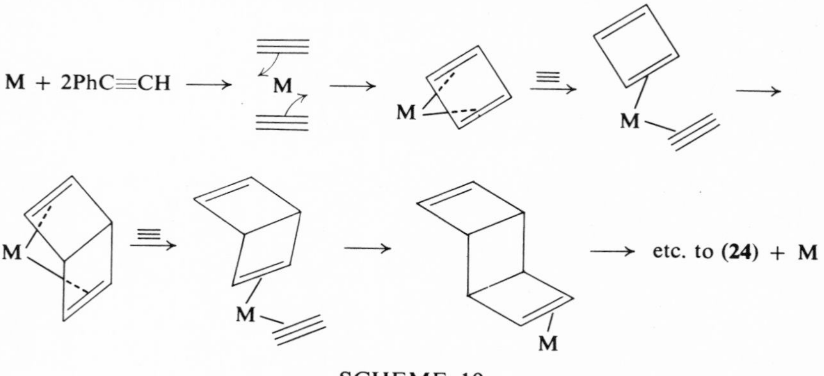

M + 2PhC≡CH ⟶ ... ⟶ ... ⟶ ... ⟶

... ⟶ ... ⟶ etc. to (24) + M

SCHEME 10

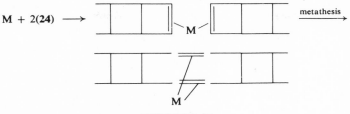

SCHEME 11

The conversion of (**24**) of molecular weight 2000 to poly(phenylacetylene) of molecular weight 12,000, was postulated to occur by a series of olefin metathesis reactions promoted by the catalyst, as shown in Scheme 11 (reproduced by permission of John Wiley & Sons, New York).

In the course of metathesis, the metal can form a five-membered ring with the olefinic bonds, and this could show fluxional activity.[126] Two possibilities for the metallocycle exist, both of which could break down electronically to form, by a free-radical mechanism, either linear, or a giant polyconjugated ring of poly(phenylacetylene). This mechanism is shown in Figure 1.

With either process, at the end of the chain an allylic radical is obtained as shown in Figure 2. Either of these can enter into the metathesis propagation as shown.

Support for the free-radical electronic breakdown to the final product has been obtained. When the polymerization of phenylacetylene was carried out in the presence of the catalyst and large amounts of free-radical traps such as galvinoxyl or 2,2-diphenyl-1-picrylhydrazyl, the system apparently was poisoned and no poly(phenylacetylene) was isolated. Obtained from the reaction, however, was the ladder compound (**24**). Apparently, the free-radical traps do not hinder the (2 + 2) cycloaddition, but poison the system to any free-radical processes. Evidence was also obtained that the process of polymerization was catalyzed throughout, and that any new chains initiated from any radicals formed in the solution were minimal.

C. Experimental Procedures

Polymerization of Phenylacetylene. The polymerization is carried out in a 50-ml, three-necked flask equipped with a thermometer, reflux condenser, and gas adapter to an oil bubbler. A solution containing 15 ml of benzene, 10 ml (9.3 g, 0.091 mol) of phenylacetylene and 0.3 g (0.001 mol) of $(tol)Mo(CO)_3$ is heated, with magnetic stirring, at 80° for 24 hr. The solution, which initially is yellow, gradually turns to a dark-brown, viscous liquid. The solvent is removed *in vacuo*, and the residue taken up in 3 ml of carbon tetrachloride and chromatographed on a neutral alumina column (2×50 cm). Elution with

(a)

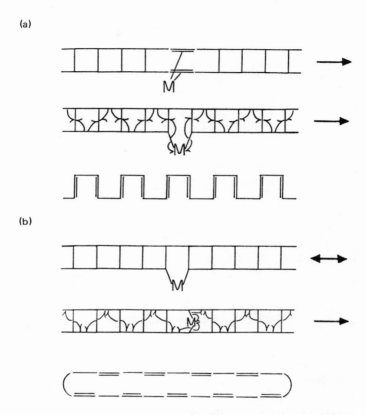

(b)

Figure 1. Electronic breakdown of the ladder compound to form (a) a linear, poly-conjugated polymer, or (b) a giant polyconjugated ring. Reproduced by permission of John Wiley & Sons, New York.

CCl_4 brings down a yellow band while the main band (also yellow) is induced to move with 2:1 v/v ether:CCl_4 eluent. The solvent is removed from the main band and the residue recrystallized twice from toluene–pentene. The yield of poly(phenylacetylene) is 8.98 g (97%) and shows a melting point of 225°. The initial band is poly(phenylacetylene) as well, but of lower molecular weight.

Isolation of the Ladder Compound. A solution of 0.5 g (0.002 mol) of (mesitylene)Cr(CO)$_3$ in 10 ml (9.3 g, 0.091 mol) of phenylacetylene (no other solvent) is heated at 75° for 1 hr. The solution is then cooled quickly with an ice bath, and the unreacted phenylacetylene (7.44 g) removed. The yellow,

(a)

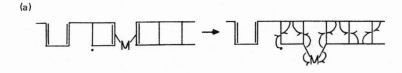

(b)

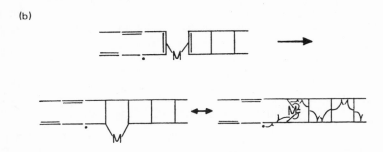

Figure 2. Reactions at the end of the chain. Reproduced by permission of John Wiley & Sons, New York.

semisolid residue is dissolved in CCl_4 and subjected to column chromatography. Two bands are eluted using 1:1 v/v ether:CCl_4; the first band contains poly(phenylacetylene). Upon removal of the solvent, a yellow solid is obtained, mp 68–70°, yield 0.71 g (7.63%), mol wt 2030.

VIII. POLYMER-BOUND SYSTEMS

The attachment of homogeneous catalysts to polymeric supports has attracted considerable interest because polymer-anchored "homogeneous" catalysts are theoretically capable of showing the advantages of both heterogeneous and homogeneous systems. Because of relatively widely separated active sites, a polymer-bound system can show high selectivity and rates of reaction, while allowing facile recovery and repeated use of the catalyst.

At least two homogeneous Group-VIB tricarbonyl catalyst systems have been "heterogenized" by attaching the $M(CO)_3$ group through the π-system of the phenyl ring on polystyrene. (Polystyrene)$Cr(CO)_3$ has been used to promote hydrogenation,[127] whereas (polystyrene)$Mo(CO)_3$ has found utility in Friedel–Crafts reactions.[128]

A. Hydrogenation of Methyl Sorbate

The catalyst was prepared by refluxing $Cr(CO)_6$ in dimethoxyethane with polystyrene crosslinked with 1% divinylbenzene. The final catalyst contained $-Cr(CO)_3$ groups bound to 20–25% of the phenyl rings. The catalyst was used to hydrogenate methyl sorbate under a variety of conditions, the results of which are shown in Table X.

Examination of the data in Table X reveals that the product distribution is a function of temperature. At 160°, the selectivity toward methyl 3-hexenoate is in the 96–98% range. At 150°, the initial selectivity toward the 3-hexenoate is considerably less, but after recycling the catalyst, the selectivity toward that

TABLE X
Hydrogenation of Methyl Sorbate[127]

Reaction No.	Conversion %	Distribution of products, %			Reaction conditions[d]
		I[a]	II[b]	III[c]	
1	100	58–65	2–7	34–40	H_2 = 34 atm; 150°, 24 hr; cyclohexane
2[e]	100	80	7	12	Same as reaction 1
3[e]	100	76	9	15	Same as reaction 1
4[e]	100	80	5	15	Same as reaction 1
5[e]	100	79	5	16	Same as reaction 1
6[e]	100	82	4	14	Same as reaction 1
7[e]	33	99.8	0.2	0	H_2 = 34 atm; 140°; 24 hr; cyclohexane
8[e]	100	96.5	2.0	0.7	H_2 = 34 atm; 160°; 24 hr; cyclohexane
9	100	97.2	2.0	0.8	Same as reaction 8
10[e]	100	97.4	1.8	0.8	Same as reaction 8
11	100	20	50	30	H_2 = 34 atm; 150°; 10 hr; DMF
12[e]	0	0	0	0	H_2 = 34 atm; 150°; 24 hr; DMF
13	100	70.7	23.6	5.7	H_2 = 34 atm; 150°; 5 hr; DMF
14	60	97.4	2.7	0	H = 34 atm; 150°; 10 hr; cyclohexane
15	100	87.4	5.6	7.0	H = 34 atm; 150°; 48 hr; cyclohexane

[a] I = (Z)-Methyl 3-hexenoate.
[b] II = (E)-Methyl 2-hexenoate.
[c] III = Methyl hexanoate.
[d] Amount of $Cr(CO)_3$ attached to resin in the 0.48–0.51 mmol range.
[e] Catalyst recycled from previous reaction.

compound stabilizes at a higher value. Selectivities at 140° are similar to those at 160°, but the conversions are much less.

The rate and product distribution of hydrogenation of methyl sorbate is also a function of the solvent. The rate of hydrogenation is considerably faster in DMF, but the catalyst cannot be recycled because extensive leaching of $Cr(CO)_3$ groups from the polystyrene support occurs as the soluble $(DMF)_3Cr(CO)_3$, rendering the support inactive.

A comparison can be made between the corresponding homogeneous and heterogeneous systems in the hydrogenation of methyl sorbate. When $(C_6H_5C_2H_5)Cr(CO)_3$ was used as the catalyst, a 95% conversion was observed after 7 hr.[82,87] However, hydrogenation reactions using the heterogenized catalyst, while slower in rate, are more selective toward production of methyl 3-hexenoate.

B. Friedel–Crafts Reactions

A study was conducted using polystyrene-anchored $Mo(CO)_3$ groups, analogous to the use of $(tol)Mo(CO)_3$, to promote Friedel–Crafts reactions. The catalyst was prepared by refluxing $Mo(CO)_3$ with polystyrene in heptane for 3 hr; the resulting catalyst system contained about 1% Mo by weight. This catalyst was used to promote a variety of Friedel–Crafts reactions which are summarized in Table XI.

TABLE XI

Friedel–Crafts Reactions Catalyzed by (Polystyrene)$Mo(CO)_3$[128]

Aromatic substrate	Organic halide	Products and comments
Anisole	t-Butyl chloride	41.7% Yield of p-t-butylanisole
Anisole	Benzyl chloride	89.3% Yield of 4- and 2-methoxydiphenyl-methane; about 10% polybenzyl
Toluene	Benzyl chloride	68.3% Yield of 4-methyldiphenylmethane; some polybenzyl
Toluene	t-Butyl chloride	4.1% Yield of p-t-butyltoluene
t-Butylbenzene	Benzyl chloride	51% Yield of p-t-butyldiphenylmethane; some polybenzyl
t-Butylbenzene	t-Butyl chloride	No reaction
Anisole	Cyclohexyl chloride	No reaction
Benzyl chloride[a]	—	100% Polybenzyl
Anisole	Benzoyl chloride	56% Yield of p-methoxybenzophenone
Anisole	Phenylacetyl chloride	81% Yield of p-phenylacetylanisole
Anisole	Hexanoyl chloride	38% Yield of p-hexanoylanisole
Anisole	Acetyl chloride	27% Yield of p-methoxyacetophenone; 8.5% o-methoxyacetophenone
Toluene	Benzoyl chloride	No reaction

[a] Neat.

A comparison of the rates of Friedel–Crafts reactions promoted by the homogeneous and heterogenized counterparts shows that the polymer-anchored catalyst is considerably more sluggish. Alkylations and acylations, in general, occur more slowly and in lower yields with the supported than with homogeneous systems. The supported catalyst is only effective where tertiary or stabilized carbonium ions can be generated, and also where good electron-releasing groups are present on the aromatic host.

An interesting feature of the heterogeneous catalyst is that it can be recycled repeatedly without noticeable loss of activity. This observation tends to indicate that the $Mo(CO)_3$ groups remain attached to the phenyl rings on polystyrene throughout the course of catalysis; otherwise, the catalyst would be leached from the support and activity would decrease rapidly from run to run.

Leaching experiments carried out on the catalyst showed that the amount of exchange of $Mo(CO)_3$ groups with the solvent was negligible, and that the active catalyst is attached to the resin.

C. Experimental Procedures

Preparation of $(Polystyrene)Cr(CO)_3$. Into a Strohmeier reactor equipped with a 250-ml reaction flask containing a magnetic stirring bar are placed 4.0 g of the cross-linked polystyrene beads, 4.0 g of $Cr(CO)_6$, and 150 ml of dimethoxyethane. The mixture is refluxed under nitrogen for 48 hr; after that time, the reaction mixture is allowed to cool to room temperature and the beads collected by filtration. The beads are repeatedly swollen with benzene and collected by filtration to remove unreacted $Cr(CO)_6$. The supported catalyst, after vacuum drying, shows carbonyl stretching frequencies at 1965 and 1880 cm^{-1}, and contains 8.9% Cr.

Hydrogenation of Methyl Sorbate. A Hoke bomb is charged with the catalyst, 1.9 g (15.1 mmol) of methyl sorbate, and 15 ml of cyclohexane. The mixture is degassed by two freeze–thaw cycles, and the bomb is pressurized with 500 psi of hydrogen and placed in an oil bath at 105° where it is shaken for 24 hr. After that time the bomb is allowed to cool to room temperature and the hydrogen vented. The catalyst is removed by filtration, the liquid contents concentrated, and the products analyzed by gas chromatography.

Preparation of $(Polystyrene)Mo(CO)_3$. Into a 100-ml, three-neck flask equipped with a nitrogen inlet and reflux condenser connected to an oil bubbler are placed 60 ml of freshly distilled heptane, 5 g of polystyrene beads, and 2 g of $Mo(CO)_6$. The system is flushed well with nitrogen and the contents are heated to reflux with magnetic stirring. The heating is continued for 3 hr, whereupon the mixture is filtered hot, and washed with hot heptane to remove unreacted $Mo(CO)_6$. The beads are now yellow and show carbonyl stretching bands at 1965 and 1880 cm^{-1}.

Alkylation of Anisole with Benzyl Chloride. In a 250-ml, three-neck flask, equipped as described above for the preparation of (polystyrene)$Mo(CO)_3$, are placed 100 ml anisole, 10.2 g (0.08 mol) of benzyl chloride, and 0.3 g of (polystyrene)$Mo(CO)_3$ catalyst. The mixture is refluxed with magnetic stirring, and after the evolution of HCl ceases (ca. 2 hr), the mixture is allowed to cool to room temperature. The catalyst is removed by filtration and unreacted anisole is removed by rotary evaporation. The viscous residue is distilled at 117–119° (0.5 torr) to yield 14.2 g (89.3%) 4- and 2-methoxydiphenylmethane. The remaining solid residue is polybenzyl.

REFERENCES

1. J. Halpern, *Homogeneous Catalysis*, Advances in Chemistry Series, No. 70, American Chemical Society, Washington, D.C., 1968, pp. 1–24.
2. F. Calderazzo, R. Ercoli, and G. Natta in *Organic Synthesis via Metal Carbonyls*, I. Wender and P. Pino, Ed., Interscience Publishers, New York, 1968, p. 158.
3. H. Zeiss, P. J. Wheatley, and H. S. J. Winkler, *Benzenoid Metal Complexes*, The Ronald Press Co., New York, 1966, p. 85.
4. R. B. King, *Transition-Metal Organometallic Chemistry*, Academic Press, New York, 1969, p. 68.
5. A. Pidcock and B. W. Taylor, *J. Chem. Soc.*, A, 877 (1967).
6. B. Nicholls and M. C. Whiting, *J. Chem. Soc.*, 551 (1959).
7. T. Kruck, *Chem. Ber.*, **97**, 2018 (1964).
8. T. A. Magee, C. N. Matthews, T. S. Wang, and J. H. Wotiz, *J. Amer. Chem. Soc.*, **83**, 3200 (1961).
9. C. N. Matthews, T. A. Magee, and J. H. Wotiz, *J. Amer. Chem. Soc.*, **81**, 2273 (1959).
10. C. E. Jones and K. J. Coskran, *Inorg. Chem.*, **10**, 55 (1971).
11. A. Pidcock, J. D. Smith, and B. W. Taylor, *J. Chem. Soc.*, A, 872 (1967).
12. F. Zingales, A. Chiesa, and F. Basolo, *J. Amer. Chem. Soc.*, **88**, 2707 (1966).
13. R. W. Harrill and H. D. Kaesz, *J. Amer. Chem. Soc.*, **90**, 1449 (1968).
14. A. Pidcock, J. D. Smith, and B. W. Taylor, *J. Chem. Soc.*, A, 1604 (1969).
14a. A. Pidcock, J. D. Smith, and B. W. Taylor, *Inorg. Chem.*, **9**, 638 (1970).
15. M. C. Ganorkar and M. H. B. Stiddard, *J. Chem. Soc.*, 5346 (1965).
16. J. F. White and M. F. Farona, *J. Organometal. Chem.*, **37**, 119 (1972).
17. R. B. King and T. F. Korenowski, *Inorg. Chem.*, **10**, 1188 (1971).
18. R. B. King, P. N. Kapoor, and R. N. Kapoor, *Inorg. Chem.*, **10**, 1841 (1971).
19. R. B. King, R. N. Kapoor, M. S. Saran, and P. N. Kapoor, *Inorg. Chem.*, **10**, 1851 (1971).
20. W. Strohmeier and H. Mittnacht, *Chem. Ber.*, **93**, 2085 (1960).
21. W. R. Jackson, B. Nicholls, and M. C. Whiting, *J. Chem. Soc.*, 469 (1960).
22. T. A. Manuel and F. G. A. Stone, *Chem. Ind.* (*London*), 231 (1960).
23. W. Strohmeier and D. von Hobe, *Z. Naturforsch.*, **18b**, 981 (1963).
24. G. Natta, R. Ercoli, and F. Calderazzo, *Chim. Ind.* (*Milan*), **40**, 287 (1958).

25. G. Natta, F. Calderazzo, and E. Santambrogio, *Chim. Ind. (Milan)*, **40**, 1003 (1958).
26. R. Ercoli, F. Calderazzo, and A. Alberola, *Chim. Ind. (Milan)*, **41**, 975 (1959).
27. W. Strohmeier and H. Mittnacht, *Z. Phys. Chem. (Frankfurt)*, **29**, 339 (1961).
28. W. Strohmeier and R. Müller, *Z. Phys. Chem. (Frankfurt)*, **40**, 86 (1964).
29. W. Strohmeier and E. H. Staricco, *Z. Phys. Chem. (Frankfurt)*, **38**, 315 (1963).
30. V. N. Setkina, N. K. Baranetskaya, K. N. Anisimov, and D. N. Kuisanov, *Izvest. Akad. Nauk SSSR, Ser. Khim.*, **10**, 1873 (1964).
31. D. N. Kursanov, V. N. Setkina, N. K. Baranetskaya, E. I. Fedin, K. N. Anisimov, and V. M. Urinyuk, *Dokl. Akad. Nauk SSSR*, **183**, 1340 (1968).
32. M. Ashraf, *Can. J. Chem.*, **50**, 118 (1972).
33. R. Riemschneider, O. Becker, and K. Franz, *Montash. Chem.*, **90**, 571 (1959).
34. R. Ercoli, F. Calderazzo, and E. Mantica, *Chim. Ind. (Milan)*, **41**, 404 (1959).
35. G. E. Herberich and E. O. Fischer, *Chem. Ber.*, **95**, 2803 (1962).
36. W. R. Jackson and W. B. Jennings, *J. Chem. Soc., B*, 1222 (1969).
37. M. C. Whiting and B. Nicholls, *Proc. Chem. Soc.*, 152 (1958).
38. E. O. Fischer, K. Ofele, H. Essler, W. Fröhlich, J. P. Mortensen, and W. Semmlinger, *Chem. Ber.*, **91**, 2763 (1958).
39. D. A. Brown, *J. Chem. Soc.*, 4389 (1963).
40. D. A. Brown and J. R. Raju, *J. Chem. Soc., A*, 40 (1966).
41. J. D. Holmes, D. A. K. Jones, and R. Pettit, *J. Organometal. Chem.*, **4**, 324 (1965).
42. A. Ceccon, *J. Organometal. Chem.*, **29**, C19 (1971).
43. A. Mandelbaum, Z. Neuwirth, and M. Cais, *Inorg. Chem.*, **2**, 902 (1963).
44. C. U. Pittman, R. L. Voges, and J. Elder, *Macromolecules*, **4**, 302 (1971).
45. G. Drehfall, H. H. Hörhold, and K. Kühne, *Chem. Ber.*, **98**, 1826 (1965).
46. W. S. Trahanovsky and R. J. Card, *J. Amer. Chem. Soc.*, **94**, 2897 (1972).
47. A. Ceccon and G. S. Biserni, *J. Organometal. Chem.*, **39**, 313 (1972).
48. G. Jaouen and R. Dabard, *Compt. Rend., Ser. C*, **271**, 1610 (1970).
49. G. Jaouen and R. Dabard, *Tetrahedron Lett.*, 1015 (1971).
50. G. Pajaro, F. Calderazzo, and R. Ercoli, *Gazz. Chim. Ital.*, **90**, 1486 (1960).
51. W. Strohmeier, *Angew. Chem. Intern. Ed.*, **3**, 730 (1964).
52. W. Strohmeier and H. Hellmann, *Z. Naturforsch.*, **18b**, 769 (1963).
53. W. Strohmeier and H. Hellmann, *Chem. Ber.*, **96**, 2859 (1963).
54. W. Strohmeier and H. Hellmann, *Chem. Ber.*, **97**, 1877 (1964).
55. W. Strohmeier, G. Popp, and J. F. Guttenberger, *Chem. Ber.*, **99**, 165 (1966).
56. E. O. Fischer and P. Kuzel, *Z. Naturforsch.*, **16b**, 475 (1961).
57. W. Strohmeier and H. Hellmann, *Chem. Ber.*, **98**, 1598 (1965).
58. D. Sellmann and G. Maisel, *Z. Naturforsch.*, **27b**, 465 (1972).
59. D. Sellmann and G. Maisel, *Z. Naturforsch.*, **27b**, 718 (1972).
60. A. N. Nesmeyanov, D. N. Kursanov, V. N. Setkin, V. D. Vil'chevskaya, N. K. Baranetskaya, A. I. Krylova, and L. A. Glushchenko, *Dokl. Akad. Nauk SSSR*, **199**, 1336 (1971).
61. W. P. Anderson, W. G. Blenderman, and K. A. Drews, *J. Organometal. Chem.*, **42**, 139 (1972).
62. H. Behrens, K. Meyer, and A. Müller, *Z. Naturforsch.*, **20b**, 74 (1965).
63. D. A. Brown, D. Cunningham, and W. K. Glass, *J. Chem. Soc., A*, 1563 (1968).
64. N. G. Connelly and L. F. Dahl, *Chem. Commun.*, 880 (1970).
65. B. V. Lokshin, V. I. Zdanovich, N. K. Baranetskaya, V. N. Setkina, and D. N. Kursanov, *J. Organometal. Chem.*, **37**, 331 (1972).
66. D. N. Kursanov, V. N. Setkina, P. V. Petrovskii, V. I. Zdanovich, N. K. Baranetskaya, and I. D. Rubin, *J. Organometal. Chem.*, **37**, 339 (1972).

67. C. P. Lillya and R. A. Sahatjian, *Inorg. Chem.*, **11**, 889 (1972).
68. W. Jetz and W. A. G. Graham, *Inorg. Chem.*, **10**, 4 (1971).
69. R. W. Turner and E. L. Amma, *J. Amer. Chem. Soc.*, **88**, 1877 (1966).
70. M. F. Farona and J. F. White, *J. Amer. Chem. Soc.*, **93**, 2826 (1971).
71. J. F. White and M. F. Farona, *J. Organometal. Chem.*, **63**, 329 (1973).
72. S. N. Massie, U.S. Patent 3,705,201; *Chem. Abstr.*, **78**, 57987s (1973).
73. S. N. Massie, U.S. Patent 3,705,202; *Chem. Abstr.*, **78**, 57988t (1973).
74. M. F. Farona and D. Thomas, unpublished results.
75. C. H. Bamford, G. C. Eastmond, and F. J. T. Fildes, *Chem. Commun.*, 144, 146 (1970), and references contained therein.
76. C. H. Bamford, G. C. Eastmond, and D. Whittle, *J. Organometal. Chem.*, **17**, 33 (1969).
77. K. Kusada, R. West, and V. N. M. Rao, *J. Amer. Chem. Soc.*, **93**, 3627 (1971).
78. J. Korenz and M. F. Farona, unpublished results.
79. H. Alper and C. C. Huang, *J. Org. Chem.*, **38**, 64 (1973).
80. A. Miyake and H. Kondo, *Angew. Chem. Intern. Ed.*, **7**, 631 (1968).
81. J. Nasielski, P. Kirsch and L. Wilputte-Steinert, *J. Organometal. Chem.*, **27**, C13 (1971).
82. E. N. Frankel and R. O. Butterfield, *J. Org. Chem.*, **34**, 3930 (1969).
83. A. Miyake and H. Kondo, *Angew. Chem. Intern. Ed.*, **7**, 880 (1968).
84. W. R. Kroll, U.S. Patent 3,644,445; *Chem. Abstr.*, **76**, 154452e (1972).
85. E. N. Frankel, *J. Org. Chem.*, **37**, 1549 (1972).
86. L. W. Gosser, U.S. Patent 3,673,270; *Chem. Abstr.*, **77**, 89083q (1972).
87. M. Cais, E. N. Frankel, and A. Rejoan, *Tetrahedron Lett.*, 1919 (1968).
88. M. Cais, N. Maoz, and A. Rejoan, *Proc. Second Intern. Symp.*, Venice, 25 (1969).
89. E. N. Frankel and F. L. Little, *J. Amer. Oil Chemists' Soc.*, **46**, 256 (1969).
90. G. Ben-et, A. Dolev, M. Schimmel, and R. Stern, *J. Amer. Oil Chemists' Soc.*, **49**, 205 (1972).
91. E. N. Frankel, F. L. Thomas, and J. C. Cowan, *J. Amer. Oil Chemists' Soc.*, **47**, 497 (1970).
92. E. N. Frankel and F. L. Thomas, *J. Amer. Oil Chemists' Soc.*, **49**, 70 (1972).
93. E. N. Frankel, E. Selke, and C. A. Glass, *J. Amer. Chemists' Soc.*, **90**, 2446 (1968).
94. E. N. Frankel, E. Selke, and C. A. Glass, *J. Org. Chem.*, **34**, 3936 (1969).
95. M. Cais and A. Rejoan, *Inorg. Chim. Acta* , **4**, 509 (1970).
96. E. N. Frankel, private communication.
97. E. O. Fischer, W. Hafner, and H. O. Stahl, *Z. Anorg. Allgem. Chem.*, **282**, 47 (1955).
98. R. L. Banks, U.S. Patent 3,463,827; *Chem. Abstr.*, **75**, 151339s (1971).
99. J. L. Wang and H. R. Menapace, *J. Catal.*, **23**, 144 (1971).
100. J. L. Wang, U.S. Patent 3,634,540; *Chem. Abstr.*, **76**, 71975b (1972).
101. W. E. Breckoff, U.S. Patent 3,391,216; *Chem. Abstr.*, **69**, 51514c (1968).
102. E. N. Frankel, *J. Amer. Oil Chemists Soc.*, **47**, 33 (1970).
103. E. N. Frankel, *J. Catal.*, **24**, 358 (1972).
104. G. S. Lewandos and R. Pettit, *J. Amer. Chem. Soc.*, **93**, 7087 (1971); *Tetrahedron Lett.*, 789 (1971).
105. J. L. Herrison, Y. Chauvin, N. H. Phung, and G. Lefebure, *Compt. Rend.*, *Ser. C*, **269**, 661 (1969).
106. L. Ramain and Y. Trambouze, *Compt. Rend.*, *Ser. C*, **273**, 1409 (1971).
107. H. W. Ruhle, Ger. Offen. 2,062,448; *Chem. Abstr.*, **75**, 151341 (1971).
108. E. A. Zuech, Fr. 1,561,026; *Chem. Abstr.*, **72**, 33017a (1970).
109. L. Bencze and L. Marko, *J. Organometal. Chem.*, **28**, 271 (1971).

110. W. R. Kroll and G. Doyle, *J. Catal.*, **24**, 356 (1972).
111. W. R. Kroll and G. Doyle, *Chem. Commun.*, 839 (1971).
112. W. R. Kroll and G. Doyle, U.S. Patent 3,689,433; *Chem. Abstr.*, **77**, 151431g (1972).
113. G. Doyle, *J. Catal.*, **30**, 118 (1973).
114. R. L. Banks and G. R. Bailey, *Ind. Eng. Chem.*, *Prod. Res. Develop.*, **3**, 170 (1964).
115. British Petroleum Co., Ltd., Fr. 1,554,287; *Chem. Abstr.*, **71**, 60655t (1969).
116. L. F. Heckelsburg, R. L. Banks, and G. C. Bailey, *Ind. Eng. Chem.*, *Prod. Res. Develop.*, **8**, 259 (1969).
117. P. P. O'Neill and J. J. Rooney, *J. Amer. Chem. Soc.*, **94**, 4383 (1972).
118. E. S. Davie, D. A. Whan, and C. Kemball, *Chem. Commun.*, 1430 (1969).
119. R. F. Howe, D. E. Davidson, and D. A. Whan, *J. Chem. Soc., Faraday Trans.*, 2266 (1972).
120. E. S. Davie, D. A. Whan, and C. Kemball, *Chem. Commun.*, 1202 (1971).
121. C. W. Bird, *Transition Metal Intermediates in Organic Synthesis*, Academic Press, New York, 1967, Chapter 1.
122. W. Strohmeier and C. Barbeau, *Z. Naturforsch.*, **19b**, 262 (1964).
123. W. Hubel and C. Hoogzand, *Chem. Ber.*, **93**, 103 (1960).
124. M. F. Farona, P. A. Lofgren, and P. S. Woon, *Chem. Commun.*, 246 (1974).
125. P. S. Woon and M. F. Farona, *J. Polymer Sci. Chem. Ed.*, **12**, 1749 (1974).
126. R. H. Grubbs and T. K. Brunk, *J. Amer. Chem. Soc.*, **94**, 2538 (1972).
127. C. U. Pittman, Jr., B. T. Kim, and W. M. Douglas, *J. Org. Chem.*, **40**, 590 (1975).
128. C. P. Tsonis and M. F. Farona, *J. Organometal. Chem.*, **114**, 293 (1976).

Index